GENETIC MAPS

Locus Maps of Complex Genomes

FIFTH EDITION

Stephen J. O'Brien, Editor

Laboratory of Viral Carcinogenesis, National Cancer Institute

BOOK 1

Viruses

COLD SPRING HARBOR LABORATORY PRESS 1990

GENETIC MAPS

First edition, March 1980
Second edition, June 1982
Third edition, June 1984
Fourth edition, July 1987
Fifth edition, February 1990

GENETIC MAPS
Locus Maps of Complex Genomes
Fifth Edition
BOOK 1 Viruses

Printed in the United States of America

Library of Congress Catalog Card Number 84-644938
ISBN 0-87969-342-8
ISSN 0738-5269

Cover design by Leon Bolognese

All Cold Spring Harbor Laboratory Press publications may be ordered directly from Cold Spring Harbor Laboratory, Box 100, Cold Spring Harbor, New York 11724. Phone 1-800-843-4388. In New York (516) 367-8423.

PREFACE

"The map is a sensitive indicator of the changing thought of man, and few of his works seem to be such an excellent mirror of culture and civilization."

Norman J.W. Thrower
Maps and Men, 1972

The early geographic explorers are revered by their countrymen and descendants throughout the world for pioneering the discovery and charting of foreign lands. The precision and attention to detail accorded to their charts and maps are almost always taken for granted today. Astronauts, and then those who saw their photographs from space, have marveled at the resemblance of these real images of the Florida peninsula, the boot of Italy, and the British Isles to the maps we had grown up with. John Noble Wilford, in his fascinating monograph entitled *The Mapmakers,* reminds us that it is only within the last quarter of the twentieth century that it could be said that the earth has been mapped. It is humbling to consider how important these maps have been to our history, our culture, and our way of life today; in this context, the early cartographers were indeed intrepid.

The new explorers of the eukaryote and prokaryote genome will certainly one day be considered with equivalent veneration, since the topography of their charts and maps may prove as daunting as the earth's surface. The animal and plant genomes are products of several hundred million years of biological evolution, and there are clearly far more secrets to be deciphered in the genomes that can be anticipated in future generations. However, genetic detectives have begun to unravel some of the mysteries of genetic organization, and as we enter the final decade of the century, the scientific community (as well as the enlightened public) has begun to grasp the enormous value of generating and expanding genetic maps of man, of agriculturally significant plants and animals, and of model systems that allow us a glimpse of how genes are organized, replicated, and regulated.

The first genetic map was formulated by A.H. Sturtevant in 1913 and consisted of five genes arranged in a linear fashion along the X chromosome of the fruit fly, *Drosophila melanogaster.* In the ensuing decades, gene mapping of numerous species has proceeded deliberately and cumulatively in organisms as diverse as flies, corn, wheat, mink, apes, man, and bacteria. All of these maps, whether based on recombination, restriction, physical or DNA sequence, are predicated on Sturtevant's logical notion that gene order on a chromosome could be displayed as a linear array of genetic markers. The results of these efforts in more than 100 genetically studied organisms are the basis of *GENETIC MAPS: Locus Maps of Complex Genomes.*

During the preparation of the previous edition, it occurred to me that the rate of growth of the gene mapping effort was so rapid that we should prepare to publish future editions in multiple volumes. The fifth edition is a realization of that idea. *Genetic Maps* now consists of six smaller books, each based on an arbitrary subdivision of biological organisms. These are BOOK 1 Viruses; BOOK 2 Bacteria, Algae, and Protozoa; BOOK 3 Lower Eukaryotes; BOOK 4 Nonhuman Vertebrates; BOOK 5 Human Maps; BOOK 6 Plants. Each of these is available in paperback at a modest cost from Cold Spring Harbor Laboratory Press. The entire compendium can be purchased as a hardback volume of 1098 map pages suitable for libraries and research institutions.

Our original intent was to publish in one volume complete, referenced genetic maps of every organism with a substantive group of assigned loci. We intentionally excluded DNA sequences, since these are easily available in several computer databanks (Genbank, EMBL, and others). Text was to be kept at a minimum, and the maps would be both comprehensive and concise. The collection was to be updated every 2–3 years, and each new volume would contain a complete revision, rendering the previous volume obsolete.

The original publication of *Genetic Maps* was an experiment, which I believe now can be judged a success. The execution of such a venture depended heavily on the support and cooperation of thousands of geneticists throughout the worldwide scientific community. This cooperation was graciously extended, and the result was an enormously valuable and accessible collection. On behalf of my colleagues in the many fields that we call genetic biology, I gratefully acknowledge the numerous geneticists, scientists, and readers who have contributed to or corrected these maps. To ensure the continuation of these heroic efforts in the future, readers are encouraged to send to me any suggestions for improvement, particularly suggestions for new maps to be included, as well as constructive criticisms of the present maps. When new organisms are recommended, I would also appreciate names and addresses of prospective authors. In addition, readers are encouraged to supply corrections, reprints, and new mapping data to the appropriate authors who may be included in future editions.

The compilation of each genetic map, like the drafting of the first geographic maps, is an extremely tedious, yet important, assignment. All of the authors deserve special thanks for the large efforts they have expended in contributing their maps. Even in this computer age, we often feel like we are proofreading the telephone book, but I, for one, believe that the final product makes the effort worthwhile. Finally, I acknowledge specifically my editorial assistants, Patricia Johnson and Virginia Frye, who have cheerfully and expeditiously carried the bulk of the editorial activities for the present volume, and Annette Kirk, Lee Martin, and John Inglis of Cold Spring Harbor Laboratory Press for support and advice on the preparation of this edition.

Stephen J. O'Brien, *Editor*

CONTENTS

Preface, iii

BOOK 1 Viruses

BACTERIOPHAGES

Lambda	D.L. Daniels, J.L. Schroeder, W. Szybalski, F. Sanger, and F.R. Blattner	**1.3**
T4	E. Kutter, B. Guttman, G. Mosig, and W. Rüger	**1.24**
P1	M. Yarmolinsky	**1.52**
P2	E. Håggard-Ljungquist	**1.63**
P4	C. Halling and R. Calendar	**1.70**
ϕ29	V. Pačes and C. Vlček	**1.74**
ϕX174	P. Weisbeek	**1.79**

DNA VIRUSES

Simian virus 40	J. Remenick and J. Brady	**1.84**
Polyomavirus	Y. Ito and B.E. Griffin	**1.90**
Adenovirus	G. Akusjärvi and G. Wadell	**1.98**
Epstein-Barr virus	P.J. Farrell	**1.102**
Herpes simplex virus	D.J. McGeoch, V.G. Preston, S.K. Weller, and P.A. Schaffer	**1.115**
Hepatitis B virus	H. Farza and P. Tiollais	**1.121**
Papillomaviruses	C.C. Baker and L.M. Cowsert	**1.128**
bovine (BPV) types 1 and 2		1.129
deer (DPV)		1.129
European elk (EEPV)		1.129
cotton tail rabbit (CRPV)		1.129
human (HPV) types 1a, 5, 6b,		1.130
8, 11, 16, 18, 31, and 33		1.131
Vaccinia virus	P.L. Earl and B. Moss	**1.138**

RNA VIRUSES

Human and simian immunodeficiency viruses (HIV-I, HIV-II, and SIVmac)	S.F. Josephs and F. Wong-Staal	**1.149**
Human T-lymphotropic retroviruses (HTLV-I and HTLV-II)	L. Ratner and F. Wong-Staal	**1.160**

Non-human lentiviruses	T. Huet and S. Wain-Hobson	**1.162**
simian immunodeficiency virus (SIV)		1.163
feline immunodeficiency virus (FIV)		1.165
Ungulate lentiviruses	T. Huet and S. Wain-Hobson	**1.166**
visna,		1.166
ovine pulmonary adenomatosis virus (OPAV),		1.166
caprine arthritis encephalitis virus (CAEV),		1.166
bovine immunodeficiency virus (BIV)		1.166
equine infectious anemia virus (EIAV)		1.167
Type C retroviruses	J.M. Coffin	**1.168**
AKV		1.168
Mo-MLV		1.169
Pr-RSV (*src*)		1.170
SSV (*sis*)		1.171
Mo-MSV (*mos*)		1.172
Ab-MLV (*abl*)		1.173
Ra-MSV (*ras*)		1.173
Fr-$SFFV_p$		1.174
SK770 (*ski*)		1.174
FeSV (*fms*)		1.175
MSV-3611 (*raf*)		1.175
E26 (*ets*)		1.176
Y73 (*yes*)		1.176
FuSV (*fps*)		1.177
AMV (*myb*)		1.177
AEV (*erbB*)		1.178
MC29 (*myc*)		1.178
GA-FeSV (*fes*)		1.179
GR-FeSV (*fgr*)		1.179
FBJ-OV (*fos*)		1.180
FBR		1.180
ARV (*rel*)		1.181
BaEV		1.181
RD114		1.181

Index, I.1

COMPLETE CONTENTS OF THE FIFTH EDITION

BOOK 1 Viruses

BACTERIOPHAGES

Lambda	D.L. Daniels, J.L. Schroeder, W. Szybalski, F. Sanger, and F.R. Blattner
T4	E. Kutter, B. Guttman, G. Mosig, and W. Rüger
P1	M. Yarmolinsky
P2	E. Hågard-Ljungquist
P4	C. Halling and R. Calendar
ϕ29	V. Pačes and C. Vlček
ϕX174	P. Weisbeek

DNA VIRUSES

Simian virus 40	J. Remenick and J. Brady
Polyomavirus	Y. Ito and B.E. Griffin
Adenovirus	G. Akusjärvi and G. Wadell
Epstein-Barr virus	P.J. Farrell
Herpes simplex virus	D.J. McGeoch, V.G. Preston, S.K. Weller, and P.A. Schaffer
Hepatitis B virus	H. Farza and P. Tiollais
Papillomaviruses bovine (BPV) types 1 and 2 deer (DPV) European elk (EEPV) cotton tail rabbit (CRPV) human (HPV) types 1a, 5, 6b, 8, 11, 16, 18, 31, and 33	C.C. Baker and L.M. Cowsert
Vaccinia virus	P.L. Earl and B. Moss

RNA VIRUSES

Human and simian immunodeficiency viruses (HIV-I, HIV-II, and SIVmac)	S.F. Josephs and F. Wong-Staal
Human T-lymphotropic retroviruses (HTLV-I and HTLV-II)	L. Ratner and F. Wong-Staal
Non-human lentiviruses simian immunodeficiency virus (SIV) feline immunodeficiency virus (FIV)	T. Huet and S. Wain-Hobson
Ungulate lentiviruses visna, ovine pulmonary adenomatosis virus (OPAV), caprine arthritis encephalitis virus (CAEV), bovine immunodeficiency virus (BIV) equine infectious anemia virus (EIAV)	T. Huet and S. Wain-Hobson
Type C retroviruses AKV Mo-MLV Pr-RSV (*src*) SSV (*sis*) Mo-MSV (*mos*) Ab-MLV (*abl*) Ra-MSV (*ras*) Fr-SFFV$_p$ SK770 (*ski*) FeSV (*fms*) MSV-3611 (*raf*) E26 (*ets*) Y73 (*yes*) FuSV (*fps*) AMV (*myb*) AEV (*erbB*) MC29 (*myc*) GA-FeSV (*fes*) GR-FeSV (*fgr*) FBJ-OV (*fos*) FBR ARV (*rel*) BaEV RD114	J.M. Coffin

BOOK 2 Bacteria, Algae, and Protozoa

BACTERIA

Salmonella typhimurium	K.E. Sanderson and D. Douey
Staphylococcus aureus	P.A. Pattee
Bacillus subtilis	D.R. Zeigler
Escherichia coli	B.J. Bachmann
Pseudomonas aeruginosa	B.W. Holloway and C. Zhang
Pseudomonas putida	D.G. Strom and A.F. Morgan
Proteus mirabilis	J.N. Coetzee, M.C. Van Dijken, and W.F. Coetzee
Proteus morganii	Y. Beck
Neisseria gonorrhoeae	J.H. Jackson and T.E. Shockley
Acinetobacter calcoaceticus	A. Vivian
Caulobacter crescentus	B. Ely
Rhizobium meliloti and *Rhizobium leguminosarum* biovars: *phaseoli* *trifolii* *viciae*	J.E. Beringer, J.P.W. Young, and A.W.B. Johnston
Haemophilus influenzae Rd	R.J. Redfield and J.J. Lee

ALGAE

Chlamydomonas reinhardtii nuclear gene loci mitochondrial genome chloroplast genome	E.H. Harris

PROTOZOA

Paramecium tetraurelia	K.J. Aufderheide and D. Nyberg
Tetrahymena thermophila	P.J. Bruns

BOOK 3 Lower Eukaryotes

FUNGI

Dictyostelium discoideum	P.C. Newell
Neurospora crassa	
nuclear genes	D.D. Perkins
mitochondrial genes	R.A. Collins
restriction polymorphism	R.L. Metzenberg and J. Grotelueschen
Saccharomyces cerevisiae	
nuclear genes	*R.K. Mortimer, D. Schild, C.R. Contopoulou, and J.A. Kans*
mitochondrial DNA	*L.A. Grivell*
Podospora anserina	D. Marcou, M. Picard-Bennoun, and J.-M. Simonet
Sordaria macrospora	G. Leblon, V. Haedens, A.D. Huynh, L. Le Chevanton, P.J.F. Moreau, and D. Zickler
Coprinus cinereus	J. North
Magnaporthe grisea (blast fungus)	D. Z. Skinner, H. Leung, and S.A. Leong
Phycomyces blakesleeanus	M. Orejas, J.M. Diaz-Minguez, M.I. Alvarez, and A.P. Eslava
Schizophyllum commune	C.A. Raper
Ustilago maydis	A.D. Budde and S.A. Leong
Ustilago violacea	A.W. Day and E.D. Garber
Aspergillus nidulans	
nuclear genes	A.J. Clutterbuck
mitochondrial genome	T.A. Brown

NEMATODE

Caenorhabditis elegans	M.L. Edgley and D.L. Riddle

INSECTS

Drosophila melanogaster	
biochemical loci	G.E. Collier
in situ hybridization data	E. Kubli and S. Schwendener
cloned genes	J. Merriam, S. Adams, G. Lee, and D. Krieger
Aedes (Stegomyia) aegypti	L.E. Munstermann
Aedes (Protomacleaya) triseriatus	L.E. Munstermann
Drosophila pseudoobscura	W.W. Anderson
Anopheles albimanus	S.K. Narang and J.A. Seawright
Anopheles quadrimaculatus	S.K. Narang, J.A. Seawright, and S.E. Mitchell
Nasonia vitripennis	G.B. Saul, 2nd

BOOK 4 Nonhuman Vertebrates

RODENTS

Mus musculus (mouse)	
nuclear genes	M.T. Davisson, T.H. Roderick, D.P. Doolittle, A.L. Hillyard, and J.N. Guidi
DNA clones and probes, RFLPs	J.T. Eppig
retroviral and cancer-related genes	C.A. Kozak
Rattus norvegicus (rat)	G. Levan, K. Klinga, C. Szpirer, and J. Szpirer
Cricetulus griseus (Chinese hamster)	
nuclear genes	R.L. Stallings, G.M. Adair, and M.J. Siciliano
CHO cells	G.M. Adair, R.L. Stallings, and M.J. Siciliano
Peromyscus maniculatus (deermouse)	W.D. Dawson
Mesocricetus auratus (Syrian hamster)	R. Robinson
Meriones unquiculatus (Mongolian gerbil)	R. Robinson

OTHER MAMMALS

Felis catus (cat)	S.J. O'Brien
Canis familiaris (dog)	P. Meera Khan, C Brahe, and L.M.M. Wijnen
Equus caballus (horse)	L.R. Weitkamp and K. Sandberg
Sus scrofa domestica L. (pig)	G. Echard
Oryctolagus cuniculus (rabbit)	R.R. Fox
Ovis aries (sheep)	G. Echard
Bos taurus (cow)	J.E.Womack
Mustela vison (American mink)	O.L. Serov and S.D. Pack
Marsupials and Monotremes	J.A. Marshall Graves

PRIMATES

Primate Genetic Maps	N. Creau-Goldberg, C. Cochet, C. Turleau, and J. de Grouchy
Pan troglodytes (chimpanzee)	
Gorilla gorilla (gorilla)	
Pongo pygmaeus (orangutan)	
Hylobates (Nomascus) concolor (gibbon)	
Macaca mulatta (rhesus monkey)	
Papio papio, hamadryas, cynocephalus (baboon)	
Cercopithecus aethiops (African green monkey)	
Cebus capucinus (capuchin monkey)	
Microcebus murinus (mouse lemur)	
Saquinus oedipus (cotton-topped marmoset)	P.A. Lalley
Aotus trivirgatus (owl monkey)	N.S.-F. Ma

FISH

Salmonid fishes *Salvelinus, Salmo, Oncorhynchus*	B. May and K.R. Johnson
Non-Salmonid fishes *Xiphophorus, Poeciliopsis, Fundulus, Lepomis*	D.C. Morizot

AMPHIBIAN

Rana pipiens (leopard frog)	D.A. Wright and C.M. Richards

BIRD

Gallus gallus (chicken)	R.G. Somes, Jr., and J.J. Bitgood

BOOK 5 Human Maps

HUMAN MAPS (Homo sapiens)	
Human gene map	I.H. Cohen, H.S. Chan, R.K. Track, and K.K. Kidd
Human gene map nuclear genes morbid anatomy mitochondrial chromosome	V.A. McKusick
Human genetic linkage maps	R. White, J.-M. Lalouel, M. Lathrop, M. Leppert, Y. Nakamura, and P. O'Connell
Human linkage map	H. Donis-Keller and C. Helms
Human genetic linkage map (chromosome 21)	J.F. Gusella, J.L. Haines, and R.E. Tanzi
Chromosome rearrangements in human neoplasia	C.D. Bloomfield and J.L. Frestedt
Human fragile sites	G.R. Sutherland
An abridged human map proto-oncogenes endogenous retroviral sequences cell surface receptors growth factors and lymphokines	S.J. O'Brien and P.A. Johnson
Human DNA restriction fragment length polymorphisms	K.K. Kidd, R.K. Track, A.M. Bowcock, F. Ricciuti, G. Hutchings, and H.S. Chan
Human mitochondrial DNA	D.C. Wallace

BOOK 6 Plants

PLANTS	
Lycopersicon esculentum (tomato)	S.D. Tanksley and M.A. Mutschler,
Triticum aestivum (wheat)	
linkage map	D.L. Milne and R.A. McIntosh
biochemical/molecular loci	G.E. Hart and M.D. Gale
Zea mays L. (corn)	E.H. Coe, Jr., D.A. Hoisington, and M.G. Neuffer
Glycine max L. Merr. (soybean)	R.G. Palmer and Y.T. Kiang
Arabidopsis thaliana	M. Koornneef
RFLP	E.M. Meyerowitz, J.L. Bowman, C. Chang, and S. Kempin
Lactuca sativa (lettuce)	R.V. Kesseli, I. Paran, and R.W. Michelmore
Brassica oleracea (broccoli) RFLP	M.K. Slocum, S.S. Figdore, W. Kennard, J.Y. Suzuki, and T.C. Osborn
Pisum sativum (garden pea)	N.F. Weeden and B. Wolko
Petunia hybrida (petunia) linkage map	A. Cornu, E. Farcy, D. Maizonnier, M. Haring, W. Veerman, and
chloroplast DNA	A.G.M. Gerats
Hordeum vulgare L. (barley)	P. von Wettstein-Knowles
RFLP	T. Blake
Secale cereale L. (rye)	G. Melz and R. Schlegel

BOOK 1

VIRUSES

A MOLECULAR MAP OF COLIPHAGE LAMBDA

Compiled by Donna L. Daniels*, John L. Schroeder*, Waclaw Szybalski&, Fred Sanger#, and Frederick R. Blattner*
*Laboratory of Genetics, and &McArdle Laboratory, University of Wisconsin, Madison, Wisconsin 53706 and #MRC Laboratory of Molecular Biology, Cambridge, England

The molecular map of λ is presented in the following sections:

Figure 1 is a scale drawing of the λ map (reprinted from Daniels et al. 1983).

Table 1 presents our tabulation of all the sites on the λ map. Listed are the names of all mapped sites (columns A and E), their base pair coordinates on the 5′ - 3′ *l* strand of λ DNA (column C), four-character symbols used by the computer for sequence annotation in Appendix II (column D), a description of the site (columns F and H) and the method of localization of the map (column G). Restriction sites that have been mapped experimentally are shown in column A. All other sites are specified in column E.

Table 2 lists the published sources of λ DNA sequence. Table 3 shows the sources of mapping information for sites other than genes. Table 4 shows the sources of mapping information for protein coding genes and open reading frames. Table 5 summarizes the locations of restriction enzyme cutsites for enzymes which cut fewer than 20 times as predicted from the complete sequence.

SOURCES

The sources of information used to compile the molecular map of phage λ DNA are:

1. The complete nucleotide sequence of λ. Various groups have sequenced regions of various strains of λ. This information has been collated and sequences have been compared from all sources to determine overlaps and discrepancies. This information forms eight "sequence blocks", which are indicated in Table 1 (column B) by ditto marks enclosed by the symbols ▽ and △. Within these blocks are only four unresolved discrepancies (single base-pair changes). Sanger et al. (1982) presented the complete sequence of λ*cI*ind1ts857*S*7 based on sequence determined in their laboratory and on compilation and comparison with published lambda sequences (see Table 2). The sequence that has been annotated is that compiled by Sanger et al. (1982) changed to "wild type". These changes are *ind*1 to *ind*$^+$ at 37589; *c*Its857 to *c*I$^+$ at 37742; *S*am7 to *S*$^+$ at 45352; and C to T correction of typographical error in Sanger et al. (1982) at 44374.

2. A map of 171 restriction sites. This map of 20 enzymes was determined by gel electrophoresis and is presented in Column A of Table 1. It was constructed by merging the published 67 site map for 15 enzymes (Daniels and Blattner 1982) with the individual enzyme maps of *Pvu*II of L. MacHattie (pers. comm.), *Hin*dII (*Hin*cII) of Robinson and Landy (1977a,b), *Pst*I (Smith et al. 1976; Legerski et al. 1978), *Sph*I, *Mst*II (D. L. Daniels, unpubl.) and *Cla*I (Mayer et al. 1981, and D. L. Daniels, unpubl.). The restriction map predicted from the DNA sequence agrees with the experimental map, confirming both.

3. Compilation of electron micrographic mapping information by Szybalski and Szybalski (1979).

4. The λ gene map by Echols and Murialdo (1978).

5. Sequence changes of a large number of mutations, both from published sequence of λ variants and personal communications from λ workers (see Table 3).

NUMBERING SYSTEM

The λ sequence as presented includes both the sticky ends. The numbering system is as follows: Base 1 is the first (5′) base of the left sticky end of the *l* strand of the λ DNA molecule as isolated from the capsid. This base is denoted LEND (left end). The first base of the double stranded portion (denoted CLND - complementary strand left end) is 13; the last base of the double stranded portion (denoted REND - [right end]) is 48502; the last base of the right sticky end (denoted CRND - complementary strand right end) is 48514. The length of the circular form of λ DNA is 48502 base pairs and runs from LEND to REND.

ANNOTATION CONVENTIONS

Annotations may refer to one of three features of sequence; a particular base, internucleotide space or a map interval.

1. The particular base form is used for point mutations and a few other sites. The number refers to the affected base.

2. Internucleotide spaces are annotated for restriction cutsites, hybrid (hy) phage junctions and insertion mutations (i). The space is located between the numbered base and the one following it. Restriction sites were obtained by computer search for the recognition sites of the enzymes and labeled at the cut site on the *l* strand. Predicted lengths of *l* strand fragments can be obtained by subtraction. Lengths of *r* strand fragments may need to be corrected depending on the position of the 3′ cutsite for the particular enzymes used.

3. Map intervals are annotated for genes, open reading frames, deletions, substitutions, sequence blocks, λdv plasmids, and transcriptional units. These annotations occur in left-right pairs. The numbers refer to the first and last bases of the interval, respectively. For genes, these are the first base of the ATG or GTC start codon and the third base of the last amino acid, not including the termination codon. For deletions and substitutions the numbers refer to the first and last bases of λ that are deleted. Sequence block markers refer to the first and last bases sequenced in the particular reference. For λ-dv plasmids the annotated interval is the region of λ that is present in the plasmid (not the region of λ deleted to create the plasmid). For transcription units the two ends of the RNA molecule are denoted.

ACCURACY

Most sites have been mapped precisely by DNA sequencing. A few amber mutations could be assigned exactly by fine-structure recombinational mapping combined with patterns of suppression.

Sites that have been mapped less precisely are also indicated. Deletion, substitution and insertion endpoints determined by heteroduplex analysis were translated from the %λ map of Szybalski and Szybalski (1979) to the base-pair coordinate system by linear interpolation between landmark sites, as described by Daniels et al. (1980). Coordinates for such sites are preceeded by a tilde ~. The method used to determine each site is indicated in Table 1 (column G).

Gene endpoints are indicated on the map in italics (column D). Their most probable locations were determined from the sequence based on the presence of an open reading frame starting with an AUG or GUG, presence of a possible ribosome binding site (RBS; Shine and Delgarno 1974; Stormo 1982a,b), similarity to the codon usage of other λ genes (Staden and McLachlan 1982), and consistency with the %λ gene maps of Echols and Murialdo (1978) and Szybalski and Szybalski (1979). In several cases the gene assignments were confirmed by determination of the first few aminoterminal amino acids (see Table 4).

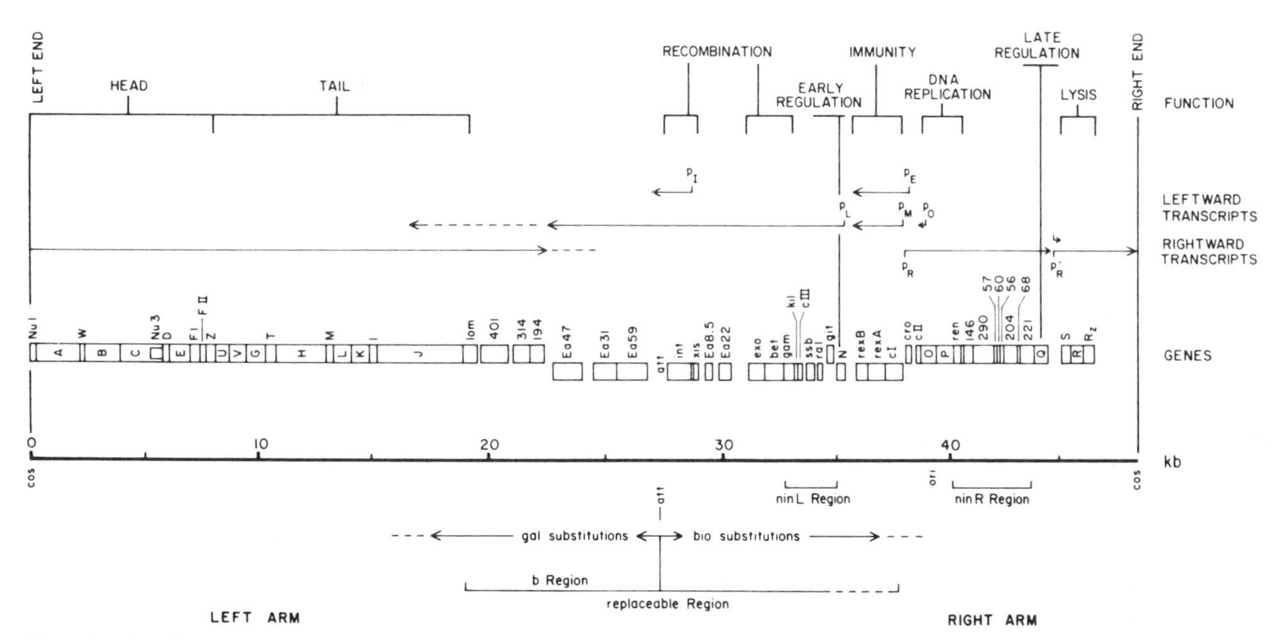

Figure 1. Map of bacteriophage λ DNA. A scale drawing of the molecular map presented in Table 1 is presented above the kilobase scale, beginning and ending at the choesive end site (*cos*). Gene clusters with related functions are indicated above the brackets, and the regulatory genes, *N* (involved in earlier regulation) and *Q* (involved in late regulation) are indicated by vertical lines. Known promoters are denoted by *p* with a subscript to indicate their unique points of origin. (p_I) *int* protein promoter; (p_E) establishment promoter for *c*I; (p_M) maintenance promoter for *c*I; (p_L) major leftward promoter; (p_R) major rightward promoter; (p_O) *oop* promoter; (p_R') late promoter. (→) Extent and direction of transcription; (----) readthrough. Map positions of genes as determined by analysis of open reading frames (ORFs) in the DNA sequence are indicated. Known genes (identified either genetically by mutation analysis or functionally by SDS-gel electrophoresis of protein product, or both) are indicated with letter names, whereas ORFs, presumed from the sequence to code for protein products but with no previously known gene assignments, are given numbers corresponding to the coding capacity of the ORF. Major areas of substitution and deletion mutations are indicated below the scale. (*att*) attachment site; (*ori*) origin of replication.

Key to Table 1:

Columns:

(A) Restriction sites
(B) Sequenced by more than one laboratory (see Table 2) (between arrowheads - ∇ and Δ)
(C) Base-pair coordinates (a tilde (~) means the site is approximate
(D) Gene or site abbreviation (gene end points are in italics.)
(E) Name of the site
(F) Site description (see below)
(G) Methods of mapping (see below)
(H) Comments

Site description (F):

dl	:Left end of a deletion
dr	:Right end of a deletion
sl	:Left end of a substitution
sr	:Right end of a substitution
dl-*att*	:Left end of a deletion whose right end is at *att*
att-dr	:The right end of a deletion whose left end is at *att*
sl-*att*	:Left end of a substitution whose right end is at *att*
att-sr	:Right end of a substitution whose left end is at *att*
i	:Insertion
5'	:First base (startpoint) of an RNA transcript
3'	:Last base (endpoint) of an RNA transcript
bs	:Protein binding site
pm	:Point mutation
M-NH3	:Amino terminus of an orf beginning with AUG
V-NH3	:Amino terminus of an orf beginning with GUG
COOH	:Carboxy terminus of an orf; corresponds to the last base of the coding sequence
hy	:λ::lambdoid junction in hybrid phages; the genotype of the hybrid and the particular novel junction mapped (::) are specified in Comments
plasmid-*l*	:Left endpoint on the λ map for the λdv plasmid
plasmid-*r*	:Right endpoint on the λ map for the λdv plasmid
f(*l*)	:Left end of a site possible significance for biological function
f(c)	:Center of a site of possible significance for biological function

Methods of Mapping (G):

sq	Wild-type or mutant site sequenced
RNA	RNA end determined by sequencing or fine structure S1 mapping of the RNA as well as promoter or terminator-like characteristics in the sequence.
AA	Amino acids at N terminus of a protein were sequenced
G	Genetic analysis - gene assigned to a particular orf or start point based on genetic evidence that gene spans a sequenced site (e.g., restriction site, point mutation, insertion site, etc.); genetic analysis of point mutations, fine structure mapping or suppression phenotypes, which allow accurate assignment to the sequence; genetic mapping of a biologically significant site to within a limited region on the sequence
RES	Restriction mapping of mutations which modify the restriction pattern
EM	Mapping by electron microscopy of heteroduplexes
f	Site assumed from features of the sequence such as homology to other such sites and/or mapping of a biological function to the vicinity of the sequence feature; position of the presumed left end (*l*) or center (c) of the sequence is specified.

REFERENCES

Abraham, J., D. Mascarenhas, R. Fisher, M. Benedik, A. Campbell and H. Echols. 1980. *Proc. Natl. Acad. Sci. USA* 77:2477-1481.

Bailone, A., and R. Devoret. 1978. *Virology* 84:547-.

Blattner, F. R. and J. E. Dahlberg. 1972. *Nature, New Biol* . 237:227-232.

Bienkowska-Szewczyk, K., B. Lipinska and A. Taylor. 1981. *Mol. Gen. Genet.* 198:111-114.

Calva, E., and R. R. Burgess. 1980. *J. Biol. Chem.* 225:11017-11022.

Court, D., C. Brady, M. Rosenberg, D. L. Wulff, M. Behr, M. Mahoney, and S. Izumi. 1980. *J. Mol. Biol.* 138:231-254.

Court, D., U. Schmeissner, M. Rosenberg, A. Oppenheim, G. Guarneros, and C. Montanez. 1984. *J. Mol. Biol.* (in press)

Dahlberg, J. E., and F. R. Blattner. 1973. *Proceedings of 1973 ICN-UCLA Symposium on Molecular Biology* (Ed. C. F. Fox), Academic Press, New York, NY pp. 533-544.

Dahlberg, J. E., and F. R. Blattner. 1975. *Nuc. Acids Res.* 2:1441-1458.

Daniels, D. L. 1981. *PhD. Thesis*, University of Wisconsin, Madison, WI U.S.A.

Daniels, D. L., and F. R. Blattner. 1982. *Virology* . 117:81-92.

Daniels, D. L., A. R. Coulson, and F. Sanger, 1983. *Cold Spring Harbor Symp. Quant. Biol.* . 47:1009-1024.

Daniels, D. L., J. R. deWet and F. R. Blattner. 1980. *J. Virol.* 33:390-400.

Daniels, D. L., J. L. Schroeder, W. Szybalski and F. R. Blattner. 1982b. In *Genetic Maps* S.J.O'Brien (Ed.) Vol. 2. p. 1 Cold Spring Harbor Laboratory, New York.

Daniels, D. L., J. L. Schroeder, W. Szybalski, F. Sanger and F. R. Blattner. 1983. *Lambda II*, p 469-517.

Daniels, D. L., J. L. Schroeder, W. Szybalski, F. Sanger, A. Coulson, G-F. Hong, G. P. Peterson, and F. R. Blattner. 1983. *Lambda II*. Cold Spring Harbor Lab p 519-676.

Davies, R. W. 1980. *Nuc Acids Res* 8:1765-1782.

Davies, R. W., P. H. Schreier and D. E. Buchel. 1978. *Nuc Acids Res* 5:3209.

Davies, R. W., P. H. Schreier and D. E. Buchel. 1977. *Nature* 270:757-760.

Denniston-Thompson, K., D. D. Moore, K. E. Kruger, M. E. Furth and F. R. Blattner. 1977. *Science* 198:1051-1056.

Drahos, D., G. R. Galluppi, M. Caruthers, and W. Szybalski. 1982. *Gene* 18:343-354.

Drahos, D., and W. Szybalski. 1981. *Gene* 16:261-274.

Echols, H., and H. Murialdo. 1978. *Microbiol. Rev.* 42:577-591.

Epp, C. 1978. *PhD Thesis*, University of Toronto, Toronto, Canada.

Flashman, S. M. 1976. PhD Thesis, Harvard University, Cambridge, MA (USA)

Flashman, S. M., 1978. *Molec Gen Genet* 166:61-73.

Franklin, N. C., and G. N. Bennett. 1979. *Gene* 8:107-119.

Garrett, J. M., and R. Young. 1982. *J. of Virology* 44(3):886-892.

Grayhack, E. J. and J. W. Roberts. 1982. *Cell* 30:637-648.

Grosschedl, R., and E. Schwarz. 1979. *Nucl. Acids Res.* 6:867-881.

Guarente, L., J. S. Nye, A. Hochschild and M. Ptashne. 1982. *Proc. Natl. Acad. Sci. USA* 79:2236

Guarneros, G., C. Montanez, T. Hernandez and D. Court. 1982. *Proc. Natl. Acad. Sci. USA* 79:238-242.

Halling, S. M., R. W. Simons, J. C. Way, R. B. Walsh and N. Kleckner. 1982. *Proc. Natl. Acad. Sci. USA* 79:2680-2612.

Hawley, D. K., 1982. *PhD Thesis*, Harvard University, Cambridge, MA (USA)

Hecht, M. H., H. C. M. Nelson, and R. T. Sauer. 1983. *Proc. Natl. Acad. Sci., USA* (in press)

Ho, Y-S., Lewis, M., and Rosenberg, M. 1982. *J. Biol. Chem.* 257:9128-9134.

Ho, Y-S., and Rosenberg, M., 1982. *Ann. Microb. (Paris)* 133:215-218.

Hobom, G., R. Grosschedl, M. Lusky, G. Scherer, E. Schwarz, and H. Kossel. 1979. *Cold Spring Harbor Symp. Quant. Biol.* 43:165-178.

Hoess, R., and A. Landy. 1978. *Proc. Natl. Acad. Sci. USA* 75:5437-5441.

Hoess, R. H., C. Foeller, K. Bidwell, and A. Landy. 1980. *Proc. Natl. Acad. Sci. USA* 77:2482-2486.
Horn, G. T., and R. D. Wells. 1981. *J. Biol. Chem.* 256:1988-2202.
Hsiang, M. W., and Y. Takeda. 1977. *Nature* 270:275-277.
Hsu, P.-L., W. Ross and A. Landy. 1980. *Nature* 285:85-91.
Ikeda, H., K. Aoki, and A. Naito. 1982. *Proc. Natl. Acad. Sci. USA* 79:3724-3728.
Imada, M., and A. Tsugita. 1971. *Nature New Biol.* 233:230-231.
Ineichen, K., J. C. W. Shepherd and T. A. Bickle. 1981. *Nucl. Acids Res.* 9:4639-4653.
Kleckner, N., 1979. *Cell* 16:711-720.
Kleid, D., Z. Humayun, A. Jeffrey, and M. Ptashne. 1976. *Proc. Natl. Acad. Sci. USA* 73:293-297.
Kleid, D. G., K. L. Agarwal, and Khorana, H. G. 1975. *J. Biol. Chem.* 250:5574-5582.
Knight, D. M. and H. Echols. 1983. *J. Mol. Biol.* 163:505
Krö ger, M., and G. Hobom. 1982. *Gene* 20:25-38.
Landsmann, J., M. Krö ger and G. Hobom. 1982. *Gene* 20:11-24.
Landy, A., and W. Ross. 1977. *Science* 197:1147-1159.
Lebowitz, P., S. M. Weissman and C. M. Radding. 1971. *J. Biol. Chem.* 246:5120-5139.
Legerski, R. J., J. L. Hodnett and H. B. Gray, Jr. 1978. *Nucl. Acids Res.* 5:1445-1464.
Lieb, M. 1981. *Mol. Gen. Genetics* 184:364-371.
Luk, K.-C., P. Dobrzański and W. Szybalski. 1982. *Gene* 17 : 259-262.
Luk, K.-C., and W. Szybalski. 1982a. *Gene* 17: 247-258.
Luk, K-C. and W. Szybalski. 1982b. *Gene* 20:127-134.
Luk, K.-C., and W. Szybalski. 1983a *Virology* 125:403-418.
Luk, K.-C. and W. Szybalski. 1983b. *Gene* 21:175-191.
Maniatis, T., M. Ptashne, K. Backman, D. Kleid, S. Flashman, A. Jeffrey, and R. Maurer. 1975. *Cell* 5:109-113.
Mayer, H., R. Grosschedl, H. Schutte and G. Hobom. 1981. *Nucl. Acids Res.* 9:4833-4845.
Meyer, B. J., D. G. Kleid, and M. Ptashne. 1975. *Proc. Natl. Acad., Sci. USA* 72:4785-4789.
Meyer, B. J., R. Maurer, and M. Ptashne. 1980. *J. Mol. Biol.* 139:163-194
Moore, D. D., and Blattner, F. R., 1982. *J. Mol. Biol.* 154:81-83.
Moore, D. D., K. Denniston-Thompson, K. E. Kruger, M. E. Furth, B. G. Williams, D. L. Daniels and F. R. Blattner. 1979. *Cold Spring Harbor Symp. Quant. Biol.* 43(1):155-164.
Moore, D. D., K. Denniston-Thompson, and F. R. Blattner. 1981. *Gene* 14: 91-101.
Muster, C. J., L. A. MacHattie and J. A. Shapiro. 1983. *J. of Bact.* 153:979-990.
Nichols, B. P., and J. E. Donelson. 1978. *J. Virol.* 26:429-434.
Petrov, N. A., V. A. Karguinov, N. N. Mikriukov, O. T. Serpinski and V. V. Kravchenko. 1981. *FEBS Letters* 133:316-320.
Ptashne, M., K. Backman, M. Z. Humayun, A. Jeffrey, R. Maurer, B. Meyer and R. T. Sauer. 1976. *Science* 194:156-161.
Roberts, R. J. 1983. *Nucl. Acids Res.* 11:135-168.
Roberts, T. M., H. Shimatake, C. Brady and M. Rosenberg. 1977. *Nature* 270:274-275.
Robinson, L. H., and A. Landy. 1977. *Gene* 2:1-31.
Robinson, L.H., and A. Landy. 1977. *Gene* 2:33-54.
Rosen, E. D., J. L. Hartley, K. Matz, B. P. Nichols, K. M. Young, J. E. Donelson, and G. N. Gussin, 1980. *Gene* 11: 197-205.
Rosenberg, M., D. Court, H. Shimatake, C. Brady and D. L. Wulff. 1978. *Nature* 272:414-423.
Rosenberg, M., D. Court, H. Shimatake, C. Brady and D. L. Wulff. 1980. IN *The operon*, (J. A. Miller and W. S. Reznikoff, Eds.), Cold Spring Harbor Laboratory, Cold Spring Harbor, NY, pp 345-371.
Rosenvold, E. C., E. Calva, R. R. Burgess and W. Szybalski. 1980. *Virology* 107:476-487.
Ross, W., M. Shulman and A. Landy. 1982. *J. Mol. Biol.* 156:501-519.
Salstrom, J. S., and W. Szybalski. 1978. *J. Mol. Biol.* 124:195-221.

Sanger, F., A. R. Coulson, G.-F. Hong, D. F. Hill, and G. B. Petersen. 1982. *J. Mol. Biol.* 162:729-773.
Sauer, R. T. 1978. *Nature* 276:301-302.
Scherer, G. 1978. *Nucl. Acids Res.* 5:3141-3156.
Schmeissener, U., D. Couirt, H. Shimatake and M. Rosenberg. 1980. *Proc. Natl. Acad. Sci. USA.* 77:3191-3195.
Schmeissner, U., Court, D. McKenney, K. and Rosenberg, M., 1981. *Nature* 292: 173-175.
Schwarz, E., G. Scherer, G. Hobom and H. Kossel. 1978. *Nature* 272:410-414.
Schwarz, E., G. Scherer, G. Hobom and H. Kossel. 1980. *Biochem. Int.* 1:386-394.
Shaw, J. E., and H. Murialdo. 1980. *Nature* 283:30-35.
Shine, J., and J. Delgarno. 1974. *Proc. Nat. Acad. Sci. USA* 71:1342-1346.
Sklar, J. L. 1977. *PhD Thesis*, Yale University, New Haven, CT. USA
Smith, D. I., F. R. Blattner and J. Davies. 1976. *Nucl. Acids Res.* 3:343-353.
Smith, G. R., D. W. Schultz and J. M. Crasemann. 1980. *Cell* 19:785-793.
Smith, G. R., M. Comb, D. W. Schultz, D. L. Daniels and F. R. Blattner. 1981. *J. Virol.* 37:336-342.
Somasekhar, G., D. Drahos, J.S. Salstrom, and W. Szybalski. 1982. *Gene* 20:477-480.
Somasekhar, G. and W. Szybalski. 1983. *Feder. Proc.* 42:2038.
Sprague, K. U., D. H. Faulds, and G. R. Smith. 1978. *Proc. Natl. Acad. Sci., USA* 75:6182-6186.
Staden, R., and A. D. McLachlan. 1982. *Nucleic Acids Res.* 10:1.
Stormo, G. D., T. D. Schneider and L. M. Gold. 1982a. *Nucl. Acids. Res.* 10:2971-2996.
Stormo, G. D., T. D. Schneider, L. Gold, and A. Ehrenfeucht. 1982. *Nucl. Acids Res.* 10:2997-3011.
Szybalski, E. H. and W. Szybalski. 1979. *Gene* 7:217-270.
Toothman, P., and I. Herskowitz. 1979. *Virology* 102:133-146, 147-160, 161-171.
Tsurimoto, T., and K. Matsubara. 1981. *Nucl. Acid Res.* 9:1789-
Walker, T. E., A. D. Auffret, A. Caine, A. Garnett, P. Haniser, D. Hill. M. Saraste. 1982. *Eur. J. Biochem.* 123:253-260.
Walz, A., V. Pirrotta and K. Ineichen. 1976. *Nature* 262:665-669.
Wu, R., and E. Taylor. 1971. *J. Mol. Biol.* 57:491.
Wulff, D. L., M. Bener, S. Izumi, J. Beck, M. Mahoney, H. Shimatake, C. Brady, D. Court, and M. Rosenberg. 1980. *J. Mol. Biol.* 138: 209-230.
Zagursky, R. J., and J. B. Hayes. 1983. *Gene* 23:277-292.

ACKNOWLEDGEMENTS

This is paper No. 2825 of the Laboratory of Genetics, University of Wisconsin, Madison, supported by National Institutes of Health grants AI-18214, GM-21812, and GM-28252 (to F.R.B.) and a special grant from Cold Spring Harbor Laboratory. Figure 1 is from *Lambda II*, 1983 from Cold Spring Harbor Laboratory.

TABLE 1 A MOLECULAR MAP OF BACTERIOPHAGE LAMBDA

A	B	C	D	E	F	G	H
ZEND		0	ZEND	Zero END			used for fragment endpoints
	▽	1	SB1L	Sequence Block 1L	start		
	"	1	LEND	Left END		sq	first base of sticky end (l-strand)
	"	13	CLND	3' end of r-strand			first base of double stranded DNA
	△	72	SB1L	Sequence Block 1L	end		
		191	*NU1*	Nu1 (orf-181)	M-NH3	G	
		~194	NU16	Nu1 t16	i	EM	
HIN2		199					
PVU2		211					
BGL2		415					
		711	*A*	A (orf-641)	V-NH3		
		733	*NU1*	Nu1	COOH		
HIN2		734					
HPA1		734					
		~919	AROE	λaroE	sl	EM	
		~919	TRKA	λtrkA	sl	EM	
		~919	SPC2	λspc2	sl	EM	
		~919	TRKA	λtrkA	sl	EM	
		~919	SPC2	λspc2	sl	EM	
		~919	FUS2	λfus2	sl	EM	
		~919	SPC1	λspc1	sl	EM	
		~968	TP2	Tp2	i(Tn402)	EM	
		~1305	G258	λgal258	sl-att	EM	
		~1451	G914	λgalA914	sl-att	EM	
		~1838	GAL3	λgal3	sl-att	EM	
		~1886	G102	λgalT-N102	sl-att	EM	
PVU2		1919					
SPH1		2216					
PVU2		2387					
PVU2		2528					
PST1		2560					
		2633	*W*	W (orf-68)	M-NH3		
		2633	*A*	A	COOH		
		2642	WAM	Wam403	pm	sq	C to T
PST1		2824					
		2836	*B*	B (orf-533)	M-NH3		
		2836	*W*	W	COOH		
		2902	B*	B*		AA	start processed protein of capsid
		~2950	RF18	λrif-18	sl-att	EM	
		~2950	RF20	λrif-20	sl-att	EM	
PVU2		3060					
PST1		3629					
PVU2		3639					
PST1		3644					
		~3675	G24	λgal24	sl-att	EM	
PST1		3860					
		~4111	S104	λgalS104	sl-att	EM	
CLA1		4199					
		~4255	S101	λgalS101	sl-att	EM	
PST1		4374					
		4418	*C*	C (orf-439)	V-NH3		
		4434	*B*	B	COOH		
		~4544	GALC	λgalC	sl-att	EM	
PST1		4713					
AVA1		4720					
PST1		4913					
PST1		5124					
		5132	*NU3*	Nu3 (orf-201)	M-NH3	AA	
PST1		5218					
HIN2		5269					
HPA1		5269					
BAM1		5505					
PST1		5686					
HIN2		5710					
HPA1		5710					
		5734	*NU3*	Nu3	COOH		
		5734	*C*	C	COOH		
		5747	*D*	D (orf-110)	M-NH3	AA	
		6076	*D*	D	COOH		
		6135	*E*	E (orf-341)	M-NH3		
		~6287	KN11	kan11	i	EM	
		7157	*E*	E	COOH		
		7202	*FI*	FI (orf-132)	M-NH3		
		~7497	G730	λgalF730	sl-att	EM	
		7597	*FI*	FI	COOH		
		7612	*FII*	FII (orf-117)	V-NH3	AA	
PVU2		7833					
HIN2		7950					
HPA1		7950					
		7962	*FII*	FII	COOH		
		7977	*Z*	Z (orf-192)	M-NH3		
		~8125	GALZ	λgalZ	sl-att	EM	
HIN2		8201					
HPA1		8201					
PST1		8524					
		8552	*U*	U (orf-131)	M-NH3		
		8552	*Z*	Z	COOH		
BCL1		8844					masked by dam methylation
		8944	*U*	U	COOH		
		8955	*V*	V (orf-246)	M-NH3	AA	
HIN2		9056					
		~9335	GALG	λgalG	sl-att	EM	
BCL1		9361					masked by dam methylation
		~9577	TRKA	λtrkA	sr	EM	
PST1		9617					
HIN2		9626					
		9692	*V*	V	COOH		
		9711	*G*	G (orf-140)	M-NH3		
PST1		9781					
		10115	*T*	T (orf-144)	V-NH3		
		10130	*G*	G	COOH		
		10542	*H*	H (orf-853)	M-NH3		
		10546	*T*	T	COOH		
		~10883	H706	λgalH706	sl-att	EM	
HIN2		11585					
HPA1		11585					
PST1		11767					
PST1		11839					
PVU1		11937					
SPH1		12006					
PVU2		12101					
PVU2		12164					
		13100	*M*	M (orf-109)	M-NH3		
		13100	*H*	H	COOH		
		13426	*M*	M	COOH		
		13429	*L*	L (orf-232)	M-NH3		
		~13743	GALK	λgalK	sl-att	EM	
HIN2		13785					
BCL1		13820					masked by dam methylation
		14124	*L*	L	COOH		
		14276	*K*	K (orf-199)	M-NH3		
PST1		14298					
PST1		14385					
		14773	*I*	I (orf-223)	M-NH3		protein not identified
	▽	14869	SB2	Sequence Block 2	start		
	"	14872	*K*	K	COOH		
HIN2	"	14993					masked by EcoK modification
HPA1	"	14993					masked by EcoK modification
	"	15441	*I*	I	COOH		
	"	15505	*J*	J (orf-1132)	M-NH3		
CLA1	"	15584					masked by dam methylation
PVU2	"	16080					
PST1	"	16085					
CLA1	"	16121					
PST1	"	16235					
KPN1	"	17057					
HIN2	"	17076					
PST1	"	17394					
	"	18482	T'J4	t'j4 terminator	3'	RNA	terminates leftward transcription
KPN1	"	18560					
	"	18597	T'J3	t'j3	3'	RNA	terminates leftward transcription
	"	~18630	H434	h434	sl	EM	
	"	18637	T'J2	t'j2	3'	RNA	terminates leftward transcription
	"	18671	T'J1	t'j1	3'	RNA	terminates leftward transcription
HIN2	"	18756					
	"	18900	*J*	J	COOH		
	"	18965	*LOM*	lom (orf-206)	M-NH3		
	△	18969	SB2	Sequence Block 2	end		
		~19066	H434	h434	sr	EM	
		19369	LAC5	lac5	sl	sq	
AVA1		19397					
SMA1		19399					
		19582	*LOM*	lom	COOH		

A	B	C	D	E	F	G	H
		~19646	B221	b221	dl	EM	
		~19646	B519	b519	dl	EM	
		19650	*ORF*	orf-401	M-NH3		
		~19700	B189	b189	dl-att	RES	
PVU2		19718					
PST1		19837					
HIN2		19841					
		~19985	PGL1	λpgl1	sl-att	EM	
PVU2		20061					
PST1		20285					
SST2		20323					
		~20420	B536	b536	dl	EM	
		~20469	G130	λgal130	sl-att	EM	
SST2		20533					
HIN2		20569					
PVU2		20697					
		~20809	B538	b538	dl	EM	
		20852	*ORF*	orf-401	COOH		
AVA1		20999					
		21029	*ORF*	orf-314	M-NH3		
ECR1		21226					
		~21292	B506	b506	dl	EM	
SST2		21609					
	▽	21661	SB3	Sequence Block 3	start		
	"	21714	DIFF	strain difference		sq	A in cb2 G in cI857S7
	△	21737	SB3	Sequence Block 3	end		
		21738	B2	b2	dl-att	sq	
		~21825	REV	λrev	sl	EM	
HIN2		21904					
HPA1		21904					
		21970	*ORF*	orf-314	COOH		
		21973	*ORF*	orf-194	M-NH3		
BAM1		22346					
BGL2		22425					
PST1		22425					
		~22530	C105	λcam105	i(Tn9)	RES	dl for deletions pMS370;968;969;972
		22554	*ORF*	orf-194	COOH		
		22689	*EA47*	Ea47	COOH		
		~22840	B536	b536	dr	EM	
PVU2		22993					
		~23033	B519	b519	dr	EM	
		~23082	AROE	λaroE	sr	EM	
HIN3		23130					
	▽	23131	SB4	Sequence Block 4	start		
HIN2	"	23147					
	"	23231	PBL	pBL	5'	RNA	leftward promoter in b region
	△	23269	SB4	Sequence Block 4	end		
		~23278	LAC5	lac5	sr	EM	
		~23472	SPC2	λspc2	sr	EM	
	▽	23495	SB5	Sequence Block 5	start		
	△	23548	SB5	Sequence Block 5	end		
		23549	B511	b511	dl-att	sq	
		~23616	1007	b1007	dl	EM	
		23918	*EA47*	Ea47 (orf-410)	M-NH3		
SPH1		23946					
		~24003	B515	b515	dl	EM	
		~24020	C107	cam107	i(Tn9)	EM RES	dl for deletions pXJS; pQLC 137-144
		~24197	B506	b506	dr	EM	
SPH1		24375					
		~24390	HYF	hyF	hy	EM	λ::att80imm80 OP80 QSR80
		~24390	HY42	hy42	hy	EM	λ::att80imm80 O80 PQSRλ
		~24390	HYM	hyM	hy	EM	λ::att80immλOPλQSR80
		~24390	T60	λtrp60-3	sl	EM	
XBA1		24508					partially blocked by dam methylation
		24512	*EA31*	Ea31	COOH		
SST1		24776					
		~24924	CAM3	cam3	i(Tn9)	EM	
HIN3		25157					
		25399	*EA31*	Ea31 (orf-296)	M-NH3		
		25399	*EA59*	bEa59	COOH		
		~25843	B515	b515	dr	EM	
SST1		25881					
		~25891	CAM1	cam1	i(Tn9)	EM	
ECR1		26104					
PVU1		26258					
CLA1		26617					
		~26714	1018	b1018	dl	EM	
MST2		26718					
HIN2		26744					
PST1		26932					

A	B	C	D	E	F	G	H
		26973	*EA59*	bEa59 (orf-525)	M-NH3		
HIN2		27318					
HPA1		27318					
SPH1		27378					
PVU2		27414					
	▽	27479	SB6	Sequence Block 6	start		
HIN3	"	27479					
	"	~27488	SD	sd	i	EM	
	"	~27488	SPC1	λspc1	sr	EM	
	"	27514	D106	Δ106	dr	sq	BAL31 from Hin3; is Sib+
	"	27534	D119	Δ119	dr	sq	BAL31 from Hin3; is Sib+
	"	27537	SIB	sib3	pm	sq	C to T
	"	27538	TI	tI terminator	3'	RNA	end of int message
	"	27546	D125	Δ125	dr	sq	BAL31 from Hin3; is Sib-
	"	27547	HEF	hef13	pm	sq	G to A
	"	27562	D112	Δ112	dr	sq	BAL31 from Hin3; is Sib-
	"	27568	SIB	sib2	pm	sq	C to T
	"	27570	PH53	PHΔ53;54	dr	sq	ExoIII-S1 from Hin3; is Sib- Att+
	"	27573	SIB	sib1	pm	sq	G to T
	"	27583	INT1	P arm 1	bs		int binds (DNase protection 20bp)
	"	27615	PH55	PHΔ55	dr	sq	ExoIII-S1 from Hin3; is Att-
	"	27615	INT2	P arm 2	bs		int binds (DNase protection 20bp)
	"	27714	INT3	core site	bs		int binds (DNase protection 34bp)
	"	27727	ATT-	att 2; 6; 24	pm	sq	att minus pm deletes T
	"	27731	ATT	att	f(c)	sq f	center of common core
	"	27772	B936	bio936	att-sr	sq	
	"	27779	INT4	P' arm	bs		int binds (DNase protection 41bp)
	"	27815	*INT*	int	COOH		
AVA1	"	27887					
	"	27896	501	att501	dr	sq	att- int-
BAM1	"	27972					
	"	~28312	1007	b1007	dr	EM	
	"	~28506	B16A	bio16A	att-sr	EM	
	"	~28603	RD10	Δ red10	dl	EM	
	"	~28640	MGB	λ MGB	i(Tn9)	RES	dl for deletions pQLC 95-118
	"	~28845	2034	bio2034	att-sr	EM	
	"	28863	*XIS*	xis	COOH		
	"	28882	*INT*	int (orf-356)	M-NH3		
	"	~28894	2037	bio2037	att-sr	EM	
HIN2	"	28928					
	"	28975	T841	trp Δ 841	sl	sq	
	"	~28991	I548	int-c548	i(IS2)	EM	
	"	29063	XIS6	xis am6	pm	sq	G to A
	"	29065	SI	sI pI	5'	RNA	start of int message sq
	"	29076	INTC	int-c226;262;518	pm	sq	C to T
	"	29078	*XIS*	xis (orf-72)	M-NH3		
	"	~29088	IC57	int-c57	i(IS2)	EM	
	"	~29088	I60	int-c60	i(IS2)	EM	
	"	~29088	I508	int-c508	i(IS2)	EM	
	"	29091	C2	cII site	bs	f	cII binds -35 region of pI
	"	~29093	B538	b538	dr	EM	
	"	29094	T303	trp Δ 303	sl	sq	
	"	29126	T29	trp Δ 29	sl	sq	
	"	~29185	B386	bio386	att-dr	EM	
	"	29377	*E8.5*	Ea8.5	COOH		
	"	~29379	C842	crg 842	i(IS2)	EM	
	"	~29427	RED1	Δ red1	dl	EM	
	"	29655	*E8.5*	Ea8.5 (orf-93)	M-NH3		
	"	29711	B508	b508	att-dr	sq	
	△	29811	SB6	Sequence Block 6	end		
		~29815	BI2	bi2	i(IS2)	EM	
		29850	*EA22*	Ea22	COOH		
CLA1		30290					
		~30348	7-20	bio7-20	att-sr	EM	
		30395	*EA22*	Ea22 (orf-182)	V-NH3		
		~30397	B122	bio122	dr	EM	
		~30445	B221	b221	dr	EM	
	▽	30493	SB7	Sequence Block 7	start		
	"	~30520	RD15	Δ red15	dl	EM G	
	"	30529	CHIB	chiB	pm	sq	121 delete C; 131 C to G
	△	30569	SB7	Sequence Block 7	end		
	▽	31043	SB8	Sequence Block 8	start		
	"	31043	B522	b522	att-dr	sq	
	"	31262	TL3	tL3 terminator	3'	RNA	
	"	~31267	HY1	Hy1	hy	EM	λ::imm80 OP 80 QSR80
	"	~31267	HY5	Hy5	hy	EM	80::immλ OPλ QSR80
	"	~31267	HYM	HyM	hy	EM	λatt80::immλOPλQSR80
	"	~31267	RED1	Δ red1	dr	EM	
	"	31351	*EXO*	exo	COOH		
	"	~31412	1018	b1018	dr	EM	

A	B	C	D	E	F	G	H
	"	~31412	1451	b1451	dl	EM	
	"	~31461	KAN3	kan3	i(Tn6)	EM	
AVA1	"	31617					
SMA1	"	31619					
ECR1	"	31747					
HIN2	"	31809					
HPA1	"	31809					
	"	~31897	1451	b1451	dr	EM	
CLA1	"	31991					masked by dam methylation
	"	~31994	B72	bio72	att-sr	EM	
	"	~31994	B69	bio69	att-sr	EM	
PST1	"	32009					
	"	32028	*EXO*	exo (orf-226)	M-NH3		
	"	32028	*BET*	bet	COOH		
	"	~32042	B169	bio169	att-sr	EM	
HIN2	"	32219					
HPA1	"	32219					
	"	~32236	RD10	Δ red10	dr	EM	
PST1	"	32256					
	"	~32527	B267	bio267	att-sr	EM	
	"	~32624	B74	bio74	att-sr	EM	
BCL1	"	32729					masked by dam methylation
SAL1	"	32745					
HIN2	"	32747					
	"	32810	*BET*	bet (orf-261)	M-NH3		
	"	~32818	RD15	Δ red15	dr	EM	
	"	32819	*GAM*	gam	COOH		
	"	~32867	NL4	ninL4	dl	EM	
	"	~32915	1453	b1453	att-dr	EM	
CLA1	"	32964					masked by dam methylation
	"	~32964	B11	bio11	att-sr	EM	
	"	~33012	BIO1	bio1	att-sr	EM	
	"	33035	DIFF	difference		sq	Ineichen et al. (1981) G(not C)
	"	33100	TL2D	tL2d terminator	3'	RNA	
	"	~33109	NL30	ninL30	dl	EM	
	"	33141	TL2C	tL2c terminator	3'	RNA	RNA ends at this T and at next A
	"	33190	*KIL*	kil	COOH		
	"	33232	*GAM*	gam (orf-138)	M-NH3	G	
SAL1	"	33244					
HIN2	"	33246					
	"	33302	*CIII*	cIII	COOH		
	"	~33303	B275	bio275	att-sr	EM	
	"	~33303	1319	b1319	att-dr	EM	
	"	~33303	NL63	ninL63	dl	EM	
	"	33330	*KIL*	orf-47 (kil?)	M-NH3		protein not identified
	"	33441	C611	cIIIam611	pm	sq	C to T
	"	~33449	NL20	ninL20	dl	EM	
	"	~33449	NL44	ninL44	dl	EM	
	"	~33449	B252	bio252	att-sr	EM	
	"	33463	*CIII*	cIII (orf-54)	M-NH3	G	
	"	33494	TL2B	tL2b terminator	3'	RNA	
	"	~33497	B250	bio250	att-sr	EM	
AVA1	"	33498					
XHO1	"	33498					
	"	33539	*EA10*	Ea10	COOH		
	"	~33546	V203	dv203	plasmid-1	EM	
CLA1	"	33585					
	"	~33739	NL8	ninL8	dl	EM	
	"	~33885	NL88	ninL88	dl	EM	
	"	33904	*EA10*	Ea10 (orf-122)	M-NH3		
	"	33930	TL2A	tL2a terminator	3'	RNA	
	"	~34055	B214	bio214	att-sr	EM	
	"	34090	*RAL*	ral	COOH		
	"	34104	TP44	trp44	dr	sq	
	"	~34127	IMML	immL	sl	EM	
	"	~34224	KAN1	kan 1	i(Tn5)	EM	
	"	~34249	B10	bio10	att-sr	EM	
	"	34287	*RAL*	ral (orf-66)	M-NH3		
MST2	"	34319					
	"	~34370	REV	λrev	sr	EM	
	"	34379	I21	imm21	sl	sq	
	"	34497	*GIT*	git sieB	M-NH3		
BAM1	"	34499					
	"	34560	TL1	tL1 terminator	3'	RNA	
	"	34640	NL99	ninL99	dl	sq	N independent to the left
	"	~34661	B243	bio243	att-sr	EM	
CLA1	"	34697					
	"	~34758	B233	bio233	att-sr	EM	
	"	~34758	T60	trp60-3	sr	EM	
	"	~34855	B232	bio232	att-sr	EM	

A	B	C	D	E	F	G	H
	"	~34855	NL63	ninL63	dr	EM	
	"	~34952	NL4	ninL4	dr	EM	
	"	34969	TP48	trp48	sl	sq	
	"	35035	*GIT*	git sieB	COOH		
	"	35040	*N*	N	COOH		
	"	~35049	NL44	ninL44	dr	EM	
CLA1	"	35051					
	"	35060	NL99	ninL99	dr	sq	
	"	~35097	V154	dv 154	plasmid-1	EM	
	"	~35100	B256	bio256	att-sr	EM	
	"	~35100	Q115	pλQLC115	dr	RES	
	"	35116	AM22	Nam22	pm	sq	C to T
	"	~35120	Q108	pλQLC108	dr	RES	
	"	~35130	Q118	pλQLC118	dr	RES	
	"	~35140	Q109	pλQLC109	dr	RES	
	"	~35150	Q105	pλQLC105	dr	RES	
	"	~35160	Q101	pλQLC101	dr	RES	
	"	~35160	Q143	pλQLC143	dr	RES	
	"	35165	AM7	Nam7	pm	sq	GT to AA
	"	~35180	1974	pλXJS1974	dr	RES	
	"	35191	MAR2	Mar2	pm	sq	G to A
	"	~35194	NL20	ninL20	dr	EM	
	"	~35194	NL8	ninL8	dr	EM	
	"	~35200	Q117	pλQLC117	dr	RES	
	"	~35200	2300	pλXJS2300	dr	RES	
	"	~35210	M972	pλMS972	dr	RES	
	"	~35210	QL98	pλQLC98	dr	RES	
	"	~35220	Q104	pλQLC104	dr	RES	
	"	35245	2521	Nam2521	pm	sq	G to T
HIN2	"	35261					
HPA1	"	35261					
	"	~35270	M370	pλMS370	dr	RES	same as pλCM1
	"	35287	AM53	Nam53	pm	sq	A to T
	"	~35290	1976	pλXJS1076	dr	RES	
	"	~35291	NL88	ninL88	dr	EM	
	"	~35330	M968	pλMS968	dr	RES	
	"	35310	2524	Nam2524	pm	sq	C to T
	"	35324	219	Nam219	pm	sq	C to A
	"	~35340	N2-1	bioN2-1	att-sr	EM	
	"	~35350	Q102	pλQLC102	dr	RES	
	"	35360	*N*	N (orf-107)	M-NH3		
	"	~35420	Q144	pλQLC144	dr	RES	
	"	~35440	QL95	pλQLC95	dr	RES	
	"	35465	DV1	dv1	plasmid-1	sq	
	"	~35485	NL32	ninL32	dr	EM	
	"	35518	NUTL	nutL	f(1)	f G	N utilization (17bp stem-loop)
	"	35528	NUT-	nutL63;96;18	pm	sq	63 C to A; 96 C to G; 18 C to T
	"	~35530	Q138	pλQLC138	dr	RES	
	"	35530	NUT-	nutL3	pm	sq	delete G
	"	~35570	Q103	pλQLC103	dr	RES	
	"	35576	3H-1	bio3H-1	att-sr	sq	
	"	35582	SL	sL pL	5'	RNA	
	"	35584	I4	imm434	sl	sq	
	"	35591	OL1	oL1	f(1) bs	f	17-bp diad symmetry cI binds
	"	35596	V2	vir2;v003	pm	sq	vir2 C to A; v003 C to T
	"	35606	V101	vir101	pm	sq	T to C
	"	35613	SEX1	sex1	pm	sq	G to A
HIN2	"	35615					
	"	35615	OL2	oL2	f(1) bs	f	17 bp cI binds cro binds
	"	35619	2668	vL2668	pm	sq	
	"	35620	V169	v169	pm	sq	
	"	35622	V305	v305	pm	sq	G to T
	"	35635	OL3	oL3	f(1) bs	f	17 bp cro binds
	"	35639	L668	1668	pm	sq	
BGL2	"	35711					
PVU1	"	35791					
	"	35804	TIMM	tIMM terminator	3'	f RNA	presumed 3' from pRE pRM plit
	"	35828	*REXB*	rexB	COOH		
	"	35848	KH54	KH54	dl	sq	
	"	35887	DIFF	difference		sq	Ineichen et al. (1981) delete A
	"	~35915	NL30	ninL30	dr	EM	
	"	35940	REX-	rex209	pm	sq	G to A
	"	35947	REX-	rex111	pm	sq	G to A
	"	~36010	M969	pλ969	dr	RES	
	"	~36010	24-5	bio24-5	att-sr	EM	
	"	~36153	KH67	KH67	dl	EM	
	"	~36153	16-3	bio16-3	att-sr	EM	
	"	36259	*REXB*	rexB (orf-144)	M-NH3		
	"	~36270	Q139	pλQLC139	dr	RES	
	"	36278	*REXA*	rexA	COOH		

A	B	C	D	E	F	G	H
	"	~36320	Q137	pλQLC137	dr	RES	
	"	36322	LIT	s lit p lit	5'	f RNA	start of lit (mis) RNA
	"	~36390	T75	bio t75	att-sr	EM	
	"	36398	KH70	KH70	dl	sq	
	"	~36520	Q111	pλQLC111	dr	RES	
	"	~36560	Q141	pλQLC141	dr	RES	
	"	~36670	Q142	pλ142	dr	RES	
	"	~36770	173	tet 173	i(Tn10)	EM	
	"	~36770	174	tet 174	i(Tn10)	EM	
	"	36817	PC1	pc 1	pm	sq	
	"	~36818	KAN2	kan2	i(Tn5)	EM	
	"	~36890	2299	pλXJS2299	dr	RES	
HIN3	"	36895					
CLA1	"	36966					
PST1	"	37005					
	"	~37010	1975	pλXJS1975	dr	RES	
	"	~37100	Q112	pλQLC112	dr	RES	
	"	~37110	1973	pλXJS1973	dr	RES	
	"	37114	*REXA*	rexA (orf-279)	M-NH3		
	"	~37150	Q106	pλQLC106	dr	RES	
	"	~37160	Q110	pλQLC110	dr	RES	
	"	37182	V021	dv021	plasmid-r	sq	
	"	~37200	Q107	pλQLC107	dr	RES	
	"	37230	*CI*	cI	COOH		
	"	37255	UA72	cI UA72	pm	sq	C to A
	"	37272	IC11	cI IC11	pm	sq	insert C
	"	37281	SP31	cI SP31	i (IS1)	sq	
	"	37287	AM14	cIam14	pm	G	C to A
	"	37293	NT1	cI NT1	pm	sq	G to T
	"	37308	504	cIam504	pm	G	G to C
	"	37309	BP83	cI BP83	pm	sq	T to C
	"	37313	505	cIam505	pm	G	G to A
	"	37320	K100	KH100	i(IS5)	sq	
	"	37330	UV59	cI UV59	pm	sq	A to G
	"	37333	UV63	cI UV63	pm	sq	delete A
	"	37334	IC28	cI IC28	pm	sq	insert A
BCL1	"	37352					masked by dam methylation
	"	37376	ET26	cI ET26	pm	sq	C to A
	"	37384	AA21	cI AA21	pm	sq	insert C
	"	37385	AA2	cI AA2	pm	sq	delete C
	"	37387	NT16	cI NT16	pm	sq	insert A
	"	37396	UV76	cI UV76	pm	sq	A to T
	"	37398	SP1	cI SP1	i (IS1)	sq	
	"	37412	NT15	cI NT15	pm	sq	delete G
	"	37422	TN10	171;172;473;474	i(Tn10)	sq	
	"	37423	13	cI 13	pm	sq	delete T
HIN2	"	37433					
	"	37442	UV64	cI UV64	pm	sq	GAAT to AAA
HIN3	"	37459					
	"	37531	UV73	cI UV73	pm	sq	GC to T
	"	37538	BL46	cI BL46	pm	sq	T to C
	"	~37579	KH67	KH67	dr	EM	
	"	~37579	30-7	bio 30-7	att-sr	EM	
(HIN3)		37584					created by ind1;as in most cI857
	"	37589	IND1	ind1	pm	sq	C to T lysogen not UV inducible
	"	37604	PF17	cI PF17	pm	sq	insert C
	"	37623	AA3	cI AA3	pm	sq	insert A
	"	37624	SP48	cI SP48	pm	sq	delete A
	"	~37627	V266	dv266	plasmid-l	EM	
	"	37629	ET39	cI ET39	pm	sq	G to T
	"	37629	499	cIam499	pm	G	G to C
	"	37630	B138	cI BP138	pm	sq	T to C
	"	37635	212	cIam212	pm	G	A to C
	"	37651	ET3	cI ET3	pm	sq	G to C
	"	37680	AM34	cIam34	pm	G	C to A
	"	37682	E86	cIamE86	pm	sq	C to A
	"	37687	54-1	cI54-1	pm	sq	A to C
	"	37690	UV52	cI UV52	pm	sq	T to C
	"	37691	ET15	cI ET15	pm	sq	C to A
	"	37694	R82	cIamR82	pm	sq	T to A
	"	37702	UV96	CI UV96	pm	sq	G to A
	"	37705	UV97	cI UV97	pm	sq	G to A
	"	37706	UV79	cI uv79	pm	sq	G to A
	"	37707	PF12	cI PF12;cI AA6	pm	sq	insert G; delete G
	"	37708	11-3	cI11-3;cICP175	pm	sq	11-3 C to T;CP175 C to A
	"	37709	60-1	cI60-1	pm	sq	T to G
	"	37710	UV47	cI UV47	pm	sq	A to T
	"	37711	5-3	cI5-3	pm	sq	A to G
	"	37718	U116	cI UV116;cI UV74	pm	sq	UV116 C to T;UV77 C to A
	"	37720	BP84	cI BP84	pm	sq	A to G

A	B	C	D	E	F	G	H
	"	37726	BU30	cI BU30	pm	sq	A to G
	"	37727	UV66	cI UV66	pm	sq	AC to TT
	"	37727	KH70	KH70	dr	sq	
	"	37733	SP44	cI SP44;cI SP37	pm	sq	SP44 G to A;SP37 G to T
	"	~37740	Q113	pλQLC113	dr	RES	
	"	37742	857	cI857	pm	sq	C to T cIts
	"	37744	63-1	cI63-1	pm	sq	A to C
	"	37746	ET7	cI ET7	pm	sq	C to A
	"	37756	BL71	cI BL71	pm	sq	T to C
	"	37762	UA8	cI UA8	pm	sq	G to T
	"	37765	UV61	cI UV61	pm	sq	T to A
	"	37768	DV93	dvh93	plasmid-l	sq	
	"	37768	UV62	cI UV62	pm	sq	A to G
	"	37768	L57	cIamL57	pm	sq	A to C
	"	37772	UV41	cI UV41	pm	sq	C to T
	"	37773	34-1	cI34-1	pm	sq	A to C
	"	37775	UV44	cI UV44	pm	sq	T to C
	"	37775	U117	cI UV117	pm	sq	T to G
	"	37780	NT17	cI NT17	pm	sq	C to A
	"	37780	26-2	cI26-2	pm	sq	C to T
	"	37781	UV91	cI UV91	pm	sq	CC to TT
	"	37781	UA33	cI UA33;cI SP43	pm	sq	UA33 C to A;SP43 C to T
	"	37781	CP32	cICP32	pm	sq	C to A
	"	37784	UV46	cI UV46	pm	sq	T to C
	"	37784	53-1	cI53-1	pm	sq	T to C
	"	37787	U108	cI UV108	pm	sq	A to T
	"	37789	UV37	cI UV37	pm	sq	A to G
	"	37789	L50	cIamL50	pm	sq	A to C
	"	37791	NT3	cI NT3	pm	sq	deltete A
	"	37792	CP-9	cICP-9;cI29-6	pm	sq	CP9 G to T; 29-6 G to A
	"	37798	BL42	cI BL42	pm	sq	A to G
	"	37801	SP38	cI SP38	pm	sq	C to T
	"	37804	37-1	cI37-1	pm	sq	G to A
	"	37805	B102	cI BP102	pm	sq	A to G
	"	37806	TN10	475	i(Tn10)	sq	
	"	37806	UA68	cI UA68	pm	sq	C to A
	"	37807	BL70	cI BL70	pm	sq	T to C
	"	37807	40-3	cI40-3	pm	sq	T to A
	"	37808	282	cIam282	pm	G	G to A
	"	37808	NT25	cI NT25;cI UA77	pm	sq	NT25 G to A;UA77 G to T
	"	37808	Q44	cIamQ44	pm	sq	G to A
	"	37810	ET22	cI ET22	pm	sq	C to G
	"	37810	38-2	cI38-2	pm	sq	C to T
	"	37811	SP61	cI SP61	pm	sq	C to T
	"	37812	PF2	cI PF2;cI AA13	pm	sq	PF2 insert C;AA13 delete C
	"	37813	SP28	cI SP28	pm	sq	A to G
	"	37813	9-1	cI9-1	pm	sq	A to G
	"	37814	B120	cI BU120	pm	sq	T to C
	"	37816	SP41	cI SP41	pm	sq	C to A
	"	~37817	T124	bio t124	att-sr	EM	
	"	37817	UV86	cI UV86	pm	sq	C to T
	"	37818	AA13	cI AA13	pm	sq	delete C
	"	37819	CP7	cICP7	pm	sq	A to T
	"	37826	PC2	pc 2	pm	sq	
	"	37828	SP5	cI SP5	pm	sq	G to A
	"	37831	BU39	cI BU39	pm	sq	A to G
	"	37835	12-1	cI12-1	pm	sq	A to G
	"	37838	PC3	pc 3	pm	sq	
	"	37840	BL51	cI BL51	pm	sq	T to C
	"	37841	SP3	cI SP3	pm	sq	G to A
	"	37841	Q33	cIamQ33	pm	sq	G to A
	"	37844	UA6	cI UA6;cI BL89	pm	sq	UA6 A to Y;BL89 A to G
	"	37844	47-1	cI47-1	pm	sq	A to G
	"	37846	TN10	408	i(Tn10)	sq	
	"	37846	UV14	cI UV14	pm	sq	A to G
	"	37846	L31	cIamL31	pm	sq	A to C
	"	37846	4-3	cI4-3	pm	sq	A to G
	"	37852	BL84	cI BL84	pm	sq	A to G
	"	37852	1-1	cI1-1	pm	sq	A to G
	"	37860	SP42	cI SP42	pm	sq	delete T
	"	37861	UV80	cI UV80	pm	sq	insert TT
	"	37862	UV57	cI UV57	pm	sq	T to C
	"	37863	UV88	cI UV88	pm	sq	C to TTT
	"	37865	ET28	cI ET28	pm	sq	T to G
	"	37870	ET13	cI ET13	pm	sq	insert T
	"	37872	302	cIam302	pm	G	A to C
	"	37872	56-1	cI56-1;cI5-1	pm	sq	56-1 A to C;5-1 A to G
	"	37873	32-1	cI32-1;cI29-7	pm	sq	32-1 T to G;29-7 T to C
	"	37882	UV85	cI UV85	pm	sq	T to C
	"	37883	SP39	cI SP39;cI BL80	pm	sq	SP39 T to G;BL80 T to C

A	B	C	D	E	F	G	H
	■	37886	U112	cI UV112;cI UV93	pm	sq	UV112 G to A;UV93 G to Y
	■	37886	36-1	cI36-1	pm	sq	G to A
	■	37894	BL97	cI BL97	pm	sq	G to A
	■	37894	SP52	cI SP52	i(IS1)	sq	
	■	37894	SP60	cI SP60	i(IS1)	sq	
	■	37894	177	cI CP177	pm	sq	G to T
	■	37896	SP51	cI SP51	pm	sq	G to C
	■	37897	BP80	cI BP80	pm	sq	T to C
	■	37898	UV23	cI UV23	pm	sq	C to T
	■	37901	UV6	cI UV6	pm	sq	C to T
	■	37901	E13	cIamE13	pm	sq	C to A
	■	37903	BL60	cI BL60	pm	sq	A to G
	■	37903	29-2	cI29-2	pm	sq	A to G
	■	37904	SP53	cI SP53	i(IS5)	sq	
	■	37905	UA54	cI UA54	pm	sq	C to A
	■	37907	B100	cI BL100	pm	sq	G to A
	■	37910	UV34	cI UV34	pm	sq	C to T
	■	37915	SP27	cI SP27	pm	sq	G to A
	■	37917	UV90	cI UV90	pm	sq	T to A
	■	37918	L7	cIamL7	pm	sq	A to C
	■	37919	SP14	cI SP14	i(IS5)	sq	
	■	37919	SP35	cI SP35	i(IS5)	sq	
	■	37922	UV12	cI UV12	pm	sq	insert R
	■	37925	KH54	KH54	dr	sq	
	■	37926	SP57	cI SP57	pm	sq	C to A
	■	37928	57-1	cI57-1	pm	sq	T to G
	■	37930	SP62	cI SP62	pm	sq	T to A
	■	37932	UA60	cI UA60	pm	sq	insert T
	■	37935	UA9	cI UA9	pm	sq	G to T
	■	37938	UV28	cI UV28	pm	sq	C to T
	■	37939	BP81	cI BP81	pm	sq	A to G
	■	37940	*CI*	cI (orf-237)	M-NH3	AA	
	■	37940	SM	sM pM	5'	RNA	formerly called pRM
	■	37941	BP86	cI BP86	pm	sq	A to G
	■	37947	SP2	cI SP2	i(IS5)	sq	
	■	37950	SP46	cI SP46	i(IS1)	sq	
	■	37951	OR3	oR3	f(1) bs	f	17 bp sequence cI binds
	■	37954	E37	pRM-E37	pm	sq	C to T
	■	37955	VC3	vC3	pm	sq	A to G
	■	37957	3R1	OR3-r1	pm	sq	C to T
	■	37958	3R2	OR3-r2; OR3-r3	pm	sq	-r2 G to T; -r3 G to A
	■	37965	C12	OR3-c12	pm	sq	A to G
	■	37966	C10	OR3-c10	pm	sq	T to C
	■	37971	UP1	prm up-1	pm	sq	A to G
	■	37973	M104	pRM-M104;116;U31	pm	sq	C to T
	■	37973	UV26	cI UV26	pm	sq	C to T
	■	37974	OR2	oR2	f(1) bs	f	cI binds (17bp)
	■	37975	BU20	cI BU20	pm	sq	A to G
	■	37978	E104	pRM-E104;v3c	pm	sq	E104 A to C; v3c A to G
	■	37979	V1	v1;vir146;pRM-E93	pm	sq	v1 C to A; E93 C to T
	■	37980	UV51	cI UV51	pm	sq	CC to TT
	■	~37983	S274	spi 274	att-sr	EM	
	■	37984	BR1	BR1	pm	sq	C to T
	■	37985	VN	vN	pm	sq	G to T
	■	37986	MAC1	MAC1;MAC41	pm	sq	MAC1 T to A; MAC41 T to C
	■	37987	MCC8	MCC8;MCH31;AAH8	pm	sq	MCC8 G to C; MCH31 G to A; AAH8 G to T
HIN2	■	37989					
	■	37989	MAH4	MAH4;AAH2	pm	sq	MAH4 delete T;AAH2 T to A
	■	37990	CH20	MCH20; MCH9	pm	sq	MCH20 G to C;MCH9 delete G
	■	37991	X3	pR-x3	pm	sq	A to G
	■	37992	MCH8	MCH8	pm	sq	C to T
	■	37998	OR1	oR1	f(1) bs	f	cI binds (17bp)
	■	37999	2668	vR2668;virR18	pm	sq	A to G
	■	38000	668	VR668	pm	sq	insert C
	■	38003	V326	vs326; BC1	pm	sq	vs326 C to A;BC1 C to T
	■	38006	U119	UV119	pm	sq	delete G (clear)
	■	38007	UV8	pRM-UV8;v3	pm	sq	UV8 C to T; v3 C to A
	■	38008	U93	pRM-U93;M36	pm	sq	U93 G to A
	■	38009	V387	vs387;vC1;NR5	pm	sq	vs387 G to C; vC1 G to T
	■	38009	UV11	UV11	pm	sq	G to A (clear)
	■	38010	B143	BU143	pm	sq	T to C (clear)
	■	38023	SR	sR pR	5'	RNA	
	■	~38031	CRO2	Δ cro2	dl	EM	
	■	38041	*CRO*	cro (orf-66)	M-NH3	AA	
BGL2	■	38103					
	■	38135	TRO	tRO terminator	3'	RNA	
	■	~38197	CRO2	Δ cro2	dr	EM	
AVA1	■	38214					
	■	38238	*CRO*	cro	COOH		
	■	38245	I4	imm434	sr	sq	

A	B	C	D	E	F	G	H
	■	38264	NUTR	NutR	f(1)	f	N utilization (homology to nutL)
	■	38302	CIN1	cin-1	pm	sq	G to A
	■	38306	CNC1	CNC1	pm	sq	T to C
	■	38307	CNC8	CNC8	PM	sq	A to G
	■	38334	TR1	tR1 terminator	3'	RNA	
	■	38339	DYA3	dya3	pm	sq	T to C
	■	38341	C17	c17		sq	9-bp duplication
	■	38343	SE	sE pE	5'	RNA	formerly called pRE
	■	38344	DYA2	dya2	pm	sq	A to G
	■	38350	R32	r32	i(IS2)	sq	
	■	38350	3048	cY3048	pm	sq	A to G
	■	38354	3071	cY3071;c3073a	pm	sq	cY3071 T to C; c3073a T to A
	■	38354	2001	cY2001	pm	sq	T to G
	■	38356	3088	cII3088	pm	sq	A to G
	■	38357	3019	cY3019	pm	sq	C to T
	■	38359	CIR5	cir5	pm	sq	T to C
	■	38360	*CII*	cII (orf-97)	M-NH3	AA	
	■	38360	3086	cII3086	pm	sq	A to G
	■	38361	3067	cII3067	pm	sq	T to C
	■	38362	3105	cII3105	pm	sq	G to A
	■	38363	3073	c3073b	pm	sq	G to C
	■	38364	3059	cII3059	pm	sq	T to A
	■	38364	CAN1	can1	pm	sq	T to C
	■	38368	CTR2	ctr2;ctr3	pm	sq	T to C
	■	38369	CTR1	ctr1	pm	sq	G to A
	■	38369	C2	cII site	bs		cII binds -35 region of pRE
	■	38370	3003	cY3003	pm	sq	C to T
	■	38371	CY42	cY42	pm	sq	A to T
	■	38372	3077	cY3077	pm	sq	A to G
	■	38375	3114	cII3114;cII3109	pm	sq	cII3114 A to G;cII3109 A to T
	■	38376	844	cY844	pm	sq	A to G
	■	38377	3075	cY3075	pm	sq	AC to TT
	■	38378	3107	cY3107	pm	sq	C to T
	■	38379	3008	cY3008	pm	sq	G to A
	■	38380	3001	cY3001	pm	sq	C to T
	■	38380	3095	cY3095	pm	sq	C to A
	■	38381	3072	cY3072;cY3078	pm	sq	cY3072 A to C;cY3078 A to G
	■	38382	3098	cY3098	pm	sq	A to G
	■	38384	3623	cII3623	pm	sq	G to T
	■	38385	3085	cII3085	pm	sq	A to G
	■	38387	3104	cII3104	pm	sq	G to A
	■	38388	3091	cII3091	pm	sq	C to T
	■	38391	3056	cII3056	pm	sq	T to C
	■	38393	3099	cII3099	pm	sq	C to T
	■	38396	3081	cII3081	pm	sq	A to T
	■	38399	3097	cII3097	pm	sq	G to A
	■	38400	3111	cII3111	pm	sq	AG to T
	■	38403	3092	cII3092	pm	sq	G to A
	■	38430	2002	cII2002	pm	sq	T to C
	■	38453	3641	cII3641	pm	sq	G to T
	■	38472	3062	cII3062	pm	sq	C to T
	■	38474	AM60	cIIam60	pm	sq	C to T
	■	38476	3283	cII3283a	pm	sq	C to A
	■	38478	3283	cII3283b	pm	sq	T to A
	■	38483	CHIC	chiC	pm	sq	A to T
	■	38488	3302	cII3302	pm	sq	G to A
	■	38498	68	cII68	pm	sq	T to A
	■	38523	3258	cII3258a	pm	sq	T to G
	■	38538	AM41	cIIam41	pm	sq	G to A
	■	38543	ICE	replication ice	f(1)	f G	5bp stem;5bp loop;proposed inceptor
HIN2	■	38548					
	■	38549	PK35	pk35	i(Tn903)	sq	
	■	38564	3258	cII3258b	pm	sq	C to T
	■	38571	3468	cII3468	pm	sq	C to T
	■	38582	3386	cII3386	pm	sq	G to A
	■	38592	3139	cII3139	pm	sq	T to C
	■	38599	OOP	oop terminator	3'	RNA	
	■	38617	I21	imm21	sr	sq	
	■	38634	3639	cII3639	pm	sq	A to G
	■	38642	3638	cII3638	pm	sq	A to G
	■	38650	*CII*	cII	COOH		
	■	38651	3520	cII3520	pm	sq	T to C
	■	38675	OOP	oop start point	5'	RNA	
	■	38686	*O*	O (orf-299)	M-NH3	AA	
BGL2	■	38754					
BGL2	■	38814					
	■	39034	ITN1	ori iteron 1	f(1) bs	f	18 bp repeated seq O binds
	■	39054	ITN2	ori iteron 2	f(1) bs	f	18 bp O binds-DNase protection
	■	39078	ITN3	ori iteron 3	f(1)bs	f	18 bp O binds-DNase protection
	■	39092	R93	r93	dl	sq	ori- mutation

A	B	C	D	E	F	G	H
	▪	39101	ITN4	ori iteron 4	f(1)bs	f	18 bp O binds-DNase protection
	▪	39115	R93	r93	dr	sq	ori- mutation
	▪	39120	R99	r99	dl	sq	ori- mutation
	▪	39122	TI12	ti12(tiny 12)	pm	sq	C to A; cis replication defective
	▪	39131	R99	r99	dr	sq	ori- mutation
	▪	39138	R96	r96	dl	sq	ori- mutation
	▪	39152	R96	r96	dr	sq	ori- mutation
	▪	39158	HY42	Hy42	hy	sq	λatt80imm80 O80::PQSRλ
	▪	39165	RP82	rep82	hy	sq	λ rep82::PQSRλ
ECR1	▪	39168					
	▪	39268	RIC	ric5b	pm	sq	C to T
	▪	39292	RIC	ric5b	pm	sq	G to A
	▪	~39373	A-20	bioE5a-20	att-sr	EM	
SPH1	▪	39422					
	▪	39582	*P*	P (orf-233)	M-NH3		
	▪	39582	*O*	O	COOH		
HIN2	▪	39608					
HPA1	▪	39608					
HIN2	▪	39836					
HPA1	▪	39836					
AVA1	▪	39888					
SMA1	▪	39890					
	▪	40091	DV93	dvh93	plasmid-r	sq	
	▪	40280	*REN*	ren (orf-96)	M-NH3		
	▪	40280	*P*	P	COOH		
	▪	40335	V021	dv021	plasmid-r	sq	
	▪	~40354	PQ4	PQ Δ 4	dl	EM	
	▪	~40354	CF1	cf1	dl	EM	
SST2	▪	40389					
	▪	~40404	PQ9	PQ Δ 9	dl	EM	
	▪	~40502	P4	P4	sl	EM	
	▪	~40502	Q3	qin3	sl	EM	
	▪	~40502	A3	qinA3	sl	EM	
	▪	~40502	M3	λgalM3	sl	EM	
	▪	~40502	C3	qinC3	sl	EM	
	▪	40502	NIN5	nin5	dl	sq	N independent growth
	▪	~40551	V203	dv203	plasmid-r	EM	
	▪	~40551	PQ1	PQΔ 1	dl	EM	
	▪	40567	*REN*	ren	COOH		
	▪	~40600	NI24	nin24	dl	EM	
	▪	40624	TR2	tR2 terminator	3'	f	presumed genetically defined tR2
	▪	40644	*ORF*	nin orf-146	M-NH3		
	▪	~40796	PQ8A	PQ Δ 8a	dl	EM	
HIN2	▪	40942					
	▪	41011	TN10	627	i(Tn10)	sq	
	▪	41081	*ORF*	nin orf-290	M-NH3		
	▪	41081	*ORF*	nin orf-146	COOH		
	▪	~41090	D86	Δ 86	dl	EM	
	▪	~41139	B1	b1	i(IS2)	EM	
	▪	~41287	3000	b3000	dl	EM	
CLA1	▪	41364					
	▪	~41679	IMML	immL	sr	EM	
BAM1	▪	41732					
	▪	41781	TN10	623	i(Tn10)	sq	
	▪	~41924	PQ9	PQ Δ 9	dr	EM	
	▪	41950	*ORF*	nin orf-57	M-NH3		
	▪	41950	*ORF*	nin orf-290	COOH		
	▪	~41973	3000	b3000	dr	EM	
CLA1	▪	42021					
	▪	42090	*ORF*	nin orf-60	V-NH3		
	▪	42120	*ORF*	nin orf-57	COOH		
	▪	42269	*ORF*	nin orf-56	V-NH3		
	▪	42269	*ORF*	nin orf-60	COOH		
	▪	~42317	D86	Δ 86	dr	EM	
	▪	42429	*ORF*	nin orf-204	M-NH3		
	▪	42436	*ORF*	nin orf-56	COOH		
	▪	42645	DV1	dv1	plasmid-r	sq	
	▪	~42660	PQ8A	PQ Δ 8a	dr	EM	
	▪	~43004	NI24	nin24	dr	EM	
	▪	43040	*ORF*	nin orf-68	M-NH3		
	▪	43040	*ORF*	nin orf-204	COOH		
	▪	43082	DIFF	difference		sq	Kroger and Hobom (1982);G (not A)
HIN2	▪	43183					
	▪	43224	*ORF*	nin orf-221	M-NH3		
	▪	43243	*ORF*	nin orf-68	COOH		
	▪	43307	NIN5	nin5	dr	sq	
	▪	~43346	CF1	cf1	sr	EM	
	▪	~43394	PQ9	PQ Δ 9	dr	EM	
	▪	~43443	PQ1	PQ Δ 1	dr	EM	
BCL1	▪	43682					masked by dam methylation
	▪	~43688	V266	dv266	plasmid-r	EM	
CLA1	▪	43825					
	▪	~43836	PQ4	PQΔ 4	dr	EM	
	▪	~43885	HYM	HyM	hy	EM	λatt80immOPλ::QSR80
	▪	~43885	HY5	Hy5	hy	EM	80immλ OPλ::QSR80
	▪	43886	*Q*	Q (orf-207)	M-NH3		
	▪	43886	*ORF*	nin orf-221	COOH		
	▪	~44081	TP1	Tp1	i(Tn402)	EM	
HIN3	▪	44141					
	▪	~44326	V154	dv154	plasmid-r	EM	
	▪	~44424	M3	λgalM3	sr	EM	
	▪	44498	DIFF	difference			Daniels (1982) -G; Petrov (1981) -A
	▪	44506	*Q*	Q	COOH		
	▪	44587	6S	p6S pR'	5'	RNA	start of 6S RNA; λ late RNA (maps between 44587 and 44610)
	▪	~44600	QUT	qut	f(c)	G	
	▪	~44606	PK3	pk3	i(Tn903)	EM	
	▪	44621	*ORF*	orf-64	M-NH3		
	▪	44780	T6S	t6S	3'	RNA	end of 6S RNA
	▪	44812	*ORF*	orf-64	COOH		
ECR1	▪	44972					
	▪	45027	CHID	chiD	pm	sq	
	▪	45348	S9B	Sts9B	pm	sq	G to A
	▪	45186	*S*	S (orf-107)	M-NH3		protein not identified
	▪	~45336	422	422	i(Tn5)	EM;RES	
	▪	45352	S7	Sam7	pm	sq	G to A
	▪	45493	*R*	R (orf-158)	M-NH3	AA	
	▪	45506	*S*	S	COOH		
	▪	45853	296	λ296	sl	sq	deletes 309 bp;inserts pBR322
	▪	45966	*RZ*	Rz (orf-153)	M-NH3	G	
	▪	45966	*R*	R	COOH		
	▪	~45994	FUS2	λfus2	sr	EM	
	▪	46008	DK23	dK23	i(TN903)	sq	
	▪	~46043	Q3	qin3	sr	EM	
	▪	~46043	A3	qinA3	sr	EM	
	▪	~46043	P4	P4	sr	EM	
	▪	~46043	C3	qinC3	sr	EM	
	▪	~46141	DK6	dk 6	i(Tn903)	EM;RES	
	▪	46161	296	λ296	sl	sq	deletes 309 bp;inserts pBR322
BCL1	▪	46366					masked by dam methylation
	▪	46423	*RZ*	Rz	COOH		
CLA1	▪	46439					
	Δ	46467	SB8	Sequence Block 8	end		
		~47471	PK26	pk26	i(Tn903)	EM	
		~47520	PK22	pk22	i(Tn903)	EM	
		~47520	PK24	pk24	i(Tn903)	EM	
		~47618	DA3	Δa3	dl	EM	
HIN2		47938					
BCL1		47942					masked by dam methylation
		~48158	PK21	pk21	i(Tn903)	EM	
		~48207	DA3	Δa3	dr	EM	
HIN2		48298					
	▽	48391	SB1R	Sequence Block 1R	start		
	▪	~48403	PK25	pk25	i(Tn903)	EM	
REND	▪	48502		Right END			used for fragment endpoints
	▪	48514	CRND	5' end of r-strand			last base of sticky end
	Δ	48514	SB1R	Sequence Block 1R	end		

TABLE 2 SOURCES OF AVAILABLE λ DNA SEQUENCES

Gene Region	Source	Coordinates
λ*cI857S7*	Complete Sequence	
	Sanger et al. (1982)	1-28461
		28808-35245
		35280-37709
		38039-38197
		38499-38680
		38757-39657
		39700-44169
		44501-48514
	Sequence Block 1L [1-72]	
	Wu and Taylor 1971	1-12
cos	Nichols and Donelson (1978)	13-72
	Sequence Block 2 [14869-18969]	
J	F. Sanger, A. R. Coulson, and G.-F. Hong, (1980) as cited by Daniels et al. (1982)	
	Sequence Block 3 [21661-21737]	
b-region	Hoess and Landy (1978)	
	Sequence Block 4 [23129-23495]	
pbL	Rosenvold et al. (1980) (sequence inaccurate)	23131-23248
	K.-C. Luk and W. Szybalski (unpubl.) (sequence accurate with one typographical error at 23143 in Abstracts of Bacteriophage Meetings, Cold Spring Harbor Laboratory, 1982)	23129-23269
	Sequence Block 5 [23495-23548]	
b-region	Hoess and Landy (1978)	
	Sequence Block 6 [27479-29811]	
att	Hsu et al. (1980)	27479-27633
att	Landy and Ross (1977)	27617-27934
att	Davies et al. (1977)	27616-28935
int-xis	Hoess et al. (1980)	27725-29275
	Davies (1980)	27724-29525
	Abraham et al. (1980)	28929-29198
	R. H. Hoess, K. Bidwell and A. Landy, pers. comm. (1980)	29065-29811
xis-red	Davies et al. (1978)	29711-29811
xis-red	Hoess and Landy (1978)	29710-29778
	Sequence Block 7 [30493-30569]	
*chi*B	Smith et al. (1980)	
	Sequence Block 8 [31042-46467]	
xis-red	Davies et al. (1978)	31043-31058
xis-red	Hoess and Landy (1978)	31042-31129
tL3	Luk and Szybalski (1982a, b)	30299-31408
bet-exo	F. Sanger, A. R. Coulson, and G.-F. Hong, (1980) as cited by Daniels et al (1982)	31270-32891
bet-N	Ineichen et al. (1981)	32503-35905
N-nutL	Franklin and Bennett (1979)	34956-35615
pL mRNA start	Dalhberg and Blattner (1975)	35434-35618
pLoL	Ptashne et al. (1976)	35578-35667
	Kleid et al. (1975)	35583-35600
oL	Horn and Wells (1981)	35468-35819
rex	Landsmann et al. (1982)	35437-37348
cI	Sauer (1978)	37225-37936
pRoR	Ptashne et al. (1976)	37903-38027
pM(prm)	Walz et al. (1976)	37905-37989
cro-O	Schwarz et al. (1978)	37990-38982
cro	Roberts et al. (1977)	38041-38241
oop	Dalhberg and Blattner (1973)	38597-38672
O	Scherer (1978)	38597-39688
P	Schwarz et al. (1980)	37768-40293
ori	Denniston-Thompson et al. (1977)	39062-39170
ori-O	Moore et al. (1979)	38008-39328
nin	Kröger and Hobom (1982)	40218-43972
Q-p'_R-*qut*-6S	Daniels and Blattner (1982)	43681-45218
Q	Petrov et al. (1981)	43860-45001
Q-S	Daniels (1981)	43681-45634
6S region	Sklar (1977)	44467-44807
6S RNA	Lebowitz et al. (1971)	44588-44780
chiD	Smith et al. (1981)	44972-45057
R protein	Imada and Tsugita (1971)	45493-45963
R	D. L. Daniels (unpubl.)	45635-45977
Rz	H. Ikeda, pers. comm. (1980);	46064-46443
S-R-Rz	F. Sanger, A. R. Coulson and G.-F. Hong, (1980) as cited by Daniels et al. (1982)	44467-46467
	Sequence Block 1R [48391-48514]	
cos	Nichols and Donelson (1978)	48391-48502
	Wu and Taylor (1971)	48503-48514

TABLE 3 SOURCES OF MAPPING INFORMATION FOR SITES ASSIGNED TO EXACT POSITION IN SEQUENCE

Site name	Reference
left end	Wu and Taylor (1971); Nichols and Donelson (1978)
*W*am403 (as in λ*WES*)	G.-F. Hong and F. Sanger, unpubl. (1983)
*B**	Walker et al. (1982)
*Hpa*I, *Hinc*II sites overlap mK site	Daniels et al. (1980); A. Glasgow and M. Howe, pers. comm. (1983)
t'J4 terminator	Luk and Szybalski (1983b)
t'J3 terminator	---ibid.
t'J2 terminator	---ibid.
t'J1 terminator	---ibid.
*b*189	A. Frischauf, H. Lehrach, A. Poustka, N. Murray, pers. comm. (1983)
*lac*5 left	J. S. Salstrom and D. Newman, pers. comm. (1982)
A in cb2; *G* in cI857S7	R. Hoess and A. Landy, pers. comm. (1983)
*b*2	Hoess and Landy (1978)
λcam105	Muster et al. (1983)
pbL	Rosenvold et al. (1980); K.-C. Luk and W. Szybalski, unpubl. (1982)
*b*511	Hoess and Landy (1978)
λcam107	Muster et al. (1983)
Δ106	Court et al. (1983)
Δ119	---ibid.
*sib*3	Guarneros et al. (1982)
Δ125	Court et al. (1983)
tI terminator	Luk et al. (1982); U. Schmeissner, pers. comm. (1981);
*hef*13	Guarneros et al. (1982)
Δ112	Court et al. (1983).
*sib*2	Guarneros et al. (1982)
pHΔ53	Hsu et al. (1980)
*sib*1	Guarneros et al. (1982)
INT 1 binding site	Hsu et al. (1980).
INT 2 binding site	---ibid.
pHΔ55	---ibid.
INT 3 binding site	---ibid.
att core	Davis et al. (1977); Landy and Ross (1977)
att 2; 6; 24	Ross et al. (1982)
*bio*936	R. Weisberg, pers. comm. (1982)
INT 4 binding site	Hsu et al. (1980)
*att*501	Ross et. al. (1982)
λMGB22	L. MacHattie, pers. comm. (1983)
*trp*Δ841	Abraham et al. (1980)
xis am6	Abraham et al. (1980); Hoess et al. (1980);
sI pI	Schmeissner et al. (1981)
int-c226;262;518	Abraham et al. (1980); Hoess et al. (1980).
cII binding site	Ho and Rosenberg (1982)
*trp*Δ303	Abraham et al. (1980)
*trp*Δ29	---ibid.
*b*508	Davies et al. (1978); Hoess and Landy (1978)
Δred 15	G. Smith, pers. comm. (1983)
*chi*B	Smith et al. (1980)
*b*522	Davies et al. (1978); Hoess and Landy (1978)
*t*L3 terminator	Luk et al. (1982)
G instead of C at 33035	Ineichen et al. (1982)
*t*L2d terminator	K.-C. Luk and W. Szybalski (1983a)
*t*L2c terminator	---ibid.
*c*IIIam611	Knight and Echols (1983)
*t*L2b terminator	K.-C. Luk and W. Szybalski (1983a)
*t*L2a terminator	---ibid.
*nin*L99	Somasekhar et al. (1982)
*trp*44 right	N. Franklin, pers. comm. (1983)
*imm*21 left	---ibid.
*t*L1 terminator	Drahos and Szybalski (1981)
*trp*48	Franklin and Bennett (1979)
pλQLC115	L. MacHattie, pers. comm. (1983)
*N*am22	Franklin and Bennett (1979).

TABLE 3 (Continued)

Site name	Reference
pλQLC108	L. MacHattie, pers. comm. (1983)
pλQLC118	---ibid.
pλQLC109	---ibid.
pλQLC105	---ibid.
pλQLC101	---ibid.
pλQLC143	---ibid.
*N*am7	Franklin and Bennett (1979)
pλXJS1974	L. MacHattie, pers. comm. (1983)
*N*mar2	Franklin and Bennett (1979)
pλQLC117	L. MacHattie, pers. comm. (1983)
pλXJS2300	---ibid.
pλMS972	---ibid.
pλQLC98	---ibid.
pλQLC104	---ibid.
*N*am2521	Franklin and Bennett (1979)
pλMS370	Muster et al. (1983)
*N*am53	Franklin and Bennett (1979)
pλXJS1076	L. MacHattie, pers. comm. (1983)
pλMS968	---ibid.
*N*am2524	Franklin and Bennett (1979)
*N*am219	---ibid.
pλQLC102	L. MacHattie, pers. comm. (1983)
pλQLC144	---ibid.
pλQLC95	---ibid.
dv1 left	Dahlberg and Blattner (1975); Landsmann et al. (1982)
nutL	Salstrom and Szybalski (1978); Drahos and Szybalski (1981); Drahos et al. (1982); Somasekhar et al. (1982)
pλQLC138	L. MacHattie, pers. comm. (1983)
pλQLC103	---ibid.
*bio*3h-1	G. Somasekhar and W. Szybalski, unpubl. (1982)
sL pL	Blattner et al. (1972)
*imm*434 left	Landsmann et al. (1982); R. Pastrana, pers. comm. (1982)
*o*L1	Ptashne et al. (1976)
*vir*2,*v*003	Dahlberg and Blattner (1975); Flashman (1978)
*vir*101	Dahlberg and Blattner (1975)
*pL sex*1	Kleid et al. (1976)
*o*L2	Ptashne et al. (1976)
*v*L2668	Bailone and Devoret (1978)
*v*169	---ibid.
*v*305	Flashman (1978)
*o*L3	Ptashne et al. (1976)
1 668	Bailone and Devoret (1982)
*t*IMM terminator	Landsmann et al. (1982); K.-C. Luk and W. Szybalski, unpubl. (1983)
*KH*54	K.-C. Luk and W. Szybalski, unpubl. (1983)
delete A at 35887	Ineichen et al. (1981)
*rex*209	Landsmann et al. (1982)
*rex*111	---ibid.
pλ969	L. MacHattie, pers. comm. (1983)
pλQLC139	---ibid.
pλQLC137	---ibid.
s lit *p* lit	Landsmann et al. (1982); K.-C. Luk and W. Szybalski, in prep. (1983)
*KH*70	K.-C. Luk and W. Szybalski, unpubl. (1983)
pλQLC111	L. MacHattie, pers. comm. (1983)
pλQLC141	---ibid.
pλ142	---ibid.
pc 1	Guarente et al. (1982)
pλXJS2299	L. MacHattie, pers. comm. (1983)
pλXJS1975	---ibid.
pλQLC112	---ibid.
pλXJS1973	---ibid
pλQLC106	---ibid.

TABLE 3 (Continued)

Site name	Reference
pλQLC110	L. MacHattie, pers. comm. (1983)
pλQLC107	---ibid.
dv021 left	Landsman et al. (1982)
*c*I UA72	[a] F.Hutchinson, T.R. Skopek, R.D. Wood, pers. comm. (1983)
*c*I IC11	---ibid.
*c*I SP31	---ibid.
*c*Iam14	Lieb (1981)
*c*I NT1	F.Hutchinson, T.R. Skopek, R.D. Wood, pers. comm. (1983)
*c*Iam504	Lieb (1981)
*c*I BP83	F.Hutchinson, T.R. Skopek, R.D. Wood, pers. comm. (1983)
*c*Iam505	Lieb (1981)
KH100	Landsmann et al. (1982)
*c*I UV59	F.Hutchinson, T.R. Skopek, R.D. Wood, pers. comm. (1983)
*c*I UV63	---ibid.
*c*I IC28	---ibid.
*c*I ET26	---ibid.
*c*I AA21	---ibid.
*c*I AA2	---ibid.
*c*I NT16	---ibid.
*c*I UV76	---ibid.
*c*I SP1	---ibid.
*c*I NT15	---ibid.
*c*I171;172;473;474::Tn*10*	Kleckner (1979)
*c*I 13	F. Hutchinson, T.R. Skopek, R.D. Wood, pers. comm. (1983)
*c*I UV64	---ibid.
*c*I UV73	---ibid.
*c*I BL46	---ibid.
ind1	Sanger et al. (1982); R.T. Sauer, pers. comm. (1980)
*c*I PF17	F. Hutchinson, T.R. Skopek, R.D. Wood, pers comm. (1983)
*c*I AA3	---ibid.
*c*I SP48	---ibid.
*c*I ET39	---ibid.
*c*Iam499	Lieb (1981).
*c*I BP138	F. Hutchinson, T.R. Skopek, R.D. Wood, pers. comm. (1983)
*c*Iam212	Lieb (1981)
*c*I ET3	F.Hutchinson, T.R. Skopek, R.D. Wood, pers. comm. (1983)
*c*Iam34	Lieb (1981)
*c*IamE86	Hecht et al. (1983)
*c*I54-1	---ibid.
*c*IamR82	---ibid.
*c*I UV52	F.Hutchinson, T.R. Skopek, R.D. Wood, pers. comm. (1983)
*c*I ET15	---ibid.
*c*I UV96	---ibid.
*c*I UV97	---ibid.
*c*I UV79	---ibid.
*c*I PF12;*c*I AA6	---ibid.
*c*I11-3;*c*ICP175	Hecht et al. (1983)
*c*I60-1	---ibid.
*c*I UV47	F.Hutchinson, T.R. Skopek, R.D. Wood, pers. comm. (1883)
*c*I5-3	Hecht et al. (1983)
*c*I UV116;*c*I UV74	F.Hutchinson, T.R. Skopek, R.D. Wood, pers. comm. (1983)
*c*I BP84	---ibid.
*c*I BU30	---ibid.
*c*I UV66	---ibid.
*c*I SP44;*c*I SP37	---ibid.

TABLE 3 (Continued)

Site name	Reference
pλQLC113	L. MacHattie, pers. comm. (1983)
*c*I857	Lieb (1981); R. T. Sauer, pers. comm. (1980)
*KH*70	K.-C. Luk and W. Szybalski (unpubl.)
*c*I63-1	Hecht et al. (1983)
*c*I ET7	F.Hutchinson, T.R. Skopek, R.D. Wood, pers. comm. (1983)
*c*I BL71	---ibid.
*c*I UA8	---ibid.
*c*I UV61	---ibid.
dvh93 left	Schwarz et al. (1980)
*c*I UV62	F.Hutchinson, T.R. Skopek, R.D. Wood, pers. comm. (1983)
*c*IamL57	Hecht et al. (1983)
*c*I UV41	F.Hutchinson, T.R. Skopek, R.D. Wood, pers. comm. (1983)
*c*I34-1	Hecht et al. (1983)
*c*I UV117	F.Hutchinson, T.R. Skopek, R.D. Wood, pers. comm. (1983)
*c*I UV44	---ibid.
*c*I NT17	---ibid.
*c*I26-2	Hecht et al. (1983)
*c*I UA33;*c*I SP43	F.Hutchinson, T.R. Skopek, R.D. Wood, pers. comm. (1983)
*c*I UV91	---ibid.
*c*ICP32	Hecht et al. (1983)
*c*I UV46	F.Hutchinson, T.R. Skopek, R.D. Wood, pers. comm. (1983)
*c*I53-1	Hecht et al. (1983)
*c*I UV108	F.Hutchinson, T.R. Skopek, R.D. Wood, pers. comm. (1983)
*c*I UV37	---ibid.
*c*IamL50	Hecht et al. (1983)
*c*I NT3	F.Hutchinson, T.R. Skopek, R.D. Wood, pers. comm. (1983)
*c*ICP-9;*c*I29-6	Hecht et al. (1983)
*c*I BL42	F.Hutchinson, T.R. Skopek, R.D. Wood, pers. comm. (1983)
*c*I SP38	---ibid.
*c*I37-1	Hecht et al. (1983)
*c*I BP102	F.Hutchinson, T.R. Skopek, R.D. Wood, pers. comm. (1983)
*c*I475::Tn*10*	Kleckner (1979)
*c*I UA68	F.Hutchinson, T.R. Skopek, R.D. Wood, pers. comm. (1983)
*c*I BL70	---ibid.
*c*I40-3	Hecht et al. (1983)
*c*Iam282	Lieb (1981)
*c*I NT25;*c*I UA77	F.Hutchinson, T.R. Skopek, R.D. Wood, pers. comm. (1983)
*c*IamQ44	Hecht et al. (1983)
*c*I ET22	F.Hutchinson, T.R. Skopek, R.D. Wood, pers. comm. (1983)
*c*I38-2	Hecht et al. (1983)
*c*I SP61	F.Hutchinson, T.R. Skopek, R.D. Wood, pers. comm. (1983)
*c*I PF2;*c*I AA13	---ibid.
*c*I SP28	---ibid.
*c*I9-1	Hecht et al. (1983)
*c*I BU120	F.Hutchinson, T.R. Skopek, R.D. Wood, pers. comm. (1983)
*c*I SP41	---ibid.
*c*I UV86	---ibid.
*c*I AA13	---ibid.
*c*ICP7	Hecht et al. (1983)
pc 2	A. Hochschild, pers. comm. (1982)
*c*I SP5	F.Hutchinson, T.R. Skopek, R.D. Wood, pers. comm. (1983)

TABLE 3 (Continued)

Site name	Reference
*c*I BU39	F.Hutchinson, T.R. Skopek, R.D. Wood, pers. comm. (1983)
*c*I12-1	Hecht et al. (1983)
pc 3	A. Hochschild, pers. comm. (1982)
*c*I BL51	F.Hutchinson, T.R. Skopek, R.D. Wood, pers. comm. (1983)
*c*I SP3	---ibid.
*c*IamQ33	Hecht et al. (1983)
*c*I UA6;*c*I BL89	F.Hutchinson, T.R. Skopek, R.D. Wood, pers. comm. (1983)
*c*I47-1	Hecht et al. (1983)
*c*I408::Tn*10*	Kleckner (1979)
*c*I UV14	F.Hutchinson, T.R. Skopek, R.D. Wood, pers. comm. (1983)
*c*IamL31	Hecht et al. (1983)
*c*I4-3	---ibid.
*c*I BL84	F.Hutchinson, T.R. Skopek, R.D. Wood, pers. comm. (1983)
*c*I1-1	Hecht et al. (1983)
*c*I SP42	F.Hutchinson, T.R. Skopek, R.D. Wood, pers. comm. (1983)
*c*I UV80	---ibid.
*c*I UV57	---ibid.
*c*I UV88	---ibid.
*c*I ET28	---ibid.
*c*I ET13	---ibid.
*c*Iam302	Lieb (1981)
*c*I56-1;*c*I5-1	Hecht et al. (1983)
*c*I32-1;*c*I29-7	---ibid.
*c*I UV85	F.Hutchinson, T.R. Skopek, R.D. Wood, pers. comm. (1983)
*c*I SP39;*c*I BL80	---ibid.
*c*I UV112;*c*I UV93	---ibid.
*c*I36-1	Hecht et al. (1983)
*c*I SP60	F.Hutchinson, T.R. Skopek, R.D. Wood, pers. comm. (1983)
*c*I SP52	---ibid.
*c*I BL97	---ibid.
*c*I CP177	Hecht et al. (1983)
*c*I SP51	F.Hutchinson, T.R. Skopek, R.D. Wood, pers. comm. (1983)
*c*I BP80	---ibid.
*c*I UV23	---ibid.
*c*I UV6	---ibid.
*c*IamE13	Hecht et al. (1983)
*c*I BL60	F.Hutchinson, T.R. Skopek, R.D. Wood, pers. comm. (1983)
*c*I29-2	Hecht et al. (1983)
*c*I SP53	F.Hutchinson, T.R. Skopek, R.D. Wood, pers. comm. (1983)
*c*I UA54	---ibid.
*c*I BL100	---ibid.
*c*I UV34	---ibid.
*c*I SP27	---ibid.
*c*I UV90	---ibid.
*c*IamL7	Hecht et al. (1983)
*c*I SP35	F.Hutchinson, T.R. Skopek, R.D. Wood, pers. comm. (1983)
*c*I SP14	---ibid.
*c*I UV12	---ibid.
*KH*54	K.-C. Luk and W. Szybalski (unpubl.)
*c*I SP57	F.Hutchinson, T.R. Skopek, R.D. Wood, pers. comm. (1983)
*c*I57-1	Hecht et al. (1983)
*c*I SP62	F.Hutchinson, T.R. Skopek, R.D. Wood, pers. comm. (1983)
*c*I UA60	---ibid.

TABLE 3 (Continued)

Site name	Reference
*c*I UA9	F.Hutchinson, T.R. Skopek, R.D. Wood, pers. comm. (1983)
*c*I UV28	---ibid.
*c*I BP81	---ibid.
*s*M,*p*M	Walz et al. (1976)
*c*I BP86	F.Hutchinson, T.R. Skopek, R.D. Wood, pers. comm. (1983)
*c*I SP2	---ibid.
*c*I SP46	---ibid.
*o*R3	Ptashne et al. (1976)
prm-E37	Rosen et al. (1980)
*v*C3	Flashman (1976); Meyer et al. (1980)
oR3-r1	Meyer et al. (1980)
oR3-r2 oR3-r3	---ibid.
oR3-c12	---ibid.
oR3-c10	---ibid.
prm up-1	---ibid.
prm-M104;116;U31	Meyer et al. (1975); Kleid et al. (1976); Rosen et al. (1980)
*c*I BU20	F.Hutchinson, T.R. Skopek, R.D. Wood, pers. comm. (1983)
*o*R2	Kleid et al. (1976); Ptashne et al. (1976)
*c*I UV26	F.Hutchinson, T.R. Skopek, R.D. Wood, pers. comm. (1983)
prm-E104	G. Gussin, pers. comm. (1982)
*v*3c	Flashman (1976); Flashman (1978)
*v*1;*vir*146	Maniatis et al. (1975); Schwarz et al. (1980)
*c*I UV51	F.Hutchinson, T.R. Skopek, R.D. Wood, pers. comm. (1983)
prm-E93	Rosen et al. (1980)
BR1	J. Eliason, pers. comm. (1982)
*v*N	Maniatis et al. (1975); Flashman (1978)
MAC1 MAC41	J. Eliason, pers. comm. (1982)
MCC8,MCH31,AAH8	---ibid.
MAH4,AAH2	---ibid.
MCH20 MCH9	---ibid.
*p*R-x3	Hawley (1982)
MCH8	J. Eliason, pers. comm. (1982)
*o*R1	Ptashne et al. (1976)
VR2668	Bailone and Devoret (1978); R. Devoret, pers. comm. (1982)
*vir*R18	Flashman (1978)
VR668	Bailone and Devoret (1978); R. Devoret, pers. comm (1982)
*v*s326	Maniatis et al. (1975)
BC1	J. Eliason, pers. comm (1982)
prm-UV8	Rosen et al. (1980)
*v*3	Maniatis et al. (1975); Schwarz et al. (1980)
prm-UV93;M36	Rosen et al. (1980)
*v*s387 *v*C1 NR5	Meyer et al. (1975); J. Eliason, pers. comm. (1982)
UV119	F.Hutchinson, T.R. Skopek, R.D. Wood, pers. comm. (1983)
UV11	---ibid.
BU143	---ibid.
*s*R *p*R	Blattner and Dahlberg (1972)
*t*RO terminator	Calva and Burgess (1980)
*imm*434 right	Rosenberg et al. (1978); Schwarz et al. (1978); Grosschedl and Schwarz (1979)
*nut*R	Rosenberg et al. (1978)
Cin-1	Wulff et al. (1980)
CNC1	---ibid
CNC8	---ibid
*t*R1 terminator	Rosenberg et al. (1978); Calva and Burgess (1980).
dya3	D.L. Wulff and M.E. Mahoney, pers. comm. (1983)

TABLE 3 (Continued)

Site name	Reference
*c*17	Rosenberg et al. (1978)
*s*E,*p*E	Schmeissner et al. (1980)
dya2	D.L. Wulff and M.E. Mahoney, pers. comm. (1983)
*r*32	Rosenberg et al. (1978)
*cy*3048	Wulff et al. (1980)
*cy*3071;*cy*3073a	D.L. Wulff and M.E. Mahoney, pers. comm. (1983)
*cy*2001	Rosenberg et al. (1978); Schmeissner et al. (1980); Wulff et al. (1980)
*c*II3088	D.L. Wulff and M.E. Mahoney, pers. comm. (1983)
*cy*3019	Wulff et al. (1980)
*c*ir5	D.L. Wulff and M.E. Mahoney, pers. comm. (1983)
*c*II3086	---ibid.
*c*II3067	---ibid.
*c*II3105	---ibid.
*cy*3073b	---ibid.
*c*II3059	---ibid.
*can*1	Wulff et al. (1980)
ctr2;ctr3	D.L. Wulff and M.E. Mahoney, pers. comm. (1983)
ctr1	---ibid.
*c*II binding site	Ho and Rosenberg (1982)
*cy*3003	Wulff et al. (1980)
*cy*42	Schwarz et al. (1978); Wulff et al. (1980)
*cy*3077	D.L. Wulff and M.E. Mahoney, pers. comm. (1983)
*c*II3114;*c*II3109	---ibid.
*cy*844	Wulff et al. (1980)
*cy*3075	D.L. Wulff and M.E. Mahoney, pers. comm. (1983)
*cy*3107	---ibid.
*cy*3008	Wulff et al. (1980)
*cy*3001	---ibid
*cy*3095	D.L. Wulff and M.E. Mahoney, pers. comm. (1983)
*cy*3072;*cy*3078	---ibid.
*cy*3098	---ibid.
*c*II3623	---ibid.
*c*II3085	---ibid.
*c*II3104	---ibid.
*c*II3091	---ibid.
*c*II3056	---ibid.
*c*II3099	---ibid.
*c*II3081	---ibid.
*c*II3097	---ibid.
*c*II3111	---ibid.
*c*II3092	---ibid.
*c*II2002	Wulff et al. (1980)
*c*II3641	D.L. Wulff and M.E. Mahoney, pers. comm. (1983)
*c*II3062	---ibid.
*c*IIam60	Schwarz et al. (1978)
*c*II3283a	D.L. Wulff and M.E. Mahoney, pers. comm. (1983)
*c*II3283b	---ibid.
*c*II3302	---ibid.
*chi*C	Sprague et al. (1978)
*c*II68	Schwarz et al. (1978)
*c*II2358a	D.L. Wulff and M.E. Mahoney, pers. comm. (1983)
*c*IIam41	Schwarz et al. (1978)
ice	Hobom et al. (1978)
pk35	Moore et al. (1979)
*c*II3258b	D.L. Wulff and M.E. Mahoney, pers. comm. (1983)
*c*II3468	---ibid.
*c*II3386	---ibid.
*c*II3139	---ibid.
to (*oop* terminator)	Dahlberg and Blattner (1973)
*imm*21 right	Schwarz et al. (1978); Rosenberg et al. (1980)
*c*II3639	D.L. Wulff and M.E. Mahoney, pers. comm. (1983)
*c*II3638	---ibid.
*c*II3520	---ibid.
so (*oop* start point)	Blattner and Dahlberg (1972); Dahlberg and Blattner (1973)
ori iteron 1	Denniston-Thompson et al. (1977); Hobom et al. (1979); Moore et al. (1981); Tsurimoto and Matsubara (1981); K. Zahn, pers. comm. (1982)
ori iteron 2	---ibid.
ori iteron 3	---ibid.
*r*93	Denniston-Thompson et al. (1977)
ori iteron 4	Hobom et al. (1979); Matsubara et al. (1981); Moore et al. (1981); K. Zahn, pers comm. (1982)
*r*99	Denniston-Thompson et al. (1977)
*t*iny 12	Hobom et al. (1979); Moore et al. (1979, 1981)
*r*96	Denniston-Thompson et al. (1977)
hy42	Moore et al. (1979; 1981)
rep82	Moore et al. (1981)
ri^{c}5b	Moore and Blattner (1982)
λdvh93	Schwartz et al. (1980)
dv021 right	Kröger and Hobom (1982)
*nin*5 left	---ibid.
*t*R2 terminator	---ibid.
627	Halling et al. (1982)
623	---ibid.
dv1 right	Kröger and Hobom (1982)
G instead of A at 43082	---ibid.
*nin*5 right	---ibid.
A instead of C at 44498	Petrov et al. (1981) - A; Daniels and Blattner (1982) - G
*p*6S *p*R′	Lebowitz et al. (1971); Blattner and Dahlberg (1972); Calva and Burgess (1980); Grayhack and Roberts (1982); D. L. Daniels, H. Lozeron, and F. R. Blattner, in prep. (1983)
qut	Somasekhar and Szybalski (1983); E. Grayhack, C. Hart, X. Yang and J. Roberts, pers. comm. (1982)
6S RNA	Lebowitz et al. (1971); Calva and Burgess (1980)
*chi*D	Smith et al. (1981)
λ422	Daniels and Blattner (1982)
*S*ts9B	R. Grimaila, G. Neal, R. Young, pers. comm. (1983)
*S*am7	Daniels (1981); Sanger et al. (1982)
λ296	Ikeda et al. (1982)
dK23	Benedik, M., R. Young, A. Taylor and A. Campbell. pers. comm. (1982)
dk6	---ibid.
right end	Wu and Taylor (1971); Nichols and Donelson (1978)

The mapping information that assigns these sites exactly in the sequence is referenced. For references to restriction sites, see text. For most of the other sites see Szybalski and Szybalski (1979).

[a] For mutants isolated by R. Hutchinson, T.R. Shopek, and R. D. Wood (pers. comm.) symbols BL, BB, or BV refer to mutagenesis by 5-bromouracil; AL or AA, to 9-aminoacridine; PF, to proflavine; IC, to ICR141; SP, to spontaneous mutants; UV, to 254 nm UV light; ET or VA, to acetophenone + 313 mm UV light; NT, to only the host irradiated with 254 nm UV light.

TABLE 4 METHOD OF ASSIGNMENT OF START CODONS OF λ PROTEINS

Gene	Predicted traits*	Method of Assignment
Nu1	191-733 orf-181 20.4	gene *Nu1* inactivated by insertion *Nu1t16* mapped by restriction enzymes (M. Gould, A. Becker, and H. Murialdo pers. comm. 1982)
A	711-2633 orf-641 73.3	uncertain; probable start based on RBS; several possible other starts in same reading frames
W	2633-2836 orf-68 7.6	probable start based on RBS is established as correct by amber mutation at 2,642 (Sanger et al. 1982)
B	2836-4434 orf-533 59.5	probable start based on RBS: one of two possible before B*
B*		start of processed *B*-coded protein in capsid, first few amino acids determined (Sanger et al. 1982; Walker et al. 1982)
C	4418-5734 orf-439 45.9	probable start based on RBS, best of six possibilities: genes *C* and *Nu3* overlap in same reading frame (Shaw and Murialdo 1980)
Nu3	5132-5734 orf-201 20.8	based on measured size of *Nu3* protein and on RBS, several other possibilities
D	5747-6076 orf-110 11.6	first few amino acids sequenced (Walker et al. 1982)
E	6135-7157 orf-341 38.2	first few amino acids sequenced (Walker et al. 1982)
FI	7202-7597 orf-132 14.3	probable start of two possible based on RBS
FII	7612-7962 orf-117 12.7	probable start of two possible based on RBS
Z	7977-8552 orf-192 21.6	probable start of three possible based on RBS
U	8552-8944 orf-131 14.6	only probable start based on RBS
V	8955-9692 orf-246 25.8	first few amino acids sequenced (Walker et al. 1982)
G	9711-10130 orf-140 15.6	most probable of two possible starts; good sequence in region of this open-reading frame. Reasonable codon usage and RBS. *V* and *T* predicted sizes fit well with physiochemical data (see Szybalski and Szybalski 1979), but *G* predicted M_r is about half its measured size.
T	10115-10546 orf-144 16.1	uncertain; probable start out of several possible based on RBS
H	10542-13100 orf-853 92.3	first few amino acids sequenced (Walker et al. 1982)
M	13100-13426 orf-109 12.5	probable start of five possible based on RBS
L	13429-14124 orf-232 25.7	probable start out of several possible based on RBS
K	14276-14872 orf-199 23.0	probable start based on RBS and size of K protein
I	14773-15441 orf-223 23.1	probable start out of three possible based on RBS, protein not identified on gels so M_r information not available
J	15505-18900 orf-1132 124.4	probable start based on RBS
lom	18965-19582 orf-206 21.9	open reading frame is probably coding for product based on codon usage, RBS. The product is *lom* based on genetic map, hydrophobicity (*lom* is a membrane protein), and M_r
orf-401 *orf*-314 *orf*-194	19650-20852 39.9 21029-21970 32.0 21973-22554 21.6	three orf's probably code for protein products based on codon usage. Orf 401 and 194 have good RBS. orf 314 poor RBS. A single base phase shift could merge orf 314 with a short orf preceding it yielding an orf 335 with a good RBS. This suggests the possibility that there is a mutation in the sequenced strain, λ*c*I857*S*7
b*Ea*47	23918-22689 48.1 orf-410	based on codon usage and RBS (Epp 1978; Szybalski and Szybalski 1979; Sanger et al. 1982; Court and Oppenheim, this volume)
b*Ea*31	25399-34512 orf-296 34.6	probable start (among two possible ones) based on codon usage and RBS (Epp 1978; Szybalski and Szybalski 1979; Court and Oppenheim, this volume)
b*Ea*59	26973-25399 orf-525 59.5	only probable start based on codon usage and RBS (Epp 1978; Court and Oppenheim, this volume)
int	28882-27815 orf-356 40.3	Davies 1980; Hoess et al. 1980
xis	29078-28863 orf-72 8.6	Davies 1980; Hoess et al. 1980
*Ea*8.5	29655-29377 orf-93 10.8	only probable start based on codon usage and RBS
*Ea*22	30395-29850 orf-182 20.9	only probable start based on codon usage and RBS
exo	32028-31351 orf-226 25.9	most probable start based on codon usage and RBS
bet	32810-32028 orf-261 29.7	only probable start based on RBS (Ineichen et al. 1981)

TABLE 4 (Continued)

Gene	Predicted traits	Method of Assignment
gam	33232-32819 orf-138 16.3	start uncertain (Ineichen et al. 1981; Sanger et al. 1982; Zagursky and Hays 1983; S. Friedman and J. Hayes. pers. comm. 1983) but to left of *SalI* @ 33244 (W. Loenen and F.R. Blattner, unpubl. 1982)
orf-47	33330-33190 (kil?) 5.5	This open reading frame has reasonable codon usage and RBS, protein not firmly identified. (Ineichen et al. 1981; Sanger et al. 1982; Court and Oppenheim, this volume)
*c*III	33463-33302 orf-54 6.0	This open reading frame is identified by cIIIam611. It has reasonable codon usage and RBS (Ineichen et al. 1981; Sanger et al. 1982; Knight and Echols 1983; Court and Oppenheim, this volume)
*Ea*10	33904-33539 orf-122 13.8	has been shown to be single-stranded binding protein and in some maps is called ssb (Ineichen et al 1981; Court and Oppenheim, this volume)
ral	34287-34090 orf-66 7.6	(Ineichen et al. 1981)
git	34497-35033 orf-179 21.0	probably *sieB* (Ineichen et al. 1981; Court and Oppenheim, this volume)
N	35360-35040 orf-107 12.3	(Franklin and Bennett 1979; N. Franklin, pers. comm. 1983)
rexB	36259-35828 orf-144 16.0	probable start of two possible based on RBS (Landsmann et al. 1982)
rexA	37114-36278 orf-279 31.3	only possible start based on RBS (Landsmann et al. 1982; Sanger et al. 1982)
*c*I	37940-37230 orf-237 26.2	amino acid sequence (Ptashne et al. 1976; Sauer 1978)
cro	38041-38238 orf-66 7.4	amino acid sequence (Hsiang et al. 1977; Roberts et al. 1977; Schwarz et al. 1978
*c*II	38360-38650 orf-97 11.1	first few amino acids sequenced (Roberts et al. 1977; Schwarz et al. 1978; Ho et al. 1982)
O	38686-39582 orf-299 33.9	(Roberts et al. 1977; Schwarz et al. 1978)
P	39582-40280 orf-233 26.5	(Schwarz et al. 1980)
ren	40280-40567 orf-96 10.6	probable start based on RBS; (Toothman and Herskowitz 1979); showed *ren* point mutations map to the left of *nin*5 deletions and that *nin*-deleted phage were *ren*-. This reading frame spans the *nin*5 endpoint (Kröger and Hobom 1982)
orf-146	40644-41081 ninB 16.6	(Kröger and Hobom 1982; Sanger et al. 1982)
orf-290	41081-41950 ninC 33.6	(Kröger and Hobom 1982; Sanger et al. 1982)
orf-57	41950-42120 ninD 7.0	(Kröger and Hobom 1982; Sanger et al. 1982)
orf-60	42090-42269 ninE 7.4	(Kröger and Hobom 1982; Sanger et al. 1982)
orf-56	42269-42436 ninF 6.3	(Kröger and Hobom 1982; Sanger et al. 1982)
orf-204	42429-43040 ninG 24.1	(Kröger and Hobom 1982; Sanger et al. 1982)
orf-68	43040-43243 ninH 7.9	(Kröger and Hobom 1982; Sanger et al. 1982)
orf-221	43224-43886 ninI 25.2	(Kröger and Hobom 1982; Sanger et al. 1982)
Q	43886-44506 orf-207 22.5	(Petrov et al. 1981; Daniels and Blattner 1982; Kröger and Hobom 1982; Sanger et al. 1982)
orf-64	44621-44812 7.1	(Daniels and Blattner 1982) possible function based on similar orf's in analogous control regions of P22, 82 (J. Roberts, pers. comm.)
S	45186-45506 orf-107 11.5	(Daniels and Blattner 1982) probable start based on RBS, reading frame assigned based on sequence of λSam7 (Daniels 1981; Sanger et al. 1982)
R	45493-45966 orf-158 17.8	amino acid sequence, (Imada and Tsugita 1971; Bienkowska-Szewczyk et al 1981; Daniels 1981; Sanger et al. 1982)
Rz	45966-46423 orf-153 17.2	most probable start based on RBS, also site of *Rz* insertion mutation dK23 has been sequenced, and the BclI site at 46366 has been shown to be within *Rz* (Garrett and Young, 1982; Sanger et al 1982; Ikeda, 1981, pers. comm.)

Genes were assigned to open reading frames, as described in the text and in Sanger et al. (1982) and Daniels et al. (1983). The assignment of genetically defined genes to open reading frames is probably correct except for kil, which is uncertain. The choice of start point is often fairly arbitrary and may, in fact, be incorrect. It is not known whether the three new open reading frames in the *b* region (401, 314 and 194) or the eight new open reading frames in the *nin* region (146, 290, 57, 60, 56, 204, 68, 221), or the open reading frame in the *Q-S* intergenic region (orf 64), actually code for products or not. For further discussion of gene assignments see Sanger et al. (1982).

Abbreviations: ORF, open reading frame: RBS, ribosome-binding site.

[a]Predicted traits are the coordinates of the gene endpoints, length in amino acids and molecular weight in kilodaltons.

TABLE 5

```
 AN ALPHABETICAL  SITE  MAP OF LAMBDA15
  SITES WITH FREQUENCIES  OF 20 OR LESS ARE PRINTED
LEFT END          :     1
AATZ( 10 SITES) :  5109  9398 11247 14978 29040 40810 41117 42251 45567 45596
ACC1( 9 SITES)  :  2191 15261 18835 19474 31302 32746 33245 40202 42922
AFL2( 3 SITES)  :  6540 12618 42630
AHA3( 13 SITES) :    92  8462 16296 23112 23286 25438 26134 26667 32705 36304 36532 38835 47431
APA1( 1 SITE)   : 10090
ASU2( 7 SITES)  : 18049 25885 27981 29151 30397 34332 42638
AVA1( 8 SITES)  :  4720 19397 20999 27887 31617 33498 38214 39888
AVA3( 14 SITES) : 10324 27205 27371 28431 30341 30988 32966 33681 34207 35867 36664 36670 37768 38306
AVR2( 2 SITES)  : 24321 24395
BAL1( 18 SITES) :  1328  2208  3262  4195  6498  6879  7586  7980  8058  8861 10611 10779 13936 14905 21262 26625 28620 36042
BAM1( 5 SITES)  :  5505 22346 27972 34499 41732
BCL1( 7 SITES)  :  8844  9361 32729 37352 43682 46366 47942
BGL2( 6 SITES)  :   415 22425 35711 38103 38754 38814
BSS2( 6 SITES)  :  3522  4126  5627 14815 16649 28008
BSTE( 10 SITES) :  5687  7058  8322  9024 16012 17941 25183 30005 36374 40049
BSTX( 14 SITES) :  2862  6713  8420  8857 10922 13270 13782 14345 18036 19748 21629 34603 38299 46441
BVU1( 7 SITES)  :   585 10090 19767 21574 24776 25881 39457
CLA1( 15 SITES) :  4199 15584 16121 26617 30290 31991 32964 33585 34697 35051 36966 41364 42021 43825 46439
ECDX( 0 SITES)  :
ECOB( 8 SITES)  :  2011  3896  4467  9961 10377 24230 28590 44883
ECOK( 5 SITES)  :  6941 14980 16369 34763 47000
ECR1( 5 SITES)  : 21226 26104 31747 39168 44972
ECRV( 20 SITES) :   654  2088  6685  8088  8826 14027 17771 18389 21273 22952 26825 28202 28215 33591 39356 41277 41545 41580 42235
                  45830
HGE2( 14 SITES) :  1784  2249  5902  6554 12512 13953 15876 17432 20243 26434 35594 35638 37998 42047
HGIJ( 7 SITES)  :   585 10090 19767 21574 24776 25881 39457
HIN3( 6 SITES)  : 23130 25157 27479 36895 37459 44141
HPA1( 15 SITES) :   734  5269  5710  7950  8201 11585 12097 14993 21904 27318 31809 32219 35261 39608 39836
KPN1( 2 SITES)  : 17057 18560
MLU1( 7 SITES)  :   458  5548 15372 17791 19996 20952 22220
MST1( 16 SITES) •   465  2505  4272  5157  6981 11565 11692 12077 12204 16048 21807 21828 27951 32685 34823 42382
MST2( 2 SITES)  : 26718 34319
NAE1( 1 SITE)   : 20042
NAR1( 1 SITE)   : 45680
NCO1( 4 SITES)  : 19329 23901 27868 44248
NDE1( 7 SITES)  : 27631 29884 33680 36113 36669 38358 40132
NRU1( 5 SITES)  :  4592 28052 31705 32409 41810
PVU1( 2 SITES)  : 26257 35790
PVU2( 13 SITES) :   211  1919  2387  2528  3060  3639  7833 16080 19718 20061 20697 22993 27414
RRU1( 5 SITES)  : 16423 18686 25687 27265 32804
SAL1( 2 SITES)  : 32745 33244
SCA1( 5 SITES)  : 16420 18683 25684 27262 32801
SNA1( 3 SITES)  : 15259 18833 19472
SPH1( 5 SITES)  :  2216 23946 24375 27378 39422
SST1( 2 SITES)  : 24776 25881
SST2( 4 SITES)  : 20323 20533 21609 40389
STU1( 6 SITES)  : 12436 31480 32999 39994 40598 40616
TTH1( 2 SITES)  : 11205 36123
XBA1( 1 SITE)   : 24508
XHO1( 1 SITE)   : 33498
XMA1( 3 SITES)  : 19397 31617 39888
XMA3( 2 SITES)  : 19944 36654
RIGHT END       : 48514
```

GENOMIC MAP OF BACTERIOPHAGE T4

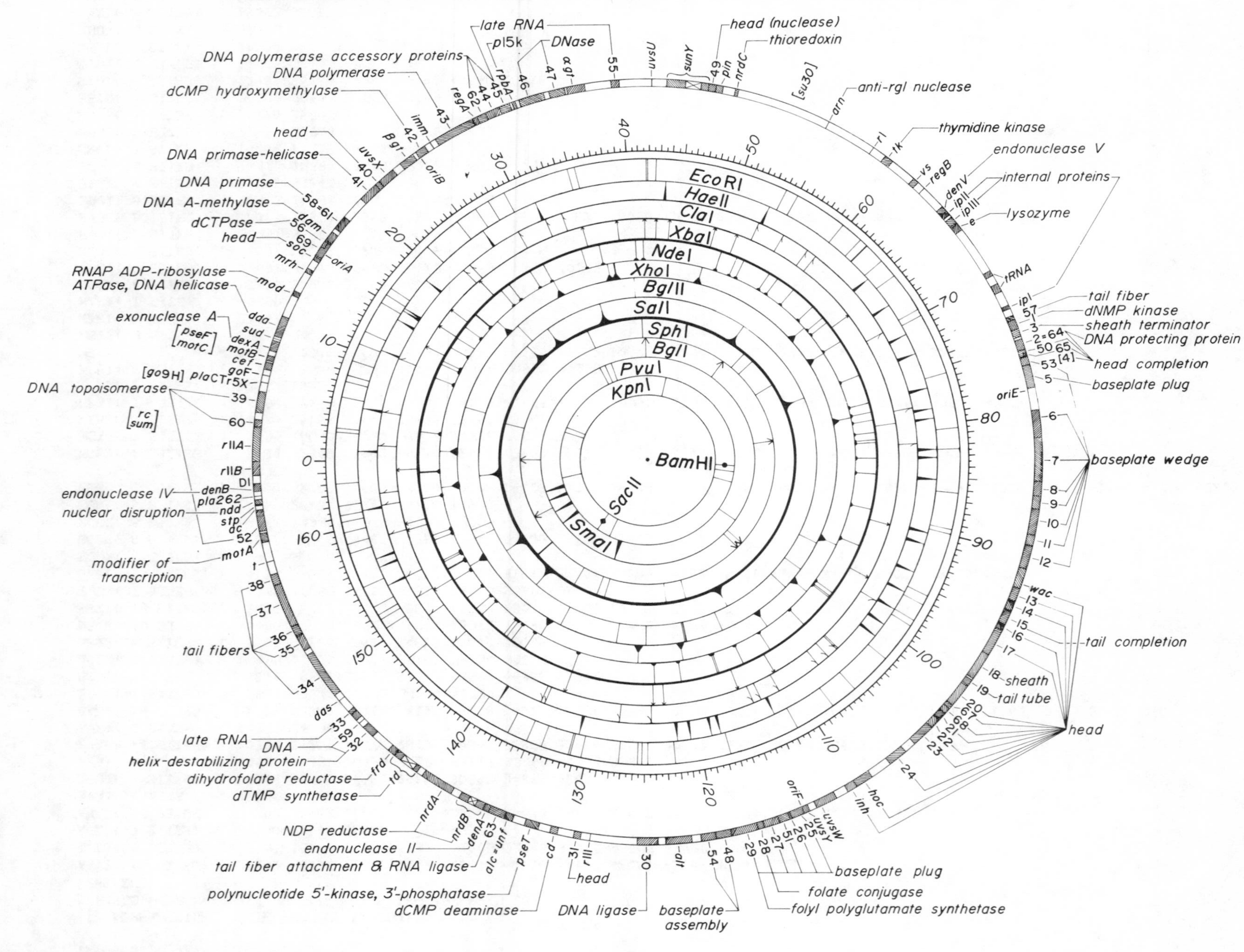

GENOMIC MAP OF BACTERIOPHAGE T4

November 20, 1989

Elizabeth Kutter and Burton Guttman, The Evergreen State College, Olympia, WA 98505; Gisela Mosig, Vanderbilt University, Nashville, TN 37235; and Wolfgang Rüger, Ruhr University, D-4630 Bochum 1, W. Germany

The genome of bacteriophage T4 contains about 166,000 base pairs (219). The mature DNA molecules are linear, but are cut with a 3% terminal redundancy from concatenated precursors and are thus circularly permuted, producing a circular genetic map (Fig. 1). About 85% of the genome has been sequenced, and the genetic and restriction maps adjusted accordingly from earlier versions. Genetic map distances agree with physical distances in most regions but are quite different in some areas around genes 43, *denV*, *uvsW*, 34 and 35.

Fig. 2 locates the cleavage sites for 20 restriction endonucleases, with genes and otherwise unidentified open reading frames (ORFs) in well-characterized regions. Those transcribed clockwise are to the left of the central line, those transcribed counterclockwise to the right; overlapping parts of genes in the same orientation are shaded. Many of the identified early, middle and late transcripts are also indicated, in that order, between the restriction map and the gene map. Regions that have been sequenced can be identified by the precision with which the cut sites are listed in the table to the right, with the delineation "hole" marking the boundaries of the sequenced regions. The expected size of the (single-enzyme-digest) fragment which starts at each cut site is also indicated, as well as its band number on gels where identified.

Table 1 gives data for all of the characterized T4 genes in standard map order, with zero, by convention, being the border between genes rIIA and rIIB. Genes that are essential on standard lab strains are designated by numbers; those whose relative positions are uncertain are in brackets. Much of the data is summarized in references 219 and 118; only more recent references are listed here. Molecular weights of protein products were determined approximately from gel data or, where underlined, were determined precisely from the DNA sequence. Tables 2 and 3 give additional data on the sequenced genes and other ORFs. By convention, each ORF is named by reference to the preceding known gene, "_.n", except that those ORFs where the only contiguous reference marker is the gene *following* are identified as "_.-n". ORFs in introns are designated "I-." Where two ORFs overlap in phase, the shorter one is designated with the same number as the longer one with a * added.

Several factors must be considered in using this map. The precise numbers will continue to shift as sequences are completed and regions joined; thus, in referring to specific coordinates, it is crucial to refer to the date of the map being used. While the sites for *Bgl*II, *Cla*I, *Eco*R1, *Kpn*I, *Sal*I and *Xho*I were mapped in several laboratories, many others have been directly mapped only relative to sites for a few other enzymes in unsequenced regions, and occasional inversions are seen, as well as discrepancies (indicated by *) which may be due to strain differences, sequencing errors, restriction-site masking or some other problem. Also, closely spaced sites may be mistaken for a single site. Furthermore, most restriction enzymes won't cut normal T4 DNA, which contains hydroxymethylcytosine instead of cytosine, so most mapping work has used cytosine-containing T4dC strains; most of these contain a deletion in the denB region, at about 165 kb on the map. *Eco*RV, *Nde*I and *Sph*I also cleave glucosylated HMC-T4 DNA.

We thank the entire T4 community for data and discussion, Eric Kutter for the computer programming, and NSF Grant DMB-8704366 (to E. K. and B. G.) and NIH grant GM-13221 (to G. M.) for support. Data cited as "personal communication" should not be used without consulting the original author.

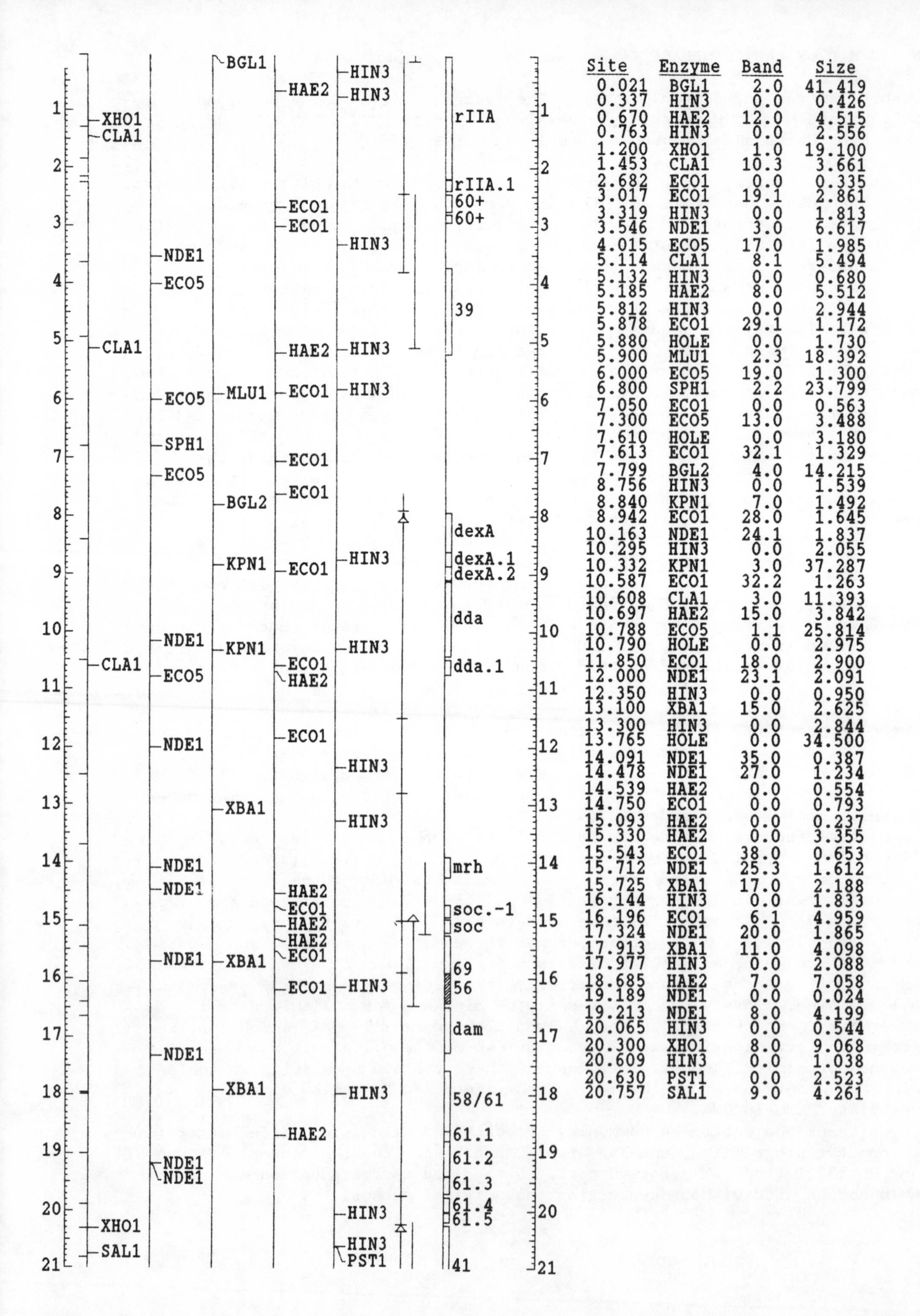

Site	Enzyme	Band	Size
0.021	BGL1	2.0	41.419
0.337	HIN3	0.0	0.426
0.670	HAE2	12.0	4.515
0.763	HIN3	0.0	2.556
1.200	XHO1	1.0	19.100
1.453	CLA1	10.3	3.661
2.682	ECO1	0.0	0.335
3.017	ECO1	19.1	2.861
3.319	HIN3	0.0	1.813
3.546	NDE1	3.0	6.617
4.015	ECO5	17.0	1.985
5.114	CLA1	8.1	5.494
5.132	HIN3	0.0	0.680
5.185	HAE2	8.0	5.512
5.812	HIN3	0.0	2.944
5.878	ECO1	29.1	1.172
5.880	HOLE	0.0	1.730
5.900	MLU1	2.3	18.392
6.000	ECO5	19.0	1.300
6.800	SPH1	2.2	23.799
7.050	ECO1	0.0	0.563
7.300	ECO5	13.0	3.488
7.610	HOLE	0.0	3.180
7.613	ECO1	32.1	1.329
7.799	BGL2	4.0	14.215
8.756	HIN3	0.0	1.539
8.840	KPN1	7.0	1.492
8.942	ECO1	28.0	1.645
10.163	NDE1	24.1	1.837
10.295	HIN3	0.0	2.055
10.332	KPN1	3.0	37.287
10.587	ECO1	32.2	1.263
10.608	CLA1	3.0	11.393
10.697	HAE2	15.0	3.842
10.788	ECO5	1.1	25.814
10.790	HOLE	0.0	2.975
11.850	ECO1	18.0	2.900
12.000	NDE1	23.1	2.091
12.350	HIN3	0.0	0.950
13.100	XBA1	15.0	2.625
13.300	HIN3	0.0	2.844
13.765	HOLE	0.0	34.500
14.091	NDE1	35.0	0.387
14.478	NDE1	27.0	1.234
14.539	HAE2	0.0	0.554
14.750	ECO1	0.0	0.793
15.093	HAE2	0.0	0.237
15.330	HAE2	0.0	3.355
15.543	ECO1	38.0	0.653
15.712	NDE1	25.3	1.612
15.725	XBA1	17.0	2.188
16.144	HIN3	0.0	1.833
16.196	ECO1	6.1	4.959
17.324	NDE1	20.0	1.865
17.913	XBA1	11.0	4.098
17.977	HIN3	0.0	2.088
18.685	HAE2	7.0	7.058
19.189	NDE1	0.0	0.024
19.213	NDE1	8.0	4.199
20.065	HIN3	0.0	0.544
20.300	XHO1	8.0	9.068
20.609	HIN3	0.0	1.038
20.630	PST1	0.0	2.523
20.757	SAL1	9.0	4.261

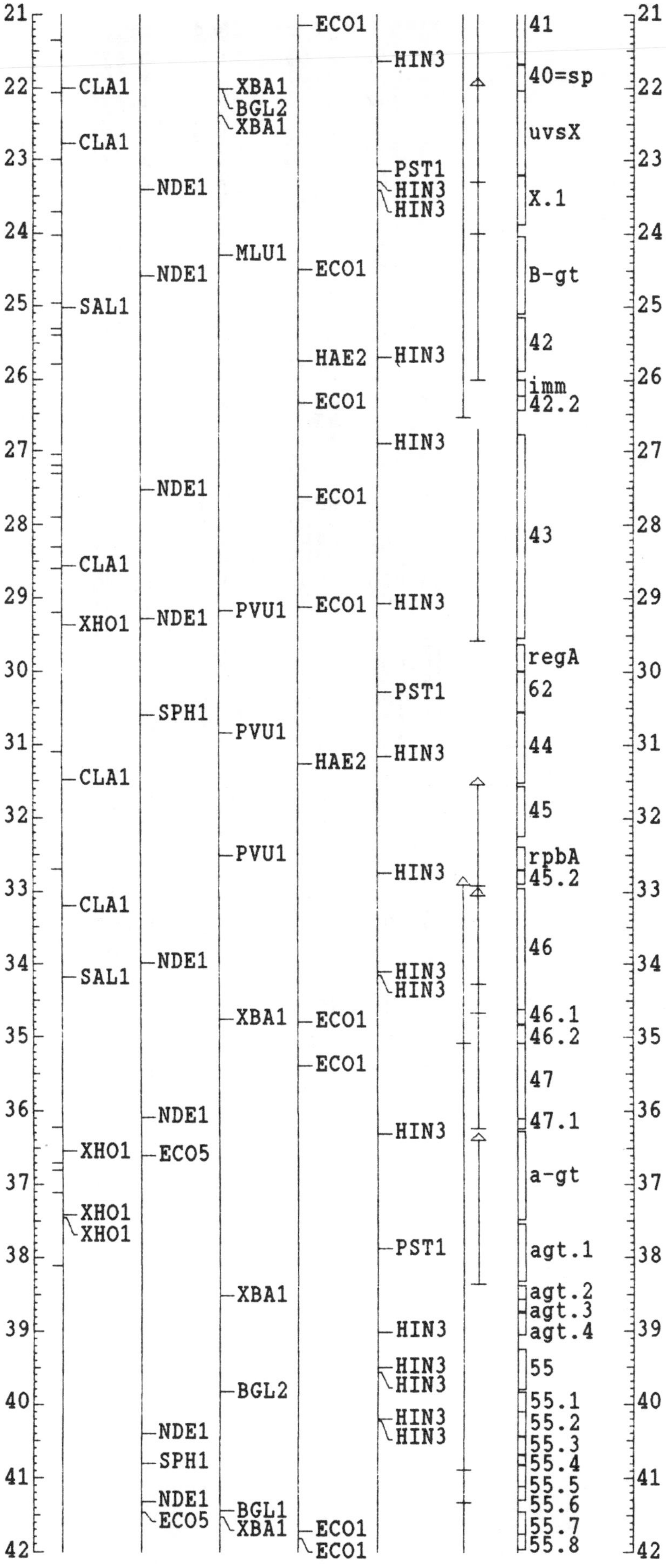

21 22 23 24 25 26 27 28 29 30 31 32 33 34 35 36 37 38 39 40 41 42

CLA1 CLA1 SAL1 CLA1 XHO1 CLA1 CLA1 SAL1 XHO1 XHO1 XHO1

NDE1 NDE1 NDE1 NDE1 SPH1 NDE1 NDE1 ECO5 NDE1 SPH1 NDE1 ECO5

XBA1 BGL2 XBA1 MLU1 PVU1 PVU1 PVU1 XBA1 XBA1 BGL2 BGL1 XBA1

ECO1 ECO1 HAE2 ECO1 ECO1 ECO1 HAE2 ECO1 ECO1 ECO1 ECO1

HIN3 PST1 HIN3 HIN3 HIN3 HIN3 HIN3 PST1 HIN3 HIN3 HIN3 HIN3 HIN3 PST1 HIN3 HIN3 HIN3 HIN3 HIN3

41 40=sp uvsX X.1 B-gt 42 imm 42.2 43 regA 62 44 45 rpbA 45.2 46 46.1 46.2 47 47.1 a-gt agt.1 agt.2 agt.3 agt.4 55 55.1 55.2 55.3 55.4 55.5 55.6 55.7 55.8

Site	Enzyme	Band	Size
21.155	ECO1	17.1	3.340
21.647	HIN3	0.0	1.656
22.001	CLA1	21.0	0.778
22.011	XBA1	0.0	0.365
22.014	BGL2	3.1	17.816
22.376	XBA1	4.0	12.389
22.779	CLA1	8.2	5.774
23.153	PST1	0.0	7.123
23.303	HIN3	0.0	0.127
23.412	NDE1	28.1	1.183
23.430	HIN3	0.0	2.260
24.292	MLU1	2.5	18.625
24.495	ECO1	26.1	1.848
24.595	NDE1	16.0	2.935
25.018	SAL1	7.0	9.164
25.690	HIN3	0.0	1.206
25.743	HAE2	9.0	5.516
26.343	ECO1	32.3	1.286
26.896	HIN3	0.0	2.173
27.530	NDE1	24.2	1.754
27.629	ECO1	30.0	1.489
28.553	CLA1	12.2	2.925
28.574	CLA*	0.0	3.696
29.069	HIN3	0.0	2.090
29.118	ECO1	4.0	5.683
29.175	PVU1	4.0	1.670
29.284	NDE1	5.4	4.698
29.368	XHO1	9.1	7.175
30.276	PST1	0.0	7.599
30.599	SPH1	4.0	10.202
30.845	PVU1	3.0	1.664
31.159	HIN3	0.0	1.585
31.259	HAE2	2.1	12.259
31.478	CLA1	16.0	1.719
32.270	CLA*	0.0	162.304
32.509	PVU1	2.0	67.045
32.744	HIN3	0.0	1.371
33.197	CLA1	1.0	19.903
33.982	NDE1	21.0	2.100
34.115	HIN3	0.0	0.048
34.163	HIN3	0.0	2.147
34.182	SAL1	3.0	23.718
34.765	XBA1	13.0	3.739
34.801	ECO1	0.0	0.581
35.382	ECO1	2.0	6.347
36.082	NDE1	7.0	4.310
36.310	HIN3	0.0	2.694
36.543	XHO1	13.0	0.867
36.602	ECO5	9.0	4.864
37.410	XHO1	14.0	0.038
37.448	XHO1	5.1	11.452
37.875	PST1	0.0	8.437
38.504	XBA1	14.0	3.022
39.004	HIN3	0.0	0.499
39.503	HIN3	0.0	0.069
39.572	HIN3	0.0	0.629
39.830	BGL2	1.0	55.828
40.201	HIN3	0.0	0.020
40.221	HIN3	0.0	3.727
40.392	NDE1	30.1	0.925
40.801	SPH1	2.3	19.935
41.317	NDE1	1.2	9.233
41.440	BGL1	5.0	17.193
41.466	ECO5	23.0	0.803
41.526	XBA1	19.0	1.206
41.729	ECO1	0.0	0.093
41.822	ECO1	0.0	0.780

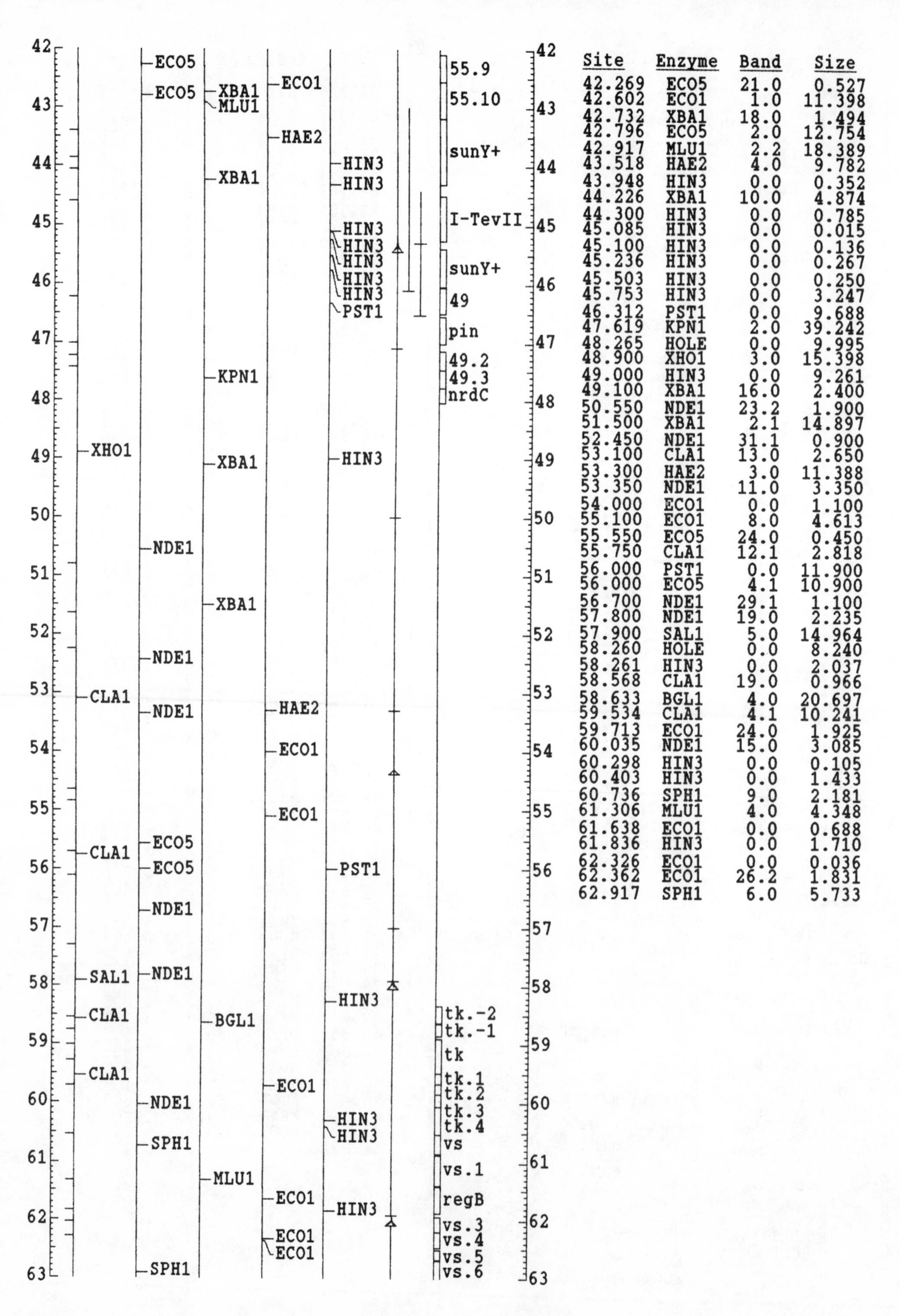

Site	Enzyme	Band	Size
42.269	ECO5	21.0	0.527
42.602	ECO1	1.0	11.398
42.732	XBA1	18.0	1.494
42.796	ECO5	2.0	12.754
42.917	MLU1	2.2	18.389
43.518	HAE2	4.0	9.782
43.948	HIN3	0.0	0.352
44.226	XBA1	10.0	4.874
44.300	HIN3	0.0	0.785
45.085	HIN3	0.0	0.015
45.100	HIN3	0.0	0.136
45.236	HIN3	0.0	0.267
45.503	HIN3	0.0	0.250
45.753	HIN3	0.0	3.247
46.312	PST1	0.0	9.688
47.619	KPN1	2.0	39.242
48.265	HOLE	0.0	9.995
48.900	XHO1	3.0	15.398
49.000	HIN3	0.0	9.261
49.100	XBA1	16.0	2.400
50.550	NDE1	23.2	1.900
51.500	XBA1	2.1	14.897
52.450	NDE1	31.1	0.900
53.100	CLA1	13.0	2.650
53.300	HAE2	3.0	11.388
53.350	NDE1	11.0	3.350
54.000	ECO1	0.0	1.100
55.100	ECO1	8.0	4.613
55.550	ECO5	24.0	0.450
55.750	CLA1	12.1	2.818
56.000	PST1	0.0	11.900
56.000	ECO5	4.1	10.900
56.700	NDE1	29.1	1.100
57.800	NDE1	19.0	2.235
57.900	SAL1	5.0	14.964
58.260	HOLE	0.0	8.240
58.261	HIN3	0.0	2.037
58.568	CLA1	19.0	0.966
58.633	BGL1	4.0	20.697
59.534	CLA1	4.1	10.241
59.713	ECO1	24.0	1.925
60.035	NDE1	15.0	3.085
60.298	HIN3	0.0	0.105
60.403	HIN3	0.0	1.433
60.736	SPH1	9.0	2.181
61.306	MLU1	4.0	4.348
61.638	ECO1	0.0	0.688
61.836	HIN3	0.0	1.710
62.326	ECO1	0.0	0.036
62.362	ECO1	26.2	1.831
62.917	SPH1	6.0	5.733

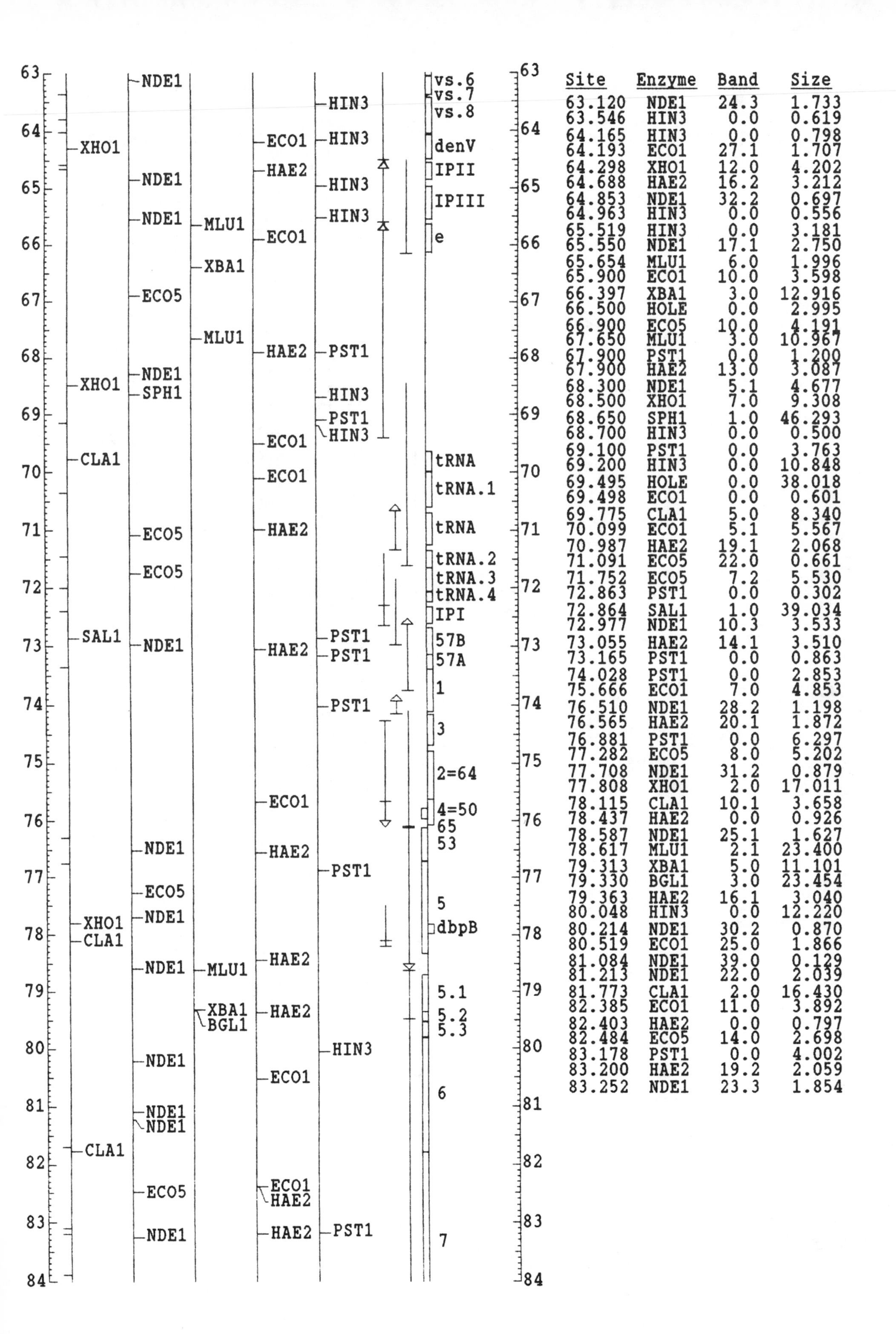

Site	Enzyme	Band	Size
63.120	NDE1	24.3	1.733
63.546	HIN3	0.0	0.619
64.165	HIN3	0.0	0.798
64.193	ECO1	27.1	1.707
64.298	XHO1	12.0	4.202
64.688	HAE2	16.2	3.212
64.853	NDE1	32.2	0.697
64.963	HIN3	0.0	0.556
65.519	HIN3	0.0	3.181
65.550	NDE1	17.1	2.750
65.654	MLU1	6.0	1.996
65.900	ECO1	10.0	3.598
66.397	XBA1	3.0	12.916
66.500	HOLE	0.0	2.995
66.900	ECO5	10.0	4.191
67.650	MLU1	3.0	10.967
67.900	PST1	0.0	1.200
67.900	HAE2	13.0	3.087
68.300	NDE1	5.1	4.677
68.500	XHO1	7.0	9.308
68.650	SPH1	1.0	46.293
68.700	HIN3	0.0	0.500
69.100	PST1	0.0	3.763
69.200	HIN3	0.0	10.848
69.495	HOLE	0.0	38.018
69.498	ECO1	0.0	0.601
69.775	CLA1	5.0	8.340
70.099	ECO1	5.1	5.567
70.987	HAE2	19.1	2.068
71.091	ECO5	22.0	0.661
71.752	ECO5	7.2	5.530
72.863	PST1	0.0	0.302
72.864	SAL1	1.0	39.034
72.977	NDE1	10.3	3.533
73.055	HAE2	14.1	3.510
73.165	PST1	0.0	0.863
74.028	PST1	0.0	2.853
75.666	ECO1	7.0	4.853
76.510	NDE1	28.2	1.198
76.565	HAE2	20.1	1.872
76.881	PST1	0.0	6.297
77.282	ECO5	8.0	5.202
77.708	NDE1	31.2	0.879
77.808	XHO1	2.0	17.011
78.115	CLA1	10.1	3.658
78.437	HAE2	0.0	0.926
78.587	NDE1	25.1	1.627
78.617	MLU1	2.1	23.400
79.313	XBA1	5.0	11.101
79.330	BGL1	3.0	23.454
79.363	HAE2	16.1	3.040
80.048	HIN3	0.0	12.220
80.214	NDE1	30.2	0.870
80.519	ECO1	25.0	1.866
81.084	NDE1	39.0	0.129
81.213	NDE1	22.0	2.039
81.773	CLA1	2.0	16.430
82.385	ECO1	11.0	3.892
82.403	HAE2	0.0	0.797
82.484	ECO5	14.0	2.698
83.178	PST1	0.0	4.002
83.200	HAE2	19.2	2.059
83.252	NDE1	23.3	1.854

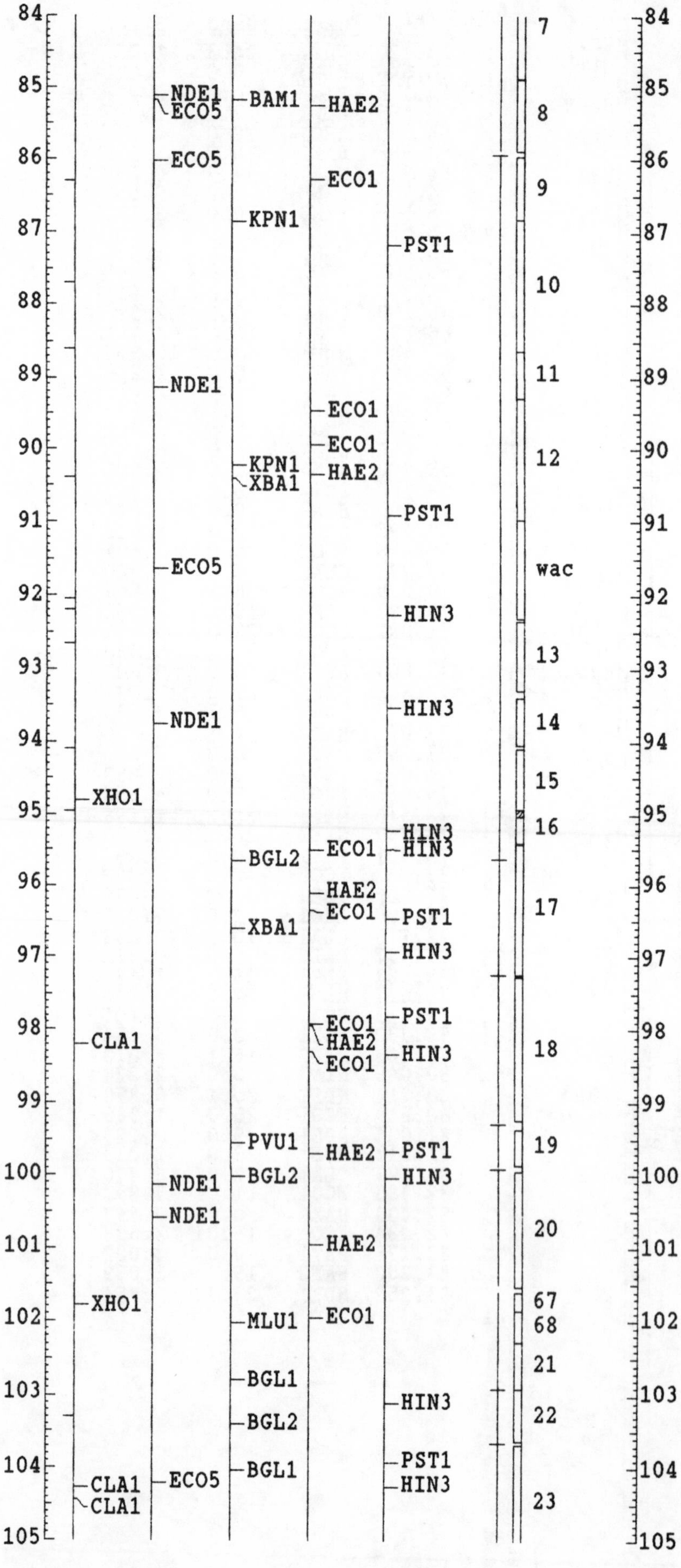

Site	Enzyme	Band	Size
85.106	NDE1	9.0	4.041
85.182	ECO5	20.0	0.837
85.189	BAM1	1.0	166.000
85.259	HAE2	10.1	5.102
86.019	ECO5	7.1	5.619
86.277	ECO1	16.1	3.176
86.861	KPN1	6.0	3.373
87.180	PST1	0.0	3.732
89.147	NDE1	6.1	4.606
89.453	ECO1	41.1	0.480
89.933	ECO1	5.2	5.579
90.234	KPN1	1.0	46.350
90.361	HAE2	0.0	5.750
90.414	XBA1	9.1	6.191
90.912	PST1	0.0	5.554
91.103	HAE*	17.2	2.800
91.638	ECO5	3.0	12.569
92.268	HIN3	0.0	1.268
93.536	HIN3	0.0	1.697
93.753	NDE1	4.1	6.381
93.903	HAE*	19.3	61.897
94.819	XHO1	9.2	6.973
95.233	HIN3	0.0	0.250
95.483	HIN3	0.0	1.456
95.512	ECO1	35.1	0.841
95.658	BGL2	8.1	4.363
96.111	HAE2	20.2	1.830
96.353	ECO1	29.1	1.572
96.466	PST1	0.0	1.342
96.605	XBA1	2.2	14.288
96.939	HIN3	0.0	1.403
97.808	PST1	0.0	1.871
97.925	ECO1	39.2	0.380
97.941	HAE2	21.1	1.755
98.203	CLA1	7.0	6.070
98.305	ECO1	13.0	3.652
98.342	HIN3	0.0	1.704
99.554	PVU1	1.0	95.621
99.679	PST1	0.0	4.241
99.696	HAE2	22.1	1.257
99.853	MLU*	5.0	166.000
100.021	BGL2	9.0	3.378
100.046	HIN3	0.0	3.061
100.134	NDE1	34.0	0.450
100.403	NDE*	37.0	16.332
100.584	NDE1	6.3	4.515
100.953	HAE2	2.2	11.869
101.792	XHO1	9.3	6.874
101.957	ECO1	12.1	3.799
102.017	MLU1	2.4	19.735
102.784	BGL1	7.0	1.255
103.107	HIN3	0.0	1.146
103.399	BGL2	5.0	10.601
103.920	PST1	0.0	2.665
104.039	BGL1	1.0	49.311
104.207	ECO5	5.0	9.231
104.253	HIN3	0.0	1.523
104.273	CLA1	23.0	0.168
104.441	CLA1	18.0	1.038

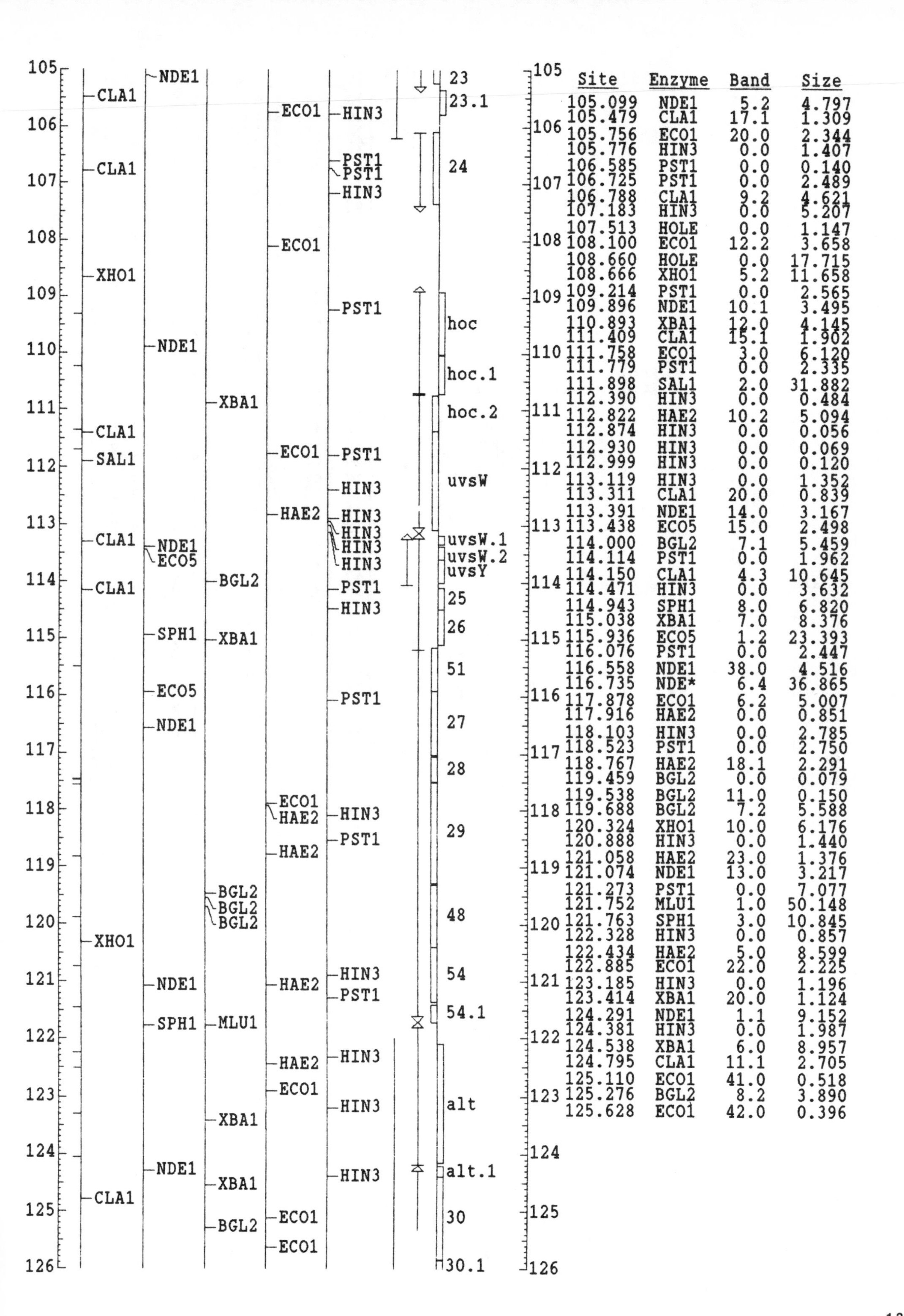

Site	Enzyme	Band	Size
105.099	NDE1	5.2	4.797
105.479	CLA1	17.1	1.309
105.756	ECO1	20.0	2.344
105.776	HIN3	0.0	1.407
106.585	PST1	0.0	0.140
106.725	PST1	0.0	2.489
106.788	CLA1	9.2	4.621
107.183	HIN3	0.0	5.207
107.513	HOLE	0.0	1.147
108.100	ECO1	12.2	3.658
108.660	HOLE	0.0	17.715
108.666	XHO1	5.2	11.658
109.214	PST1	0.0	2.565
109.896	NDE1	10.1	3.495
110.893	XBA1	12.0	4.145
111.409	CLA1	15.1	1.902
111.758	ECO1	3.0	6.120
111.779	PST1	0.0	2.335
111.898	SAL1	2.0	31.882
112.390	HIN3	0.0	0.484
112.822	HAE2	10.2	5.094
112.874	HIN3	0.0	0.056
112.930	HIN3	0.0	0.069
112.999	HIN3	0.0	0.120
113.119	HIN3	0.0	1.352
113.311	CLA1	20.0	0.839
113.391	NDE1	14.0	3.167
113.438	ECO5	15.0	2.498
114.000	BGL2	7.1	5.459
114.114	PST1	0.0	1.962
114.150	CLA1	4.3	10.645
114.471	HIN3	0.0	3.632
114.943	SPH1	8.0	6.820
115.038	XBA1	7.0	8.376
115.936	ECO5	1.2	23.393
116.076	PST1	0.0	2.447
116.558	NDE1	38.0	4.516
116.735	NDE*	6.4	36.865
117.878	ECO1	6.2	5.007
117.916	HAE2	0.0	0.851
118.103	HIN3	0.0	2.785
118.523	PST1	0.0	2.750
118.767	HAE2	18.1	2.291
119.459	BGL2	0.0	0.079
119.538	BGL2	11.0	0.150
119.688	BGL2	7.2	5.588
120.324	XHO1	10.0	6.176
120.888	HIN3	0.0	1.440
121.058	HAE2	23.0	1.376
121.074	NDE1	13.0	3.217
121.273	PST1	0.0	7.077
121.752	MLU1	1.0	50.148
121.763	SPH1	3.0	10.845
122.328	HIN3	0.0	0.857
122.434	HAE2	5.0	8.599
122.885	ECO1	22.0	2.225
123.185	HIN3	0.0	1.196
123.414	XBA1	20.0	1.124
124.291	NDE1	1.1	9.152
124.381	HIN3	0.0	1.987
124.538	XBA1	6.0	8.957
124.795	CLA1	11.1	2.705
125.110	ECO1	41.0	0.518
125.276	BGL2	8.2	3.890
125.628	ECO1	42.0	0.396

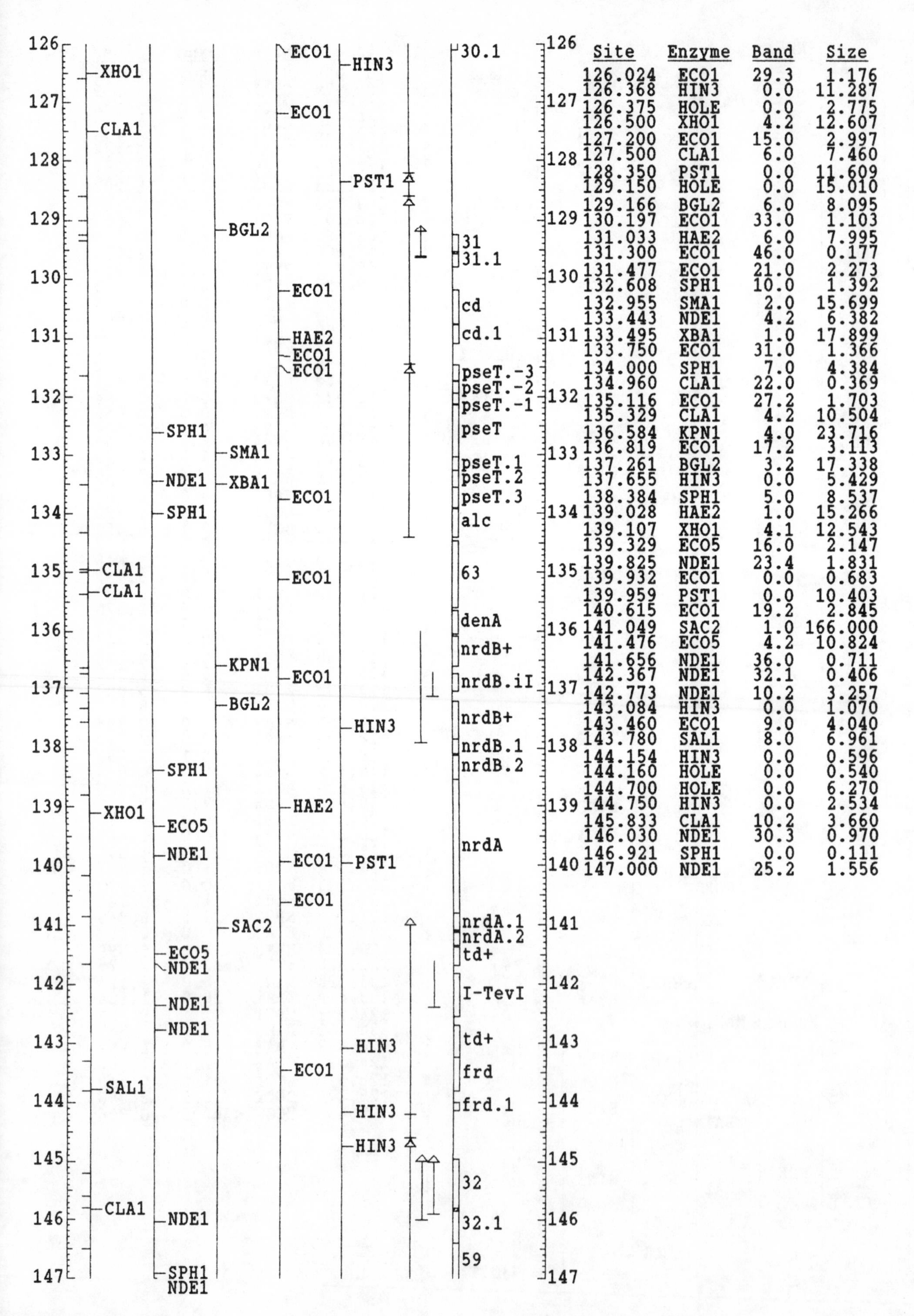

Site	Enzyme	Band	Size
126.024	ECO1	29.3	1.176
126.368	HIN3	0.0	11.287
126.375	HOLE	0.0	2.775
126.500	XHO1	4.2	12.607
127.200	ECO1	15.0	2.997
127.500	CLA1	6.0	7.460
128.350	PST1	0.0	11.609
129.150	HOLE	0.0	15.010
129.166	BGL2	6.0	8.095
130.197	ECO1	33.0	1.103
131.033	HAE2	6.0	7.995
131.300	ECO1	46.0	0.177
131.477	ECO1	21.0	2.273
132.608	SPH1	10.0	1.392
132.955	SMA1	2.0	15.699
133.443	NDE1	4.2	6.382
133.495	XBA1	1.0	17.899
133.750	ECO1	31.0	1.366
134.000	SPH1	7.0	4.384
134.960	CLA1	22.0	0.369
135.116	ECO1	27.2	1.703
135.329	CLA1	4.2	10.504
136.584	KPN1	4.0	23.716
136.819	ECO1	17.2	3.113
137.261	BGL2	3.2	17.338
137.655	HIN3	0.0	5.429
138.384	SPH1	5.0	8.537
139.028	HAE2	1.0	15.266
139.107	XHO1	4.1	12.543
139.329	ECO5	16.0	2.147
139.825	NDE1	23.4	1.831
139.932	ECO1	0.0	0.683
139.959	PST1	0.0	10.403
140.615	ECO1	19.2	2.845
141.049	SAC2	1.0	166.000
141.476	ECO5	4.2	10.824
141.656	NDE1	36.0	0.711
142.367	NDE1	32.1	0.406
142.773	NDE1	10.2	3.257
143.084	HIN3	0.0	1.070
143.460	ECO1	9.0	4.040
143.780	SAL1	8.0	6.961
144.154	HIN3	0.0	0.596
144.160	HOLE	0.0	0.540
144.700	HOLE	0.0	6.270
144.750	HIN3	0.0	2.534
145.833	CLA1	10.2	3.660
146.030	NDE1	30.3	0.970
146.921	SPH1	0.0	0.111
147.000	NDE1	25.2	1.556

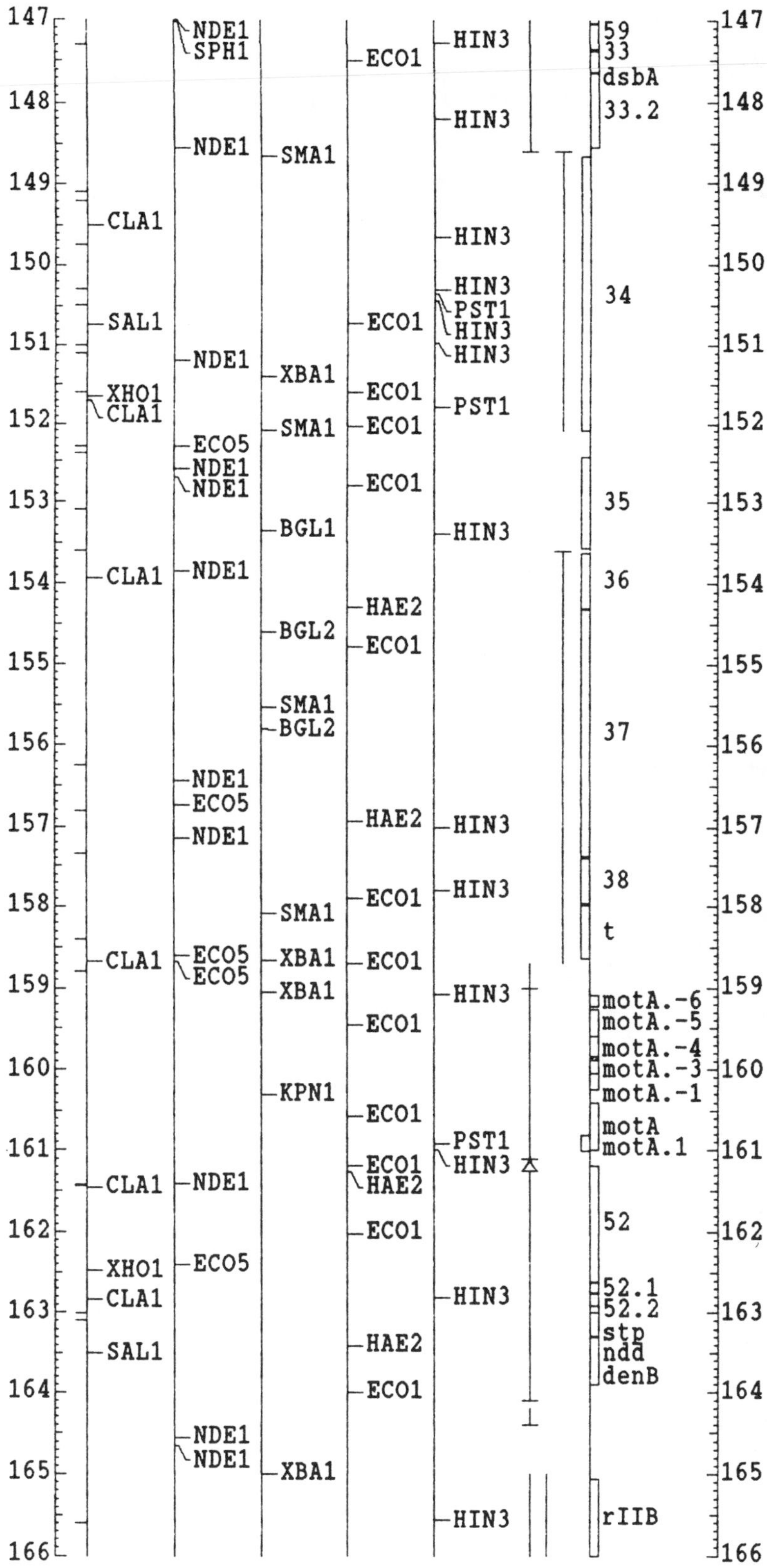

Site	Enzyme	Band	Size
147.000	NDE1	25.2	1.556
147.032	SPH1	2.1	25.768
147.284	HIN3	0.0	0.923
147.500	ECO1	14.0	3.227
148.207	HIN3	0.0	1.455
148.556	NDE1	17.2	2.644
148.654	SMA1	3.1	3.446
149.493	CLA1	14.0	2.207
149.662	HIN3	0.0	0.642
150.304	HIN3	0.0	0.145
150.362	PST1	0.0	1.438
150.449	HIN3	0.0	0.520
150.727	ECO1	0.0	0.873
150.741	SAL1	6.0	12.759
150.969	HIN3	0.0	2.422
150.970	HOLE	0.0	2.420
151.200	NDE1	26.0	1.400
151.394	XBA1	8.0	7.269
151.600	ECO1	0.0	0.450
151.650	XHO1	6.0	10.828
151.700	CLA1	15.2	2.228
151.800	PST1	0.0	9.116
152.050	ECO1	0.0	0.750
152.100	SMA1	3.2	3.431
152.300	ECO5	11.0	4.429
152.600	NDE1	41.0	0.100
152.700	NDE1	29.2	1.152
152.800	ECO1	23.0	1.976
153.350	BGL1	6.0	12.671
153.390	HOLE	0.0	5.310
153.391	HIN3	0.0	3.616
153.600	NDE*	33.0	112.803
153.852	NDE1	18.0	2.590
153.928	CLA1	9.1	4.742
154.294	HAE2	21.2	2.633
154.599	BGL2	10.0	1.194
154.776	ECO1	16.2	3.122
155.531	SMA1	4.0	2.553
155.793	BGL2	2.0	18.006
155.800	HAE*	0.0	101.303
156.442	NDE1	33.0	0.708
156.729	ECO5	18.0	1.871
156.927	HAE2	11.0	4.341
157.007	HIN3	0.0	0.799
157.150	NDE1	6.2	4.264
157.806	HIN3	0.0	1.271
157.898	ECO1	0.0	0.800
158.084	SMA1	1.0	140.871
158.600	ECO5	12.0	0.067
158.663	XBA1	21.0	0.387
158.667	ECO5	25.0	3.733
158.670	CLA1	11.2	2.795
158.698	ECO1	0.0	0.758
158.700	HOLE	0.0	0.374
159.050	XBA1	9.2	5.950
159.074	HOLE	0.0	4.356
159.077	HIN3	0.0	1.914
159.456	ECO1	0.0	1.118
160.300	KPN1	5.0	14.540
160.574	ECO1	0.0	0.612
160.916	PST1	0.0	25.714
160.991	HIN3	0.0	1.818
161.186	ECO1	0.0	0.834
161.268	HAE2	18.2	2.140
161.414	NDE1	12.0	3.136
161.465	CLA1	17.2	1.369
162.020	ECO1	0.0	1.980
162.400	ECO5	6.0	7.615
162.478	XHO1	11.0	4.722
162.809	HIN3	0.0	2.754
162.834	CLA1	9.3	4.619
163.408	HAE2	14.2	3.262
163.430	HOLE	0.0	1.610
163.500	SAL1	4.0	23.257
164.000	ECO1	0.0	4.682
164.550	NDE1	40.0	0.100
164.650	NDE1	5.3	4.896
165.000	XBA1	2.3	14.100
165.040	HOLE	0.0	6.840
165.563	HIN3	0.0	0.774

GENES OF BACTERIOPHAGE T4

Gene	Mutant Phenotype	Restrictive Host	Function and Comments	Mol Wt (x 1000)	References
rIIA	Rapid lysis; suppresses lig⁻ and some 32⁻; nonessential	rex+ lysogens, tabR	Membrane protein; affects membrane ATPase of E. coli	72, 74, 83, 95, 82.8	24, 118, 181a, 26a
[m = sum]	Suppresses lig⁻; nonessential				118
[rc]	Acriflavine resistance				118
60	DNA delay	S/6, 25°	Membrane protein, DNA topoisomerase subunit	16, 18, 18.6; gene contains untranslated sequence	71, 72
39	DNA delay	S/6, 25°	Membrane protein, DNA topoisomerase subunit, DNA-dependent ATPase	63 → 22 58.5	70, 71, 73, 118
plaCTr5x	Nonessential	CTr5x			118
[goF = go9H = comCα ≈ motC]	Nonessential		Mutations overcome block in HDF (rho) hosts		118, 155, 190
cef = mb = M1 ≈ motC	Nonessential	roc⁻, CT439	Modifier of suppressor tRNAs and of species 1, 2 and 3 RNA	18	118, 150, 155
[del(39-56)$_4$ ≈ motB]	Nonessential		Modifier of transcription?	12	118, 155
[pseF]	Nonessential		5'-phosphatase		118
dexA ≈ sud	Nonessential; suppresses 32⁻	optA	Exonuclease A	26.0	41, 111, 118
dda	Nonessential		DNA-dependent ATPase, DNA helicase	56	41, 58, 75, 88, 111, 118
[mod]	Nonessential		Adenylribosylation of RNA polymerase		118
mrh	Nonessential		Inhibits growth on rpoH mutants	13.4	37
soc	Nonessential		Small outer capsid protein	9	9, 112, 113, 118
{oriA in 69}					118, 134
69, 69* (overlaps 56)	Does not use oriA	25°		44, 26; has untranslated sequence	17, 112, 118, 133, 134

Gene	Mutant Phenotype	Restrictive Host	Function and Comments	Mol wt (x 1000)	References
56 (overlaps 69)	DNA negative		dCTPase, dUTPase, dCDPase, dUDPase	20	40, 112, 133
dam	Nonessential		DNA adenine methylase	30	112, 118, 173, 174
58 = 61	DNA delay	S/6, 25°	Primase subunit	40, 39.5	11, 14, 20, 118, 136
41	DNA arrest; single-stranded DNA, UV-sensitive		GTPase, dGTPase-, ATPase, dATPase; helicase-primase subunit	57, 59, 63, 66, 53.8	14, 65, 66, 67, 136, 163, 176
40 = sp = rIV	Polyheads; suppresses e^-		genetic exclus.; helps head vertex assembly; lysis	14, 18, 13.3	38, 65, 67, 118, 126, 139, 140
uvsX = fdsA	UV-sensitive, recombination deficient; suppresses 49^-		RecA-like recombination protein; DNA-dependent ATPase	40, 43.8	25, 38, 51, 65, 88, 118, 223, 224, 225
X.1	Nonessential		DNA-binding protein	25	35
βgt	No β-glucosylation of HMC-DNA		β-glucosyltransferase	40.6	118, 198
42	DNA negative		dCMP-Hydroxymethylase	28.4	101, 102, 110, 118, 197
imm	Immunity to superinfection exclusion		Plasma membrane; inhibits injec. of superinf. DNA	9.3	102, 110, 118, 139
43	DNA negative		DNA polymerase	112, 103.6	118, 187
dsd (in 43)	DNA delay	optA	DNA polymerase		26
regA			Translational regulation of (early) protein synthesis	14.6	1, 118, 129, 200
62	DNA negative		DNA-polymerase accessory protein; ssDNA-dependent ATPase, dATPase	21.4	118
44				35.9	118
45	DNA negative, no late mRNA		Accessory protein of DNA- and RNA-polymerases	24.7	118, 189
rpbA	Nonessential		RNA polymerase binding, "15K", protein	11.4	68, 216

Gene	Mutant Phenotype	Restrictive Host	Function and Comments	Mol wt (x 1000)	References
46	DNA arrest; recombination		Recombination nuclease	63.5	49, 118
47	deficient; reduced host DNA degradation			39.1	
αgt	No α-glucosylation of HMC-DNA		α-Glucosyl transferase	46.6	49, 118, 198
[gor]	Suppresses RNA polymerase defect				118
55	No late RNA synthesis		RNA polymerase σ factor for late T4 promoters	21.5	34, 49, 118
sunY			Unknown	67.9; contains intron	47, 179, 197a, 199
I-TevII			In sunY intron; intron mobility endonuclease	30.4	7b, 156b
49, 49*	Partially filled heads, highly branched DNA		Endonuclease VII; cleaves recombination junctions	18.1 12	6, 86, 118, 197a, 199
pin	Nonessential		Inhibitor of E. coli protease	18.8	180, 183, 197a, 199
nrdC	Nonessential		Thioredoxin	10.0103, 118, 197a, 199	
arn	DNA degradation under certain conditions		Antirestriction nuclease (vs. E. coli rglB = mcrB)		27, 118
[Su30]	Nonessential; suppresses lig^-				118
rI	Nonessential; rapid lysis				118
tk	Nonessential		Thymidine kinase	28, 21.6	118, 128, 207
vs	Nonessential		Modifies valyl-tRNA synthetase	13.1	116, 118, 128, 142, 143, 207
regB	Nonessential; folate analogue resistant		Regulation of gene expression; site-specific ribonuclease	18.0	118, 168a, 203, 207
[stI]	Nonessential "star"				118
[stIII]	Nonessential; suppresses e^- and t^-; star				118
denV	UV-sensitive		Endonuclease V, N-glycosidase	16.1	118, 157, 162, 205, 206, 207
ipII	Nonessential		Internal protein	11.1 → 9.9	118, 205, 207

Gene	Mutant Phenotype	Restrictive Host	Function and Comments	Mol wt (x 1000)	References
ipIII	Semi-essential		Internal protein	21.7 → 20.4	118, 207
e	No lysis		Endolysin	18.7	118, 124, 145
goF3	Grow on HDF hosts				118, 124
Stable species 1 (= C) RNA			unknown		10, 118, 150
Stable species 2 (= D) RNA			unknown		10, 118, 150
tRNAs	Nonessential (nonsense suppressors)	CT439	tRNA precursors		10, 56, 59, 118, 120
arg	psu_4, op				
ile					
thr					
ser	psu_a^+, psu_b^+, psu_1^+				
pro					
gly					
leu	psu_3^+				
gln	psu^+, psu_2SB				
ipI		CT596	Internal protein	10 → 8.5	10, 118
57B	Poor tail fiber assembly, bypassed in some host mutants		Morphogenetic catalyst of long and short tail fiber assembly	16	10, 118
57A			DNA-binding protein	7	10, 36, 118
1	DNA negative		dHMP-kinase	22, 25, 27.3	10, 109, 118
3	Unstable tails		Tail tube, proximal tip	23.3, 19.6	10, 87a, 109, 118
2 = 64	Inactive filled heads, noninfectious particles		Head completion; terminal DNA-protecting protein	25, 27, 31.6	87a, 118, 109,
4 = 50 = [65]	Inactive filled heads, noninfectious particles		Head completion; function unknown	17.6	87a, 105, 118
65	Encoded on complementary strand of 4		DNA-binding protein	7.8	105, 169
53	Defective tails		Baseplate, (wedge)	23, 22.9	87a, 118, 132

Gene	Mutant Phenotype	Restrictive Host	Function and Comments	Mol wt (x 1000)	References
5	Defective tails		Baseplate hub, baseplate lysozyme	63.0 → 44,	118, 132, 135
dbpB	Encoded on complementary strand of 5		DNA-binding protein	6.1	105
{oriE}					105, 106, 118
6	Defective tails, permit fiberless plating		Baseplate, 1/6 arm	74.4, 78, 86	118, 177
7	Defective tails, permit fiberless plating		Baseplate, 1/6 arm	127, 140, 119.2	117, 118, 177
8	Defective tails		Baseplate, 1/6 arm	39, 46, 38.0	99, 117, 118, 177
9	Defective tails fiberless particles		Baseplate, long tail fiber attachment	30, 34, 31.0	99, 118, 153
10	Defective tails		Baseplate, 1/6 arm	88, 90	118, 149, 153
11	Defective tails		Baseplate, 1/6 arm	24, 25, 26, 23.7	118, 149, 153
12	Defective tails		Baseplate, short tail fibers	54, 56, 57, 56.2	118, 178
wac	Nonessential		Whisker antigen	52, 51.9	118, 153
13			Head completion	33, 34.7	118, 178a
14			Head completion	30, 29.6	118, 178a
15	Defective tails		Proximal tail sheath stabilizer	32, 35, 31.4	118, 178a
16	DNA packaging defective; empty heads		Head filling packasome component	18.3	118, 151, 160
17 = q	DNA packaging defective; suppresses some 32^- or 20^- mutations; quinacrine resistant		Head filling packasome	69, 70.0	118, 151, 160
18	Defective tails		Tail sheath monomer	70, 80, 71.4	3, 4, 118
19	Defective tails		Tail tube monomer	20, 21, 18.5	3, 4, 118
20	Polyheads		Head plug protein (connector to neck)	63, 65, 67, 61.0	118, 161, 117a

Gene	Mutant Phenotype	Restrictive Host	Function and Comments	Mol wt (x 1000)	References
67 = pip	Head defect		Core protein, precursor to internal peptides	9.1	83, 84a, 118, 199a
68	Isometric heads		Prohead core protein	17	82, 83, 84, 199a
21	Faulty heads		Head assembly core; maturation protease	23.2 → 18.4 cleaved to small peptides	81, 118
22	Faulty heads		Head assembly core (later degraded)	29.8	83, 118, 147
23	No or faulty heads		Major head subunit; cleaved to a packaging-related DNA-dependent ATPase-endonuclease	56.0 → 48.7 → 43	118, 147, 220a
gol (in 23)	grow on lit hosts (CTr5x)		Disrupt general translation and gene-23 transcription		118, 186a
24 = os	Faulty heads, osmotic shock resistance		Vertex head subunit	48.4 → 46	118, 221
hoc = eph	Nonessential		Minor capsid protein	39.1, 40	20, 77, 118, 195
inh			Inhibitor of gene-21 protease	35	77, 118, 195
dar = uvsW	Suppresses 59⁻, 46⁻; UV-sensitive				25, 28, 77, 107, 118, 220
uvsY = fdsB	UV-sensitive; DNA synthesis reduced, recombination deficient; suppresses 49⁻		Recombination protein	15.8	25, 28, 30, 55, 61, 87, 118, 194. 220, 225
{oriF}					56, 93, 94, 95, 118, 222
25	Tail defects		Baseplate, 1/6 arm; lysozyme	15	54, 55, 61, 107, 118, 135, 225
26	Tail defects		Baseplate central hub	23.8 → 42	13, 18, 89, 118, 132
51	Tail defects		Baseplate central hub protease	29.3 → 16.5	89, 118, 172, 193
27	Tail defects, permit fiberless plating		Baseplate central hub	47, 49	13, 118, 172

Gene	Mutant Phenotype	Restrictive Host	Function and Comments	Mol wt (x 1000)	References
28	Tail defects		Baseplate, distal surface of central hub, gamma glutamyl hydrolase	24, 25, 20.1	13, 118
29	Tail defects		Baseplate, tail "bulge", folyl-poly-glutamyl-synthetase	77, 64.5	31, 74, 118
48	Tail defects		Baseplate, tail tube fibre, length "measure"?	37, 44, 39.7	31, 32, 74, 118
54	Tail defects		Baseplate, tail tube; polymerization initiator?	36, 35.0	31, 32, 74, 118
alt	Nonessential		Adenylribosylation of RNA polymerase, (packaged with DNA)	76.8 → 75.9	63a, 118, 169
30 = lig	DNA arrest, hyper-rec		DNA ligase	55.2	5, 118
rIII	Nonessential, rapid lysis				118
31	Capsid protein lumps; suppresses mutations in host-defective (e.g., groE) E. coli genes		Organizes head protein assembly and DNA topoisomerase complex	12.0	29, 84b, 118, 137, 138, 181
cd	Nonessential		dCMP deaminase	21	118, 114
pseT	Nonessential	CTr5x (lit$^-$)	Deoxyribonucleotide 3' phosphatase, 5' polynucleotide kinase	30, 34	76, 79, 80, 118, 127
alc = unf	Allows transcription of cytosine-containing T4 DNA	E. coli (pR386)	Abolishes transcript elongation on dC-DNA; unfolding of host DNA; DNA and RNA polymerase-binding protein	19.1	62, 63, 98, 118, 159, 185a
63	Poor tail fiber attachment		RNA ligase; helps tail fiber attachment	43.5	15, 79, 80, 118, 159
denA	Nonessential; defective in host DNA degradation		Endonuclease II	15.8	118, 159

Gene	Mutant Phenotype	Restrictive Host	Function and Comments	Mol wt (x 1000)	References
nrdB	Nonessential		Ribonucleotide reductase B subunit (split gene)	35, 40, 45.3	46, 47, 118, 182
nrdA	Nonessential		Ribonucleotide reductase A subunit	80, 85	118, 119, 201
td	Nonessential		Thymidylate synthase; baseplate hub component (split gene)	32, 33.0	21, 22, 33, 47, 118, 156
I-TevI	Nonessential		In td intron; intron mobility endonuclease	28.1	7a, 156b, 214
frd			Dihydrofolate reductase; baseplate wedge component	21.6	156
32	DNA arrest; recombination and excision repair deficient; UV-sensitive	tab-32	ssDNA-binding (= helix-destabilizing) protein	33.5	118, 131, 42
59	DNA arrest		positions gp41 on ssDNA	26.0	59, 118
33	No late RNA synthesis		RNA-polymerase-binding-protein	12.8	59, 118
dbpA			dsDNA-binding protein	10.4	59, 60
das-suα	Suppresses 46⁻, uvsX⁻		das, sur, and/or dbpA may be the same gene		59, 118, 169
[sur]	Suppresses 46⁻, 47⁻, uvsX⁻				209, 210
34	Fiberless particles		Proximal tail fiber subunit (A antigen)	145	45, 59, 118
{oriG in 34}					93, 94, 95, 118
35	Fiberless particles		Hinge tail fiber subunit	39, 40	45, 118, 130b
36	Fiberless particles		Small distal tail fiber subunit	24.3	118, 130b
37	Fiberless particles, host range		Large distal tail fiber subunit	112.8	118, 130b, 165
38	Fiberless particles		Assembly catalyst of gp37	26, 27, 28	118, 164, 165b, 130b

Gene	Mutant Phenotype	Restrictive Host	Function and Comments	Mol wt (x 1000)	References
t = stII	Lysis defective; suppresses 63^- and rII mutations			25.2	118, 130a, 165a
motA = sip	Suppresses rII$^-$in K(\) hosts		Regulates middle gene expression, activates middle promoters	24, 23.6	118, 202, 203, 204
[rV]	Temperature-dependent rapid lysis				118
52	DNA delay	S/6, 25°	Membrane protein, DNA topoisomerase subunit	52, 50.6	69, 118, 168
ac	acriflavine resistant			5.0	118, 15
[ama] [rs]	Nonessential, acriflavine resistant			5.4	118, 15
stp	suppresses pseT mutations			3.2	118, 15
ndd = D2b	Nuclear disruption defective	CT447		15, 11.1	118, 184, 185, 15
pla262	Nonessential	CT262			118, 15
denB	Nonessential; allows production of T4 with dC in DNA		Endonuclease IV	22	118
D1	Nonessential				118
rIIB	Nonessential, rapid lysis; suppresses $30^-$and some 32^- mutations	rex$^+$ \ lysogens, tabR	Membrane protein; affects membrane ATPase of E. coli	33, 41, 35.5	24, 69, 118

Note: Genes originally given different names and now known to be identical are signified by =; those that are probably identical are signified by ≈. Genes within the introns of other genes are denoted by I-. Genes 69 and 49 also encode smaller proteins, denoted 69* and 49* that begin within the gene and end at the same site as the longer sequence. Origins of replication, ori, are listed in braces.

GENES

start	finish	gene	start	finish	gene	start	finish	gene
0.014	2.189	*rIIA*	64.970	65.549	IPIII	113.603	114.014	*uvsY*
2.460	2.753	60+	65.639	66.131	*e*	114.093	114.489	25
2.802	2.940	60+	69.642	69.981	tRNA	114.491	115.114	26
3.722	5.212	39	70.693	71.263	tRNA	115.166	115.913	-51
7.934	8.617	*dexA*	72.326	72.611	IPI	115.915	117.029	-27
9.119	10.436	*dda*	72.687	73.143	57B	117.051	117.498	-28
13.902	14.250	*mrh*	73.142	73.382	57A	117.506	119.276	-29
14.981	15.210	*soc*	73.384	74.107	1	119.296	120.388	-48
15.255	16.430	69	74.153	74.683	3	120.390	121.350	-54
15.907	16.430	56	74.790	75.615	2=64	122.096	124.153	*alt*
16.494	17.270	*dam*	75.615	76.067	4=50	124.395	125.855	30
17.582	18.611	58/61	75.782	75.965	-65	129.247	129.580	31
20.260	21.684	41	76.120	76.708	-53	130.186	130.765	*cd*
21.698	22.040	40=*sp*	76.694	78.319	-5	132.130	133.033	*pseT*
22.032	23.205	*uvsX*	77.820	77.972	*dbpB*	133.903	134.403	*alc*
23.215	23.878	X.1	79.816	81.794	-6	134.471	135.593	63
24.042	25.098	*ßgt*	81.793	84.889	-7	135.649	136.056	*denA*
25.150	25.887	42	84.884	85.888	-8	136.087	136.586	*nrdB+*
26.000	26.250	*imm*	85.955	86.819	-9	137.185	137.848	*nrdB+*
26.775	29.542	43	86.821	88.627	-10	138.546	140.808	*nrdA*
29.622	29.987	*regA*	88.629	89.286	-11	141.366	141.675	*td+*
29.992	30.552	62	89.285	90.966	-12	141.817	142.552	I-TevI
30.557	31.513	44	90.965	92.326	-*wac*	142.692	143.241	*td+*
31.568	32.248	45	92.359	93.287	-13	143.240	143.801	*frd*
32.392	32.694	*rpbA*	93.391	94.052	-14	144.950	145.850	32
32.951	34.630	46	94.103	95.019	-15	146.401	147.049	59
35.077	36.093	47	94.929	95.421	-16	147.051	147.384	33
36.273	37.472	*αgt*	95.407	97.247	-17	147.367	147.631	*dsbA*
39.249	39.803	55	97.271	99.248	-18	148.664	152.100	-34
43.182	44.300	*sunY+*	99.367	99.856	-19	152.450	153.560	-35
44.483	45.256	I-TevII	99.942	101.529	-20	153.630	154.300	-36
45.400	46.049	*sunY+*	101.610	101.860	-67	154.310	157.380	-37
46.031	46.501	49	101.860	102.280	-68	157.413	157.962	-38
46.544	47.026	*pin*	102.280	102.900	-21	157.985	158.639	-*t*
47.777	48.037	*nrdC*	102.900	103.630	-22	160.387	161.019	*motA*
58.899	59.478	*tk*	103.680	105.240	-23	161.186	162.617	52
60.522	60.865	*vs*	106.081	107.362	-24	162.908	162.995	*stp*
61.415	61.874	*regB*	108.917	110.044	*hoc*	162.997	163.288	*ndd*
64.075	64.489	*denV*	111.366	113.088	-*uvsW*	163.300	163.900	*denB*
64.552	64.852	IPII				165.064	166.000	*rIIB*

UNIDENTIFIED OPEN READING FRAMES

start	finish	gene	references	MW
2.202	2.403	*rIIA*.1	26a	8.1
7.748	7.869	*dexA*.-1	41	5
8.617	8.859	*dexA*.1	41	9
8.855	9.097	*dexA*.2	58	
10.495	10.746	*dda*.1	58	
14.732	14.936	*soc*.-1	37	
18.614	18.778	61.1	136	5.9
18.780	19.401	61.2	136	24.3
19.404	19.694	61.3	136	11.0
19.760	20.014	61.4	136	11.8
20.019	20.198	61.5	136	7.0
26.250	26.450	42.2	102	
32.707	32.892	45.2	68	
34.630	34.833	46.1	49	8.1
34.817	35.077	46.2	49	10.3
36.093	36.230	47.1	49	5.3
37.534	38.325	*αgt*.1	49	30.3
38.380	38.556	*αgt*.2	49	6.9
38.556	38.756	*αgt*.3	49	7.9
38.728	39.042	*αgt*.4	49	12.4
39.840	40.100	55.1	199	9.8
40.106	40.429	55.2	199	12.7
40.446	40.682	55.3	199	9.1
40.686	40.814	55.4	199	5.7
40.825	41.115	55.5	199	11.8
41.111	41.290	55.6	199	7.0
41.452	41.757	55.7	199	11.7
41.763	41.972	55.8	199	7.9
42.093	42.560	55.9	199	18.2
42.553	43.182	55.10	199	23.9
47.153	47.470	49.2	199	12.6
47.470	47.775	49.3	103,199	11.9
58.338	58.629	*tk*.-2	207	11.1
58.644	58.854	*tk*.-1	207	8.3
59.482	59.668	*tk*.1	207	7.2
59.667	59.850	*tk*.2	207	7.1
59.849	60.059	*tk*.3	207	8.5
60.058	60.523	*tk*.4	207	17.5
60.862	61.403	*vs*.1	207	20.7
61.936	62.212	*vs*.3	207	10.9
62.214	62.478	*vs*.4	207	10.2
62.517	62.691	*vs*.5	207	6.6

start	finish	gene	references	MW
62.693	63.053	*vs*.6	207	13.8
63.063	63.441	*vs*.7	207	15.0
63.389	64.046	*vs*.8	207	25.0
69.989	70.595	tRNA.1	11,208	22
71.363	71.648	tRNA.2	11,208	11
71.653	72.061	tRNA.3	11,208	16
72.064	72.247	tRNA.4	11,208	7
78.700	79.350	-5.1	132,177	
79.355	79.523	-5.2	177	
79.513	79.804	-5.3	177	
105.368	105.787	23.1	221	
110.052	110.729	*hoc*.1	77,195	
110.752	111.366	-*hoc*.2	77,195	
113.185	113.349	*uvsW*.1	107	6.0
113.379	113.601	*uvsW*.2	107	9.0
121.402	121.711	-54.1	169	10.8
124.210	124.396	*alt*.1	169	7.2
125.844	126.109	30.1	5,118	10
129.545	129.795	31.1	84b	
130.764	131.100	*cd*.1	114	
131.461	131.734	*pseT*.-3	127	11
131.739	131.937	*pseT*.-2	127	8
131.932	132.127	*pseT*.-1	127	8
133.036	133.260	*pseT*.1	127	9
133.259	133.558	*pseT*.2	127,159	12
133.557	133.910	*pseT*.3	127,159	13.1
136.725	137.015	*nrdB*.iI	47,182	11.3
137.846	138.095	*nrdB*.1	182	9.4
138.121	138.544	*nrdB*.2	182,201	16.4
140.801	141.125	*nrdA*.1	201	12.4
141.081	141.342	*nrdA*.2	201	10.1
143.996	144.136	*frd*.1	156	9
145.800	146.400	32.1	59	24.5
147.645	148.557	33.2	59	35.5
159.092	159.220	*motA*.-6	204	5.2
159.265	159.588	*motA*.-5	204	12.4
159.591	159.887	*motA*.-4	204	11.4
159.838	160.047	*motA*.-3	204	8.5
160.047	160.250	*motA*.-1	204	7.9
160.808	161.003	-*motA*.1	204	7.1
162.623	162.761	52.1	15	
162.756	162.909	52.2	15	

REFERENCES

1. Adari, H.Y., K. Rose, K.R. Williams, W.H. Konigsberg, T.C. Lin and E.K. Spicer. Proc. Natl. Acad. Sci. USA **82**: 1901-1905 (1985).
2. Adari, H.Y., and E. K. Spicer. In Proteins: Structure, Function, and Genetics (1986).
3. Arisaka, F., L. Ishimoti, G. Kassavetis, T. Kumazaki, S.-I. Ishii and F. Eiserling. J. Virol. **62**: 882-886 (1988).
4. Arisaka, F., T. Nakako, H. Takahashi and S.-I. Ishii. J. Virol. **62**: 1186-1193 (1988).
5. Armstrong J., R.S. Brown and A. Tsugita. Nucl. Acids Res. **11**: 7145-7155 (1983).
6. Barth, K.A., D. Powell, M. Trupin and G. Mosig. Genetics **120**: 329-343 (1988).
7. Beck, P.J., J.P. Condreay and I.J. Molineux. J. Bact. **167**: 251-256 (1986).

7a. Bell-Pedersen, D., S.M. Quirk, M. Aubrey and M. Belfort. Gene, in press.

7b. Bell-Pedersen, D., S.M. Quirk and M. Belfort, manuscript in preparation.

8. Berget, P., personal communication.
9. Bijlenga, K.L., T. Ishii and A. Tusigata. J. Mol. Biol. **120**: 249-263 (1978).
10. Broida, J., and J. Abelson. J. Mol. Biol. **185**: 545-563 (1985).
11. Burke, R.L., M. Munn, J. Barry and B.M. Alberts. J. Biol. Chem. **260**: 1711-1722 (1985).
12. Carpousis, A.J., E.A. Mudd, H.M. Krisch. J. Mol. Biol., 1-37 (1988).
13. Cascino, A., personal communication (1989).
14. Cha, T., and B.M. Alberts. J. Biol. Chem. **261**: 7001-7010 (1986).
15. Chapman, D., I. Morad, G. Kaufmann, M.M. Gait, L. Jorissen and L. Snyder. J. Mol. Biol. **199**: 373-378 (1988).
16. Champness, W.C., and L. Snyder. J. Virol. **50**: 555-562 (1984).
17. Chang, A., R. Weiss and G. Mosig, personal communication (1989).
18. Chen, C., M. Gruidl, S. Gargano, A. Cascino and G. Mosig, manuscript in preparation (1989).
19. Christensen, A.C., and E.T. Young. Nature **299**: 369-372 (1982).
20. Childs, J.D., and R. Pilon. J. Virol. **46**: 629-631 (1983).
21. Chu, F.K., G.F. Maley, F. Maley and M. Belfort. Proc. Natl. Acad. Sci. USA **81**: 3049-3053 (1984).
22. Chu, F.K., G.G. Maley, D.K. West, M. Belfort and F. Maley. Cell **45**: 157-166 (1986).
23. Chu, F.K., G.F. Maley, F. Maley. FASEB J. **2**: 216-223, (1988)
24. Colowick, M.S., and S.P. Colowick. Trans. N. Y. Acad. Sci. **41**: 35-40 (1983).
25. Conkling, M.A., and J.W. Drake. Genetics **107**: 505-523 (1984).
26. Cunningham, R.P., and H. Berger. J. Virol. **79**: 320-329 (1977).

26a. Daegelen, P. and E. Brody. Submitted to Genetics (1989)

27. Dharmalingam, K., H.R. Revel and E.B. Goldberg. J. Bact. **149**: 694-699 (1982).
28. de Vries, J.K., and S.S. Wallace. J. Virol. **47**: 406-412 (1983).
29. Doermann, A.H., and L.D. Simon. J. Virol. **51**: 315-320 (1984).
30. Drake, J.W. J. Bact. **162**: 1311-1313 (1985).
31. Duda, R.L., M. Gingery and F.A. Eiserling. Virology **151**: 296-314 (1986).
32. Duda, R.L., J.S. Wall, J.F. Hainfeld, M. Sweet, F.A. Eiserling. Proc. Natl. Acad. Sci. USA **82**: 5550-5554 (1985).

33. Ehrenman, K., J. Pedersen-Lane, D. West, R. Herman, F. Maley and M. Belfort. Proc. Natl. Acad. Sci. USA **83**: 5875-5879 (1986).
34. Elliot, T., and E.P. Geiduschek. Cell **36**: 211-219 (1984).
35. Ellis, R., and D. Hinton. J. Cell. Biochem. supp. 13D: 123 (1989).
36. Franklin, J., J. Rech and G. Mosig, manuscript in preparation (1989).
37. Frazier, M., and G. Mosig, Gene, in press (1989).
38. Fujisawa, H., T. Yonesaki and T. Minagawa. Nucl. Acids Res. **13**: 7473-7481 (1985).
39. Gargano, S., and A. Cascino, personal communication, 1989.
40. Gary, T., and G. Mosig, manuscript in preparation (1989).
41. Gauss, P., M. Gayle, R.B. Winter, L. Gold. Mol. Gen. Genet. **206**: 24-34 (1987).
42. Gauss, P., K.B. Krassa, D.S. McPheeters, M.A. Nelson, L. Gold. Proc. Natl. Acad. Sci. USA **84**: 8515-8519 (1987).
43. Gavrilenko, I.V., N.P. Kuzmin and V.I. Tanyashin. Mol. Gen., Micro. and Virol. (Russian), in press (1987).
44. Gerald, W.L., and Karam, J.D. Genetics **107**: 537 (1984).
45. Goldberg, E., personal communication (1987).
46. Gott, J., D. Shub and M. Belfort. Cell **47**: 81-87 (1986)
47. Gott, J.M., A. Zeeh, D. Bell-Pedersen, K. Ehrenman, M. Belfort and D.A. Shub. Genes and Devel. **2**: 1791-1799 (1989).
48. Gram, H., H-D. Liebig, A. Hack, E. Niggemann and W. Rüger. Mol. Gen. Genet. **194**: 232-240 (1984).
49. Gram, H., and W. Rüger. EMBO J. **4**: 257-264 (1985).
50. Gray, T.M., B.W. Matthews. J. Biol. Chem. **262**: 16858-16864 (1987).
51. Griffith, J., and T. Formosa. J. Biol. Chem. **260**: 4484-4491 (1985).
52. Grossi, G.F., M.F. Macchiato and G. Gialanella. Z. Naturforsch. **38c**: 294 (1983).
53. Gruber, H., G. Kern, P. Gauss, L. Gold. J. Bact. **170**: 5830-5836 (1988).
54. Gruidl, M.E., N.C. Canan and G. Mosig. Nucl. Acids Res. **16**: 9862 (1988).
55. Gruidl, M.E., and G. Mosig. Genetics **114**: 1061-1079 (1986).
56. Guerrier-Takada, C., W.H. McClain and S. Altman. Cell **38**: 219-224 (1984).
57. Guild, N., Thesis, U. of Colorado (198?).
58. Hacker, K., and B. Alberts, personal communication.
59. Hahn, S., V. Kruse and W. Rüger. Nucl. Acids Res. **14**: 9311-9327 (1986).
60. Hahn, S., and W. Rüger. Nucl. Acids Res. **17**: 6729 (1989).
61. Harris, L.D., J.D. Griffith. J. Mol. Biol. **206**: 19-28 (1989).
62. Herman, R.E., N. Haas and D.P. Snustad. Genetics **108**: 305-317 (1984).
63. Herman, R.E., and D.P. Snustad. J. Virol. **53**: 430-439 (1985).
63a. Hilse, D., T. Koch and W. Rüger. Nucl. Acids Res. **17**: 6731 (1989).
64. Hinton, D.M., and N.G. Nossal. J. Biol. Chem. **260**: 12858-12865 (1985).
65. Hinton, D.M., and N.G. Nossal. J. Biol. Chem. **261**: 5663-5673 (1986).
66. Hinton, D.M., L.L. Silver and N.G. Nossal. J. Biol. Chem. **260**: 12851-12857 (1985).
67. Hinton, D.M. J. Biol. Chem. 264, in press (1989).
68. Hsu, T., R. Wei, M. Dawson and J. Karam. J. Virol. **68**: 366-374 (1987)
69. Huang, W.M. Nucl. Acids Res. **14**: 7379-7390 (1986).
70. Huang, W.M. Nucl. Acids Res. **14**: 7751-7765 (1986).
71. Huang, W.M., L.S. Wei and S. Casjens. J. Biol. Chem. **260**: 8973-8977 (1985).
72. Huang, W.M., S-Z. Ao, S. Casjens, R. Orlandi, R. Zeikus, R. Weiss, D. Winge and M. Fang. Science **239**: 1005-1012 (1988).
73. Huff, A.C., J.K. Leatherwood and K.N. Kreuzer. Proc. Natl. Acad. Sci. USA **86**: 1307-1311 (1989).

74. Ishimoto, L.K., K.S. Ishimoto, A. Cascino, M. Cipollaro and F.A. Eiserling. Virology **164**: 81-90 (1988).
75. Jongeneel, C.V., T. Formosa and M. Alberts. J. Biol. Chem. **259**: 12925-12932 (1984).
76. Jabbar, M.A., and L. Snyder. J. Virol. **51**: 522-529 (1984).
77. Kaliman, A.V., A.A. Zimin, N.N. Nazipova, V.M. Kryukov, V.I. Tanyashin, A.S. Kraev, M.V. Mironova, K.G. Skryabin and A.A. Baev. Dokl. Akad. Nauk. SSSR **299**: 737-742 (1988).
78. Kalinska, A., and L.W. Black. J. Virol. **58**: 951-954 (1986).
79. Kaufmann, G., and M. Amitsur. Nucl. Acids Res. **13**: 4333-4341 (1985).
80. Kaufmann, G., M. David, G.D. Borasio, A. Teichmann, A. Paz and M. Amitsur. J. Mol. Biol. **188**: 15-22 (1986).
81. Keller, B., and T.A. Bickle. Gene **49**: 245-251 (1986).
82. Keller, B., C. Sengstag, E. Kellenberger and T.A. Bickle. J. Mol. Biol. **179**: 415-430 (1984).
83. Keller, B., J. Dubochet, M. Adrian, M. Maeder, M. Wurtz, E. Kellenberger. J. Virol. **62**: 2960-2969 (1988).
84. Keller, B., E. Kellenberger, T.A. Bickle and A. Tsugita. J. Mol. Biol. **186**: 665-667 (1985).
84a. Keller, B., M. Maeder, C. Becker-Laburte, E. Kellenberger and T.A. Bickle. J. Mol. Biol. **190**: 83-95 (1986).
84b. Keppel, F., B. Lipinska, D. Ang and C. Georgopoulos. Gene, in press (1989).
85. Klausa, V.I., and R.G. Nivinskas. Genetika **24**: 42-52 (1988).
86. Kleff, S. and B. Kemper. EMBO J. **7**: 1527-1535 (1988).
87. Kobayashi, M., H. Saito and H. Takahashi. Nucl. Acids Res. **16**: 7729 (1988).
87a. Koch, T., N. Lamm and W. Rüger. Nucl. Acids Res. **17**: 4392 (1989).
88. Kodadek, T., M.L. Wong and B.M. Alberts. J. Biol. Chem. **263**: 9427-9436 (1988).
89. Kozloff, L.M., and M. Lute. J. Virol. **52**: 344-349 (1984).
90. Krayev, A.S., A.A. Zimin, M.V. Mironova, A.A. Tamalaytia, V.I. Tanyashin, K.G. Skeyalin and A.A. Bayer. Dokl. Acad. Sci. USSR **270**: 1495-1500 (1983).
91. Kraev, M.V., K.G. Mironova, Skryabin and A.A. Baev. Dokl. Akad. Nauk. SSSR (1988).
92. Kreuzer, K., personal communication (1989).
93. Kreuzer, K., and B. Alberts. Proc. Natl. Acad. Sci. USA **82**: 3345-3349 (1985).
94. Kreuzer, K.N., and B.M. Alberts. J. Mol. Biol. **188**: 185-198 (1986).
95. Kreuzer, K.N., H.W. Engman and W.Y. Yap. J. Biol. Chem. **263**: 11348-11357 (1988).
96. Kricker, M., personal communication (1989).
97. Krisch, H.M., and B. Allet. DNA **89**: 4937-4941 (1982).
98. Kutter, E., R. Drivdahl and K. Rand. Genetics **108**: 291-304 (1984).
98a. Kutter, E., K. d'Acci, J. McKinney, S. Provost, S. Peterson, J. Gleckler and B. Guttman, manuscript in preparation for J. Virol. (1989)
99. Kuzmin, N.P., V.M. Kryukov, V.I. Tanyashin and A.A. Bayev. Dokl. Acad. Sci. USSR **269**: 995-999 (1983).
100. Kzayer, A.S., A. Zimin, M.V. Mironova, A.A. Tanulatis, V.I. Tanyashin, K.G. Skzyabin and A.A. Bayer. Dokl. Acad. Sci. USSR **270**: 1495-1500 (1983).
101. Lamm, N., J. Tomaschewski and W. Rüger. Nucl. Acids Res. **15**: 3920 (1987).

102. Lamm, N., Y. Wang, C.K. Mathews and W. Rüger. Eur. J. Biochem **172**: 553-564 (1988).
103. LeMaster, D.M. J. Virol. **59**: 759-760 (1986).
104. Liang, Y., R. Wei, T. Hsu, C. Alford, M. Dawson and J. Karam. Genetics **119**: 743-749 (1988).
104a. Liebig, H.D., and W. Rüger. J. Mol. Biol. **209**: in press (1989).
105. Lin, G., and G. Mosig, manuscript in preparation (1989).
106. Lin, G. Ph.D. Thesis, Vanderbilt University, Nashville, TN.
107. Lin, T.-C., and W. Konigsberg, personal communication.
108. Lin, T.-C., J. Rush, E.K. Spicer and W.H. Konigsberg. Proc. Natl. Acad. Sci. USA **84**: 7000-7004 (1987).
109. Lipinska, B., A.S.M.K. Rao, B.M. Bolten, R. Balakrishnan and E.B. Goldberg. J. Bact. **171**: 488-497 (1989).
110. Lu, M., and U. Henning. J. Virology **63**: 3472-3478 (1989).
111. Macdonald, P.M., and G. Mosig. Genetics **106**: 1-16 (1984).
112. Macdonald, P.M., and G. Mosig. EMBO J. **3**: 2863-2871 (1984).
113. Macdonald, P.M., E. Kutter and G. Mosig. Genetics **106**: 17-27 (1984).
114. Maley, G., personal communication (1989).
115. Malik, S., and A. Goldfarb. J. Biol. Chem. **263**: 1174-1181 (1988).
116. Marchin, G., personal communication (1989).
117. Marsh, R., personal communication (1989).
117a. Marusich, E. I. and V. V. Mesyanzhinov. Nucl. Acids Res. **17**: 7514 (1989)
118. Mathews, C.K., E.M. Kutter, G. Mosig and P. Berget, (eds.). Bacteriophage T4. ASM, Washington, D.C. (1983).
119. Mathews, C., and J. Booth, personal communication (1989).
120. [omitted in revision]
121. McClain, W.H., and K. Foss. Cell **38**: 225-231 (1984).
122. McClain, W.H., J.H. Wilson and J.G. Seidman. J. Mol. Biol. **203**: 549-553 (1988).
123. [omitted in revision]
124. McPheeters, D.S., A. Christensen, E.T. Young, G. Stormo and L. Gold. Nucl. Acids Res. **14**: 5813-5826 (1986).
125. Menkens, A.E., and K.N. Kreuzer. J. Biol. Chem. **263**: 11358-11365 (1988).
126. Michaud, G., L. Black and D. Hinton, personal communication (1989).
127. Midgley, C., and N. Murray. EMBO J. **4**: 2695-2703 (1985).
128. Mileham, A.J., N.E. Murray and H.R. Revel. J. Virol. **50**: 619-622 (1984).
129. Miller, E.S., R.B. Winter, K.M. Campbell, S.D. Power and L. Gold. J. Biol. Chem. **260**: 13053-13059 (1985).
130. Miner, Z., and S. Hattman. J. Bact. **170**: 5177-5184 (1988).
130a. Montag, D., M. Degen and U. Henning. Nucl. Acids Res. **15**: 6736.
130b. Montag, D., I. Riede, M.-L. Eschbach, M. Degen and U. Henning. J. Mol. Biol. **196**: 165-174 (1987).
131. Mosig, G. Genetics **110**: 159-171 (1985).
132. Mosig, G., G. Lin, J. Franklin and W.-H. Fan. The New Biologist **1**: in press (1989).
133. Mosig, G., and P.M. Macdonald. J. Mol. Biol. **189**: 243-248 (1986).
134. Mosig, G., P.M. Macdonald, D. Powell, M. Trupin and T. Gary, in DNA Replication and Recombination (T. Kelly and R. McMacken, eds.), Alan Liss, New York, 403-414 (1987).
135. Nakagawa, H., F. Arisaka and S.-I. Ishii. J. Virol. **54**: 460-466 (1985).
136. Nakanishi, M., and B. Alberts, personal communication (1989).
137. Nivinskas, R.G., and L.U. Blek. Mol. Biol. (Mosc) **22**: 1507-1516 (1988).
137a. Nivinskas, R.G., and L.W. Black. Gene **73**: 251-257 (1989).

138. Nivinskas, R. G., A. Randonikene and N. Gild. Mol. Biol. (Mosc) **23**:739-749 (1989)
139. Obringer, J.W. Genet. Res. **52**: 81-90 (1988).
140. Obringer, J., P. McCreary and H. Bernstein. J. Virol **62**: 3043-3045 (1988).
141. Oliver, D.B., and R.A. Crowther. J. Mol. Biol. **153**: 545-568 (1981).
142. Olson, N.J., and G.L. Marchin. J. Virol. **51**: 42-46 (1984).
143. Olson, N.J., and G.L. Marchin. J. Virol. **53**: 702-704 (1985).
144. Orsini, G., and E.N. Brody. Virology **162**: 397-405 (1988).
145. Owen, J., E. Schultz, A. Taylor and G.R. Smith. J. Mol. Biol. **165**: 229-248 (1983).
146. Parker, M., personal communication (1987).
147. Parker, M.L., A.C. Christensen, A. Boosman, J. Stockard, E.T. Young and A.H. Doermann. J. Mol. Biol. **180**: 399-416 (1984).
148. Pedersen-Lane, D., and M. Belfort. Proc. Natl. Acad. Sci. USA **85**: 1151-1155 (1988).
149. Plishker, M.F., and P.B. Berget. J. Mol. Biol. **178**: 699-709 (1984).
150. Plunkett, G., G.P. Mazzara and W.H. McClain. Arch. Biochem. Biophys. **210**: 298-306 (1981).
151. Powell, D., J. Franklin, F. Arisaka and G. Mosig, manuscript in preparation (1989).
152. Pribnow, D., D.C. Sigurdson, L. Gold, B.S. Singer, C. Napoli, J. Brosirs, T.J. Dull and H.F. Noller. J. Mol. Biol. **149**: 337-376 (1981).
153. Prilipov, A.G., N.A. Selivanov, V.P. Efimov, E.I. Marusich and V.V. Mesyanzhinov. Nucl. Acids Res. **17**: 3303 (1989).
154. Prilipov, A.G., N.A. Selivanov, L.I. Nikolaeva and V.V. Mesyanzhinov. Nucl. Acids Res. **16**: 10361 (1988).
155. Pulitzer, J.F., M. Colombo and M. Ciaramella. J. Mol. Biol. **182**: 249-263 (1985).
156. Purohit, S., and C.K. Mathews. J. Biol. Chem. **259**: 6261-6266 (1984).
156a. Quirk, S.M., D. Bell-Pedersen, J. Tomaschewski, W. Rüger and M. Belfort. Nucl. Acids Res. **17**: 301-315 (1989).
156b. Quirk, S., D. Bell-Pedersen, and M. Belfort. Cell **56**: 455-465 (1989).
157. Radany, E.H., L. Naumovski, J.D. Love, K.A. Gutekunst, D.H. Hall and E.C. Friedberg. J. Virol. **52**: 846-856 (1984).
158. Raleigh, E.A., R. Trimarchi and H. Revel. Genetics **122**: 279-296 (1989).
159. Rand, K.N., and M.J. Gait. EMBO J. **3**: 397-402 (1984) and personal communication.
160. Rao, V.B., and L.W. Black. J. Mol. Biol. **200**: 475-488 (1988).
161. Rao, V.B., and L.W. Black, personal communication (1989).
162. Recinos, III, A., M.L. Augustine, K.M. Higgins and R.S. Lloyd. J. Bact. **168**: 1014-1018 (1986).
163. Richardson, R.W., and N.G. Nossal. J. Biol. Chem. **264**: 4725-4731 (1989).
164. Riede, I., M. Degen and U. Henning. EMBO J. **4**: 2343-2346 (1985).
165. Riede, I., K. Drexler, M.-L. Eschbach and U. Henning. J. Mol. Biol. **191**: 255-266 (1986).
165a. Riede, I. J. Bact. **169**: 2956-2961 (1987).
166. Robinson, D.R., N.R.M. Watts and D.H. Coombs. J. Virol. **62**: 1723-1729 (1988).
167. Rodriguez, A.P., Ph.D. Thesis, Vanderbilt University (1976).
168. Rowe, T.C., K.M. Tewey and L.F. Lui. J. Biol. Chem. **259**: 9177-9181 (1984).
168a. Ruckman, J., D. Parma, C. Tuerk, D. Hall and L. Gold. The New Biologist, in press.

169. Rüger, W., personal communication (1989).
170. Rush, J., T.-C. Lin, M. Quinones, E.K. Spicer, I. Douglas, K.R. Williams and W. Konigsberg. J. Biol. Chem. **264**: 1-35 (1989).
171. Santoro, M., V. Scarlato, A. Franze, O. Grau, M. Cipollaro, S. Gargano, R. Bova, M.R. Micheli, A. Storlazzi and A. Cascino. Gene **72**: 241-245 (1988).
172. Scarlato, V., A. Storlazzi, S. Gargano and A. Cascino. Virology **171**: 475-483 (1989).
173. Schlagman, S., and S. Hattman. Gene **22**: 139-156 (1983).
174. Schlagman, S.L., Z. Miner, Z. Feher and S. Hattman. Gene **73**: 517-530 (1988).
175. Schoemaker, J. Trends in Biotechnology **1**: 99 (1983).
176. Selick, H.E., J. Barry, T. Cha, M. Munn, M. Nakanishi, M. Wong and B. Alberts, in DNA Replication and Recombination (T. Kelly and R. McMacken, eds.), Alan Liss, New York, (1987).
177. Selivanov, N.A., and V.V. Mesyanzhinov. Nucl. Acids Res. in press (1989).
178. Selivanov, N.A., A.G. Prilipov and V.V. Mesyanzhinov. Nucl. Acids Res. **16**: 2334 (1988).
178a. Selivanov, N.A., A.G. Prilipov and V.V. Mesyanzhinov. Nucl. Acids Res. **17**: 3583 (1989).
179. Shub, D.A., J.M. Gott, M.-Q. Xu, B.F. Lang, F. Michel, J. Tomaschewski, J. Pedersen-Lane and M. Belfort. Proc. Natl. Acad. Sci. USA. **85**: 1151-1155 (1988).
180. Simon, L.D., B. Randolph, N. Irwin and G. Binkowski. Proc. Natl. Acad. Sci. USA **80**: 2059-2062 (1983).
181. Simon, L.D., and B. Randolph. J. Virol. **51**: 321-328 (1984).
182. Sjoeberg, B.M., S. Hahne, C.Z. Mathews, C.K. Mathews, K.N. Rand and M.J. Gait. EMBO J. **5**: 2031-2036 (1986).
183. Skorupski, K., J. Tomaschewski, W. Rüger and L.D. Simon. J. Bact. **170**: 3016-3024 (1988).
184. Snustad, P., and D. Oppenheimer, personal communication (1989).
185. Snustad, P., A.C. Casey and R.E. Herman. J. Bact. **163**: 1290-1292 (1985).
185a. Snyder, L., L. Gold and E. Kutter. Proc. Natl. Acad. Sci. USA **73**: 3098-3102 (1976)
186. Snyder, M., and W.B. Wood. Genetics **122**: 471-479 (1989).
186a. Snyder, L., K. Bergsland, C. Kao and R. Gulati, manuscript submitted (1989).
187. Spicer, E.K., J. Rush, C. Fung, L.J. Reha-Krantz, J.D. Karam and W.H. Konigsberg. J. Biol, Chem. **263**: 7478-7486 (1988).
188. Spicer, E.K., N.G. Nossal and K.R. Williams. J. Biol. Chem. **259**: 15425-15432 (1984).
189. Spicer, E.K., J.A. Noble, N.G. Nossal, W.H. Konigsberg and F.R. Williams. J. Biol. Chem. **257**: 8972-8979 (1987).
190. Stitt, B., and G. Mosig. J. Bact. **171**: 3872-3880 (1989).
191. Sugino, A., and J.W. Drake. J. Mol. Biol. **176**: 239-249 (1984).
192. Szewczyk, B., K. Bienkowska-Szewczyk and L.M. Kozloff. Mol. Gen. Genet. **202**: 363-367 (1986).
193. Szewczyk, B., and J. Nieradko, personal communication (1989).
194. Takahashi, H.M., M. Kobayashi, T. Noguchi and H. Saito. Virology **147**: 349-353 (1985).
195. Tanyashin, V., personal communication (1989).
196. Thylen, C. J. Gen. Virol. **68**: 253-262 (1987).
197. Thylen, C. J. Bact. **170**: 1994-1998 (1988).
198. Tomaschewski, J., H. Gram, J. Crabb and W. Rüger. Nucl. Acids Res. **13**: 7551-7568 (1985).

198a. Tomaschewski, J., Ph.D. thesis, Ruhr Universität Bochum (1987).
199. Tomaschewski, J., and W. Rüger. Nucl. Acids Res. **15**: 3632-3633 (1987).
199a. Traub, F., B. Keller, A. Kuhn, and M. Maeder. J. Virol. **49**: 902-908 (1984).
200. Trojanowska, M., E.S. Miller, J. Karam, G. Stormo and L. Gold. Nucl. Acids Res. **12**: 5979-5993 (1984).
201. Tseng, M.-J., J.M. Hilfinger, A. Walsh, and G.R. Greenberg. J. Biol. Chem. **263**: 16242-16251 (1988).
202. Uzan, M., J. Leautey, Y. D'Aubenton-Carafa and E. Brody. EMBO J. **2**: 1207-1212 (1983).
203. Uzan, M., R. Favre and E. Brody. Proc. Natl. Acad. Sci. USA **85**: 8895-8899 (1988).
204. Uzan, M., personal communication (1989).
205. Valerie, K., E.E. Henderson and J.K. DeRiel. Nucl. Acids Res. **12**: 8085-8096 (1984).
206. Valerie, K., E.E. Henderson and J.K. DeRiel. Proc. Natl. Acad. Sci. USA **82**: 4763-4767 (1985).
207. Valerie, K., J. Stevens, M. Lynch, E. Henderson and J. DeRiel. Nucl. Acids Res. **14**: 8637-8654 (1986).
208. Volker, T.A., J. Gafner, T.A. Bickle and M.K. Showe. J. Mol. Biol. **161**: 479-489 (1982).
209. Wakem, L.P., and K. Ebisuzaki. Virology **137**: 324-330 (1984).
210. Wakem, L.P., and K. Ebisuzaki. Virology **137**: 331-337 (1984).
211. Wakem, L.P., C.L. Zahradka and K. Ebisuzaki. Virology **137**: 338-346 (1984).
212. Watts, N.R.M., and D.H. Coombs. J. Virol **63**: 2427-2436 (1989).
213. Weaver, L.H., and B.W. Matthews. J. Mol. Biol. **193**: 189-199 (1987).
214. West, D.K., L.M. Changchien, G.F. Maley and F. Maley. J. Biol. Chem. **264**: 10343-10346 (1989).
215. Wiberg, J., and M. Murtha, personal communication (1984).
216. Williams, K.P., G.A. Kassavetis, F.S. Esch and E.P. Geiduschek. J. Virol. **61**: 597-599 (1986).
217. Williams, K.P., R. Müller, W. Rüger and E.P. Geiduschek. J. Bact. **171**: 3579-3582 (1989).
218. Winter, R.B., L. Morrissey, L.P. Gauss, L. Gold, T. Hsu and J. Karam. Proc. Natl. Acad. Sci. USA **84**: 7822-7826 (1987).
219. Wood, W.B., and H.R. Revel. Bact. Rev. **40**: 847-868 (1976).
220. Wu, Y., and K. Ebisuzaki. J. Virol. **52**: 1028-1031 (1984).
221. Yasuda, G., G.A. Churchill, M. Parker and D. Mooney. J. Mol. Biol., in press (1989).
222. Yee, J.-K., and R.C. Marsh. J. Virol. **54**: 271-277 (1985).
223. Yonesaki, T., Y. Ryo, T. Minagawa and H. Takahashi. Eur. J. Biochem. **148**: 127-134 (1985).
224. Yonesaki, T., and T. Minagawa. EMBO J. **4**: 3321-3327 (1985).
225. Yonesaki, T., T. Minagawa. J. Biol. Chem. **264**: 7814-7820 (1989).
226. Zograf, Y.N., V.V. Ogryz'ko, I.A. Bass and D.I. Chernyi. Molekulyarnaya Biologiya **19**: 818-832 (1985).

Bacteriophage P1

Michael Yarmolinsky
Laboratory of Biochemistry
Bldg. 37 Room 4D-15
National Cancer Institute
N.I.H., Bethesda, MD 20892

July, 1989

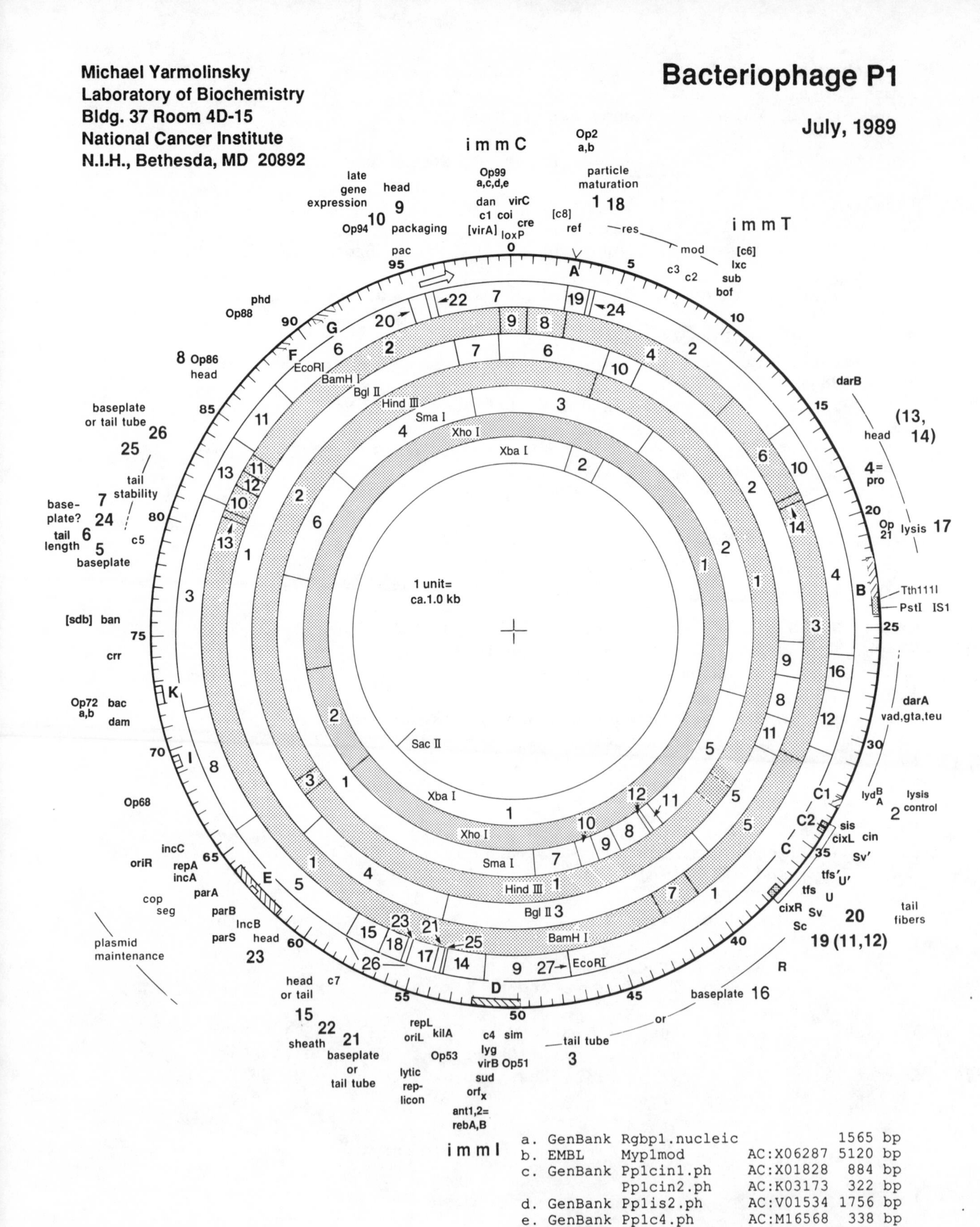

a. GenBank Rgbp1.nucleic 1565 bp
b. EMBL Myp1mod AC:X06287 5120 bp
c. GenBank Pp1cin1.ph AC:X01828 884 bp
Pp1cin2.ph AC:K03173 322 bp
d. GenBank Pp1is2.ph AC:V01534 1756 bp
e. GenBank Pp1c4.ph AC:M16568 338 bp
f. GenBank Ecop1para.ba AC:X02954 3833 bp
g. GenBank Pp1rep.ph AC:K02380 2080 bp
adjacent to Ecop1para.ba + 218 bp overlap

Genetic symbols lie outside the circles. If bracketed, they refer to genes identified only in P1's close relative, P7 (99). Several cistron designations are no longer in use, but their associated allele numbers have been reassigned in many cases (128). Cistron numbers 11 through 16 of (91) were reassigned (85) as was 17 (127). Allele numbers of several conditional mutations can be found in (139). Map coordinates have undergone a minor change. At a time when loxP (now coordinate 0/100) was not mapped precisely relative to physical markers, the unique PstI cleavage site was assigned position 20 so as to roughly align the physical and genetic maps (138). Linkage cluster boundaries (127,128) are omitted here as are coordinates of deletion prophages, some of which appeared in previous versions of the map (127,138). Each cl protein binding site or operator (Op) is assigned the integral number portion of its map position. In any one interval Op sites are lettered in alphabetical and clockwise order. Approximate positions of nonhomologies between P1 and P7 (18,57,67,79,144) are indicated by bold face letters and striped bars: insertions relative to P1 (A, the Ap^r transposon Tn902; F) and substitutions (B, C1, D, G, E); the invertible segment of P1 (and P7) is designated C. Stippled bars flanking C represent inverted repeats within one of which a deletion relative to the P1 sequence is marked as a line (C2). IS1 (part of B) is also stippled. Regions of partial non-homology between P1 and P7, within E and at positions I and K (77,79), are indicated by open bars. Restriction maps of P7 and P1/P7 hybrids can be found in (17,56). A linear denaturation map of P1 appears in (78) and is reproduced in circular form in a comprehensive review of P1 biology (142).

The cleavage sites of EcoRI, BamHI, BglII, HindIII, and the unique sites of SacII and of PstI (in the IS1) were mapped by (8), EcoRI sites that generate fragments 26 and 27 by (45), XhoI by (61), SmaI and XbaI by (76), and an additional unique site within IS1, for Tth111I by (53). No site was found for SalI (8).

An overlapping library of P1 DNA in a λ vector has been constructed and ordered with respect to the restriction map (83). The DNA sequences of selected regions of P1 that have been determined are: from a region including loxP through cre (119[a]); ref (72) and its leader sequence (136); res and mod (52[b]); bof (44,90); an 85-bp region that includes Op21 (20); IS1 (82); part of the dar operon, i.e. parts of BglII-9, -8, and -11 (47); cin and its substrates (48[c],58); from cixL through the invertible C segment to the BglII:5-3 junction (67,89); from the BglII:5-3 junction to the BamHI:5-7 junction (106[d], revised in [35]); the c4 gene (11[e]); from 176 bp upstream of c4 (including Op51) into the first gene beyond Op53 (45); from the middle of Op51 through a potential cl-binding site beyond repL (40) within which lies the repL operon, sequenced independently (113); the entire par (2[f]) and rep (3[g]) regions; an 85-bp region that includes Op68 (20); dam (v.22); a 62-bp Op72a,b sequence (43,126); a 59-bp Op86 sequence (126); an 85-bp region that includes Op88 (20); from the unique KpnI site within EcoRI-6 (at about map position 90) through the EcoRI-20 (71), including a site identified as Op94 (70); EcoRI-20,-22, and a 1.5 kb proximal and 0.3 kb distal region of EcoRI-7 (110) of which 237 bp around pac (within EcoRI-20) is published (115); and from the BglII:6-7 junction to loxP (30), within which lies a part of cl, sequenced entirely by (12). Footnotes are listed below the map.

Gene, orientation[1]	Probable function, references
loxP	Locus of cross(x)over in plasmid, cre substrate (50,108)
virC	Class of vir mutations that are presumed to confer virulence by overproduction of Coi (42,101,142)
cre (+)	Cyclization recombinase; cointegrate resolvase (7,49,103,116,117,119)
Op2a,b	Control of ref expression (30,136)
[c8]	Establishment of lysogeny (98)
ref (+)	Recombination enhancement function (72,135,136)
1 (-)	Particle maturation (62,80,91,109,131)
18	Lytic growth (91)
res (-) (=hsdR)	Restriction component of host specificity DNA system (34,41,52,62)
mod (-) (=hsdMS)	Modification and site recognition component of hsd (34,41,52,62)
c2,c3	Intracistronic complementation groups within mod; killing of host by c2 and c3 mutants leads to plaque clarity (52,84a,88,91,92)
[c6][2] (-)	Maintenance of lysogeny (98)
bof[2] (-)	Regulatory function that acts cooperatively with c1, detected as ban on function, one of several effects (44,90,109,124,135,136)
sub[2] (-)	Suppressor of bof-1 (a bof allele?) (124)
lxc[2] (-)	Mutants constitutively express P1 function that complements lexC (=ssbA) defect, one of several pleiotropic effects (66,135)
immT	Tertiary immunity region: the gene(s) of footnote 2 (142)
darB	Defense (only in cis) against a subset of type I restriction enzymes (e.g. EcoB, EcoK) (64)
13,14	Head (85,131)
pro (=4)	Protein processing, required for head morphogenesis and maturation of the DarA precursor protein (80,91,109,120,127,128,131)
17	Lysis (endolysin?) (109,127,128,129,131)
Op21	Unknown (20)
IS1	Facilitates cointegration (26,59,60)

1. Listing is clockwise from loxP. Names of sites and gene clusters are indented. Transcription: (+) = clockwise, (-) = counterclockwise.

2. bof, sub, lxc, and possibly c6 may be the same gene. The direction of transcription for each is assumed to be that of bof.

Gene, orientation	Probable function, references
darA[1] (-)	Defense (only in cis) against a subset of type I restriction enzymes (e.g. EcoA) (64,120)
vad[1] (-)	Viral architecture determination (defect increases proportion of small head particles) (55,64)
teu[2] (-)	Transduction enhancement by UV (possibly by binding to DNA and offering protection against nuclease) (64,142)
gta[1] (-)	Generalized transduction affected (frequency decreased) (64)
lydB[1] (-)	lydA,B prevent lysis delay (as does[?] λ gene S) (35,53,55,64)
lydA[1] (-)	lydA is first gene of dar operon (35,53,55,64)
2	Lysis delay; relation to lyd unclear (80,91,128,129,131)
cin (+)	C-segment inversion (and genome fusions) providing for variation in tail fibers (19,38,39,48,63,68,69)
sis	Sequence for inversion stimulation, enhances cix recombination in cis by binding to FIS (37,38,51)
cixL	C-segment inversion cross(x)-over site (left) (57,63)
S_v' (-)	
U'(tfs')(-)	
U (tfs,20)(-)	
Sv (-)	Tail fibers3, v=variable (54,58,67,80,105,125,128,131)
cixR	C-segment inversion cross(x)over site (right) (57,63)
Sc (-)	c=constant
S (19) (-)	S = S_c+Sv or Sv' (54,80,125,128,131)
11,12	Tail fibers[3] (85,131)
R (-)	Tail fiber structure or assembly. Positive regulator of translation of S (35,106)

1. The dar operon, under late promoter control (36), determines the presence of four internal proteins (120,130) which are dispensible for plaque formation (55). Relationships among members of the dar operon (darA, vad, teu, gta, and lyd) are unclear, but gta is genetically distinct from lyd (53). Unmapped mutations of (132) might be in gta or lyd.
2. Sus50 of (137) is possibly in teu (v.142).
3. Gene symbols S_c, S_v, S'_v, U, U', and tfs and tfs' (tail fiber specificity [67]) are derived from names of homologs in Mu (58). Possible synonyms of 11 and 12 are undetermined.

Gene, orientation	Probable function, references
16 (-)?	Baseplate or tail tube (128,131). Distal end of gene that is upstream of R is postulated in (35) to be 16 of (85)
3	Baseplate or tail tube (80,91,109,128,131)
sim	Unknown, confers superimmunity in high copy number (27)
Op51	immI control by immC (12,30,45,126,142)
c4 (+)	Immunity-specific prevention of ant/reb expression, hence control of plaque clarity (11,45,80,91,92,99,109)
lyg	Establishment and/or maintenance of lysogeny. lyg is possibly a class of c4 alleles that fail to confer plaque clarity (94)
virB	Class of vir mutations (e.g. vir^s [73] in c4 [45]) that confer virulence by constitutive Ant synthesis (42,91,96,109,134,142)
sud	Mutation suppresses dan-1, causes a weak constitutive expression of ant and is probably a weak virB mutation (23)
orf_x	Unknown, apparently required for expression of ant (45)
ant/reb (+) (ant1,ant2= rebA, rebB)[1]	Antagonism of cl repression (133,142)/cl-repressor bypass (102, 134); alternatives reflect differences in interpretation of function. The single orf that encodes both gene products (40,42, 45) is downstream of orf_x. Included within the ant/reb orf are the sites of reb-22 and presumably other ant/reb mutations, although ant-16 is within orf_x and ant-17 is upstream of c4 (45)
immI	Cluster of immunity functions listed above (responsible for immunity difference between P1 and P7) (16,40,45,112,133,142)
Op53	Control of repL operon (30,40,45,113,119,126)
kilA (+)	Unknown; first gene of repL operon, partially homologous to ant, expression can kill host (40,113)
repL (+)	Lytic replication; acts to initiate replication at oriL (21,40, 113); minor contributor to plasmid replication (140)
oriL	Origin of lytic replication. (Originally L referred to location "left" of EcoRI-5) (21,40,111,113)
21	Baseplate or tail tube (109,128,131)
22	Sheath (109,128,131)
15	Head or tail (85,131)

1. Unmapped mutations pla or p of (121,122) that, like ant mutations (143), permit P1 to lysogenize E. coli lon mutants, are possibly ant alleles.

Gene, orientation	Probable function, references
c7	Unclear, mutation affects plaque clarity and plasmid maintenance (80,97,98)
23	Head (109,128,131)
incB	Incompatibility determinant associated with intact parS (v.infra)
parS	Centromere analog required in cis for partioning. The intact site, but not all functional versions of it, bind IHF cooperatively with ParB (2,4,5,25,32,33,75)
parB (-)	Plasmid partitioning (2,5,142), cooperatively autoregulated (31) (v. also parS)
parA (-)	
seg[1]	Prevention of segregation of plasmid-free cells (65,107,109,118)
cop[2]	Control of plasmid copy number (9,100,118)
incA	Set of nine 19-bp iterons downstream of repA which bind RepA and control plasmid replication, conferring group Y incompatibility when it trans (1,3,5,13,14,84,118)
repA (-)	Plasmid replication; acts in trans to initiate replication from oriR and autoregulate repA transcription (1,3,5,14)
incC[3]	Set of five 19-bp iterons upstream of repA which bind RepA and participate in oriR function and autoregulation of repA. Can confer group Y incompatibility in trans (3,15)
oriR	Origin of (cis-acting region required for) plasmid replication of repressed P1 prophage. (Originally R referred to location "right" of EcoRI-5) (1,3,6,15,111)
Op68	Control of dam operon (20)
dam (+)	Methylation of adenine residues of GATC sequences (analog of E. coli gene); implicated in control of late gene expression, packaging (20,22,142)
Op72a,b	Control of ban operon (43,46,74,112)

1. seg includes par mutants (e.g. seg-101 = par-101 [4] and repA mutants (e.g. seg-103 = repA103 [6]).
2. cop includes point mutations in repA and (in principle) deletions in incA. Phenotype designations seg and cop should be replaced by genotypically more precise symbols, once the genotype is established.
3. incC is part of oriR which, in turn, is part of the basic replicon or rep region, hence the designation rep-11 for an incC mutation (6).

Gene, orientation	Probable function, references
bac	Cis-acting ban-control element; certain bac mutations render ban expression constitutive (24,43,46,81,126)
crr	P1crr prophage overproduces Ban, conferring cryoresistance on an unsuppressed dnaB266(am) P1bac-1 crr strain (29,46,86,123)
ban (+)[sdb]	DnaB analog (24,46,81,86,109,123) [suppressor of dnaB (104)]
5	Baseplate (91,109,128,131)
6	Tail length determination (91,109,128,131)
c5	Control of plaque clarity; establishment or maintenance of lysogeny (65,93,96,118)
24	Baseplate or tail stability (109,127,128,131)
7 25	Tail stability (91,109,128,131)
26	Baseplate or tail tube (109,128,131)
8	Head (91,109,128,131)
Op86	Unknown (30,126,142)
Op88	Unknown (20)
phd	Prevention of host death; antidote to product of unmapped gene(s) (doc) responsible for death on curing of P1 (141)
Op94	Control of 10 (70)
10 (+)	Late gene expression (36,42,80,91,109,110,128,129,131,142)
pac	Clustered sites from which clockwise packaging of DNA commences (8,114,115)
9 (+)	DNA cutting at pac (80,91,109,110,114,128,131,142)
c1 (-)	Repressor of lytic functions (10,28,87,92,96,98,109)
dan	dan-1 suppresses the thermolability of c1.100 repressor and is closely linked to c1.100 (23)
[virA]	Class of vir mutations, of which the only (presumed) member is vir-12, that confer virulence by a dominant c1 defect (98,99)
Op99a,c, d,e	Autoregulation of c1 (10,12,30,142). Highly degenerate site Op99b is omitted here as it fails to bind c1 repressor (110)
coi (-)	c one inactivator (96,142)
immC	Cluster of immunity genes flanking lox-cre of which c1 is the critical element (95,96,112,133,142)

1. Abeles, A.L. 1986. J. Biol. Chem. 261:3548-3555.
2. Abeles, A.L., Friedman, S.A., and Austin, S.J. 1985. J. Mol. Biol. 185:261-272. Erratum corrected in J. Mol. Biol. 189:387.
3. Abeles, A.L., Snyder, K.M., and Chattoraj, D.K. 1984. J. Mol. Biol. 173:307-324.
4. Austin, S., and Abeles, A.L. 1983. J. Mol. Biol. 169:373-387.
5. Austin, S., Hart, F., Abeles, A., and Sternberg, N. 1982. J. Bacteriol. 152:63-71.
6. Austin, S., Mural, R., Chattoraj, D., and Abeles, A. 1985. J. Mol. Biol. 83:195-202.
7. Austin, S., Ziese, M., and Sternberg, N. 1981. Cell 25:729-736.
8. Büchi, B., and Arber, W. 1977. Mol. Gen. Genet. 153:311-324.
9. Baumstark, B.R., Lowery, K., and Scott, J.R. 1984. Mol. Gen. Genet. 194:513-516.
10. Baumstark, B.R., and Scott, J.R. 1980. J. Mol. Biol. 140:471-480.
11. Baumstark, B.R., and Scott, J.R. 1987. Virology 156:197-203.
12. Baumstark, B.R., Stovall, S.R., and Ashkar, S. 1987. Virology 156:404-413.
13. Chattoraj, D., Cordes, K., and Abeles, A. 1984. Proc. Natl. Acad. Sci. USA 81:6456-6460.
14. Chattoraj, D.K., Mason, R.J., and Wickner, S.H. 1989. Cell 52:551-557.
15. Chattoraj, D.K., Snyder, K.M., and Abelels, A.L. 1985. Proc. Natl. Acad. Sci. USA 82:2588-2592.
16. Chesney, R.H., and Scott, J.R. 1975. Virology 67:375-384.
17. Chesney, R.H., Scott, J.R., and Vapnek, D. 1979. J. Mol. Biol. 130:161-173.
18. Chow, L.T., Broker, T.R., Kahmann, R., and Kamp, D. 1978. In "Microbiology - 1978," (D. Schlessinger, ed.), pp. 55-56, Amer. Soc. Microbiol., Washington, D.C.
19. Chow, L.T., and Bukhari, A.I. 1976. Virology 74:242-248.
20. Citron, M., Velleman, M., and Schuster, H. 1989. J. Biol. Chem. 264:3611-3617.
21. Cohen, G., and Sternberg, N. 1989. J. Mol. Biol. 207:99-109.
22. Coulby, J., and Sternberg, N. 1988. Gene 74:191.
23. D'Ari, R. 1977. J. Virol. 23:467-475.
24. D'Ari, R., Jaffé-Brachet, A., Touati-Schwartz, D., and Yarmolinsky, M. 1975. J. Mol. Biol. 94:341-366.
25. Davis, M.A., and Austi , S.J. 1988. EMBO J. 7:1881-1888.
26. deBruijn, F.J., and Bukhari, A.I. 1978. Gene 3:315-331.
27. Devlin, B.H., Baumstark, B.R., and Scott, J.R. 1982. Virology 120:360-375.
28. Dreiseikelmann, B., Velleman, M., and Schuster, H. 1988. J. Biol. Chem. 263:1391-1397.
29. Edelbluth, C., Lanka, E., Von der Hude, W., Mikolajczyk, M., and Schuster, H. 1979. Eur. J. Biochem. 94:427-435.
30. Eliason, J.L., and Sternberg, N. 1987. J. Mol. Biol. 198:281-293.
31. Friedman, S.A., and Austin, S.J. 1988. Plasmid 19:103-112.
32. Funnell, B. 1988. J. Bacteriol. 170:954-960.
33. Funnell, B. 1988. Proc. Natl. Acad. Sci. USA 85:6657-6661.
34. Glover, S.W., and Colson, C. 1969. Genet. Res. 13:227-240.
35. Guidolin, A., Zingg, J.-M., and Arber, W. 1989. Gene 76:239-243.

36. Guidolin, A., Zingg, J.-M., Lehnherr, H., and Arber, W. 1989. J. Mol. Biol. 209:000-000.
37. Haffter, P., and Bickle, T.A. 1987. J. Mol. Biol. 198: 579-587.
38. Haffter, P., and Bickle, T.A. 1988. EMBO J. 7:3991-3996.
39. Haffter, P., Pripfl, T., and Bickle, T.A. 1989. Mol. Gen. Genet. 215;245-249.
40. Hansen, E.B. 1989. J. Mol. Biol. 207:135-149.
41. Heilmann, H., Burkardt, H.J., Puhler, A., and Reeve, J.N. 1980. J. Mol. Biol. 144:387-396.
42. Heilmann, H., Reeve, J.N., and Puhler, A. 1980. Mol. Gen. Genet. 178:149-154.
43. Heinzel, T., Velleman, M., and Schuster, H. 1989. J. Mol. Biol. 205:127-135.
44. Heirich, M., Velleman, M., and Schuster, H. (in preparation).
45. Heisig, A., Riedel, H.-D., Dobrinski, B., Lurz, R., and Schuster, H. 1989. J. Mol. Biol. (in press).
46. Heisig, A., Severin, I., Seefluth, A.K., and Schuster, H. 1987. Mol. Gen. Genet. 206:368-376.
47. Hiestand-Nauer, R., unpublished.
48. Hiestand-Nauer, R., and Iida, S. 1983. EMBO J. 2:1733-1740.
49. Hochman, L., Segev, N., Sternberg, N., and Cohen, G. 1983. Virology 131: 11-17.
50. Hoess, R.H., Ziese, M., and Sternberg, N. 1982. Proc. Natl. Acad. Sci. USA 79:3398-3402.
51. Huber, H.E., Iida, S., Arber, W., and Bickle, T.A. 1985. Proc. Natl. Acad. Sci. USA 82:3776-3780.
52. Hümberlin, M., Suri, B., Rao, D.N., Hornby, D.P., Eberle, H., Pripfl, T., Kenel, S., and Bickle, T.A. 1988. J. Mol. Biol. 200:23-29.
53. Iida, S., personal communication.
54. Iida, S. 1984. Virology 134:421-434.
55. Iida, S., and Arber, W. 1977. Mol. Gen. Genet. 153:259-270.
56. Iida, S., and Arber, W. 1979. Mol. Gen. Genet. 173:249-261.
57. Iida, S., Hiestand-Nauer, R., Meyer, J., and Arber, W. 1985. Virology 143:347-351.
58. Iida, S., Huber, H., Hiestand-Nauer, R., Meyer, J., Bickle, T.A., and Arber, W. 1984. Cold Spring Harbor Symp. Quant. Biol. 49:769-777.
59. Iida, S., Meyer, J., and Arber, W. 1978. Plasmid 1:357-365.
60. Iida, S., Meyer, J., and Arber, W. 1981. Mol. Gen. Genet. 184:1-10.
61. Iida, S., Meyer, J., and Arber, W. 1985. J. Gen. Microbiol. 131:129-134.
62. Iida, S., Meyer, J., Bächi, B., Stålhammar-Carlemalm, M., Schrickel, S., Bickle, T.A., and Arber, W. 1983. J. Mol. Biol. 165:1-18.
63. Iida, S., Meyer, J., Kennedy, K.E., and Arber, W. 1982. EMBO J. 1:1445-1453.
64. Iida, S., Streiff, M.B., Bickle, T.A., and Arber, W. 1987. Virology 157:156-166.
65. Jaffé-Brachet, A., and D'Ari, R. 1977. J. Virol. 23:476-482.
66. Johnson, B.F. 1982. Mol. Gen. Genet. 186:122-126.
67. Kamp, D., Kardas, E., Ritthaler, W., Sandulache, R., Schmucker, R., and Stern, B. 1984. Cold Spring Harbor Symp. Quant. Biol. 49:301-311.

68. Kennedy, K.E., Iida, S., Meyer, J., Stalhammar-Carlemalm, M., Hiestand-Nauer, R., and Arber, W. 1983. Mol. Gen. Genet. 189:413-421.
69. Lee, H.-J., Ohtsubo, E., Deonier, R.C., and Davidson, N. 1974. J. Mol. Biol. 89:585-597.
70. Lehnherr, H., unpublished.
71. Lehnherr, H., and Guidolin, A., unpublished.
72. Lu, S.D., Lu, D., and Gottesman, M. 1989. J. Bacteriol. 171:3427-3432.
73. Luria, S.E., Adams, J.N., and Ting, R.C. 1960. Virology 12:348-390.
74. Lurz, R., Heisig, A., Velleman, M., Dobrinski, B., and Schuster, H. 1987. J. Biol. Chem. 262:16575-16579.
75. Martin, K.A., Friedman, S.A., and Austin, S.J. 1987. Proc. Natl. Acad. Sci. USA 84:8544-8547.
76. Mattes, R. 1985. Habilitationsschrift (Univ. Regensburg, 1985).
77. Meyer, J., Iida, S., and Arber, W. 1983. J. Mol. Biol. 165:191-195.
78. Meyer, J., Stålhammar-Carlemalm, M., and Iida, S. 1981. Virology 110:167-175.
79. Meyer, J., Stålhammar-Carlemalm, M., Streiff, M.B., Iida, S., and Arber, W. 1986. Plasmid 16:81-89.
80. Mural, R.J., Chesney, R.H., Vapnek, D., Kropf, M.M., and Scott, J.R. 1979. Virology 93:387-397.
81. Ogawa, T. 1975. J. Mol. Biol. 94:327-340.
82. Ohtsubo, H., and Ohtsubo, E. 1978. Proc. Natl. Acad. Sci. USA 75: 615-619.
83. O'Regan, G.T., Sternberg, N.L., and Cohen, G. 1987. Gene 60:129-135.
84. Pal, S.K., Mason, R.J., and Chattoraj, D.K. 1986. J. Mol. Biol. 192:275-285.
84a. Rao, D.N., Eberle, H., and Bickle, T.A. 1989. J. Bacteriol. 171:2347-2352.
85. Razza, J.B., Watkins, C.A., and Scott, J.R. 1980. Virology 105:52-59.
86. Reeve, J.N., Lanka, E., and Schuster, H. 1980. Mol. Gen. Genet. 177:193-197.
87. Rosner, J.L. 1972. Virology 48:679-689.
88. Rosner, J.L. 1973. Virology 52:213-222.
89. Sandmeier, H., and Iida, S., unpublished results.
90. Schaefer, T., and Hays, J.B., in preparation.
91. Scott, J.R. 1968. Virology 36:564-574.
92. Scott, J.R. 1970. Virology 41:66-71.
93. Scott, J.R. 1972. Virology 48:282-283.
94. Scott, J.R. 1974. Virology 62:344-349.
95. Scott, J.R. 1975. Virology 65:173-178.
96. Scott, J.R. 1980. Curr. Top. Microbiol. Immunol. 90:49-65.
97. Scott, J.R., Chesney, R.H., and Novick, R.P. 1978a. In "Microbiology - 1978," (D. Schlessinger, ed.), pp. 74-77, Amer. Soc. Microbiol., Washington, D.C.
98. Scott, J.R., and Kropf, M.M. 1977. Virology 82:362-368.
99. Scott, J.R., Kropf, M., and Mendelson, L. 1977. Virology 76:39-46.
100. Scott, J.R., Kropf, M.M., Padolsky, L., Goodspeed, J.K., Davis, R., and Vapnek, D. 1982. J. Bacteriol. 150:1329-1339.
101. Scott, J.R., Laping, J.L., and Chesney, R.H. 1977. Virology 78:346-348.
102. Scott, J.R., West, B.W., and Laping, J.L. 1978. Virology 85:587-600.
103. Segev, N., and Cohen, G. 1981. Virology 114:333-342.

104. Selvaraj, G., and Iyer, V.N. 1980. Mol. Gen. Genet. 178:561-566.
105. Sengstag, C., and Arber, W. 1983. EMBO J. 2:67-71.
106. Sengstag, C., Shepherd, J.C.W., and Arber, W. 1983. EMBO J. 2:1777-1781.
107. Som, T., Sternberg, N., and Austin, S. 1981. Plasmid 5:150-160.
108. Sternberg, N. 1978. Cold Spring Harbor Symp. Quant. Biol. 43:1143-1146.
109. Sternberg, N. 1979. Virology 96:129-142.
110. Sternberg, N., unpublished.
111. Sternberg, N., and Austin, S. 1983. J. Bacteriol. 153:800-812.
112. Sternberg, N., Austin, S., Hamilton, D., and Yarmolinsky, M. 1978. Proc. Natl. Acad. Sci. USA 75:5594-5598.
113. Sternberg, N., and Cohen, G. 1989. J. Mol. Biol. 207:111-133.
114. Sternberg, N., and Coulby, J. 1987. J. Mol. Biol. 194:453-468.
115. Sternberg, N., and Coulby, J. 1987. J. Mol. Biol. 194:469-480.
116. Sternberg, N., Hamilton, D., Austin, S., Yarmolinsky, M., and Hoess, R. 1980. Cold Spring Harbor Symp. Quant. Biol. 45:297-309.
117. Sternberg, N., Hamilton, D., and Hoess, R. 1981. J. Mol. Biol. 150: 487-507.
118. Sternberg, N., Powers, M., Yarmolinsky, M., and Austin, S. 1981. Plasmid 5:138-149.
119. Sternberg, N., Sauer, B., Hoess, R., and Abremski, K. 1986. J. Mol. Biol. 187:197-212.
120. Streiff, M.B., Iida, S., and Bickle, T.A. 1987. Virology 157:167-171.
121. Takano, T. 1971. Proc. Natl. Acad. Sci. USA 68:1469-1473.
122. Takano, T. 1977. Microbiol. Immunol. 21:573-581.
123. Touati-Schwartz, D. 1979. Mol. Gen. Genet. 174:173-188.
124. Touati-Schwartz, D. 1979. Mol. Gen. Genet. 174:189-202.
125. Toussaint, A., Lefebvre, N., Scott, J.R., Cowan, J.A., DeBruijn, F., and Bukhari, A.I. 1978. Virology 89:146-161.
126. Velleman, M., Dreiseikelmann, B., and Schuster, H. 1987. Proc. Natl. Acad. Sci. USA 84:5570-5574.
127. Walker, D.H. Jr., and Walker, J.T. 1975. J. Virol. 16:525-534.
128. Walker, D.H. Jr., and Walker, J.T. 1976. J. Virol. 20:177-187.
129. Walker, J.T., and Walker, D.H. Jr. 1980. J. Virol. 35:519-530.
130. Walker, J.T., and Walker, D.H. Jr. 1981. In "Progress in Clinical and Biological Research" (M.S. DuBow, ed.) Vol. 64, pp. 69-77, Alan R. Liss, New York.
131. Walker, J.T., and Walker, D.H. Jr. 1983. J. Virol. 45:1118-1139.
132. Wall, J.D., and Harriman, P.D. 1974. Virology 59:532-544.
133. Wandersman, C., and Yarmolinsky, M. 1977. Virology 77:386-400.
134. West, B.W., and Scott, J.R. 1977. Virology 78:267-276.
135. Windle, B.E., and Hays, J.B. 1986. Proc. Natl. Acad. Sci. USA 83:3885-3889.
136. Windle, B.E., Laufer, C.S., and Hays, J.B. 1988. J. Bacteriol. 170:4881-4889.
137. Yamamoto, Y. 1982. Virology 118:329-344.
138. Yarmolinsky, M. 1977. In "DNA Insertion Sequences, Episomes and Plasmids" (A.I. Bukhari, J.A. Shapiro, and S.L. Adhya, eds), pp. 721-732, Cold Spring Harbor Laboratory, Cold Spring Harbor, New York.
139. Yarmolinsky, M. 1984. In "Genetic Maps 1984" (S.J. O'Brien, ed), 3:42-54. Cold Spring Harbor Laboratory, Cold Spring Harbor, New York.
140. Yarmolinsky, M.B., Hansen, E.B., Jafri, S., and Chattoraj, D.K. 1989. J. Bacteriol. 171:000-000.
141. Yarmolinsky, M.B., Jafri, S., and Maguin, E., unpublished.
142. Yarmolinsky, M., and Sternberg, N. 1988. in "The Bacteriophages" (R. Calendar, ed.) Plenum Press, New York, Vol. 1, pp. 291-438.
143. Yarmolinsky, M.B., Stevens, E. 1983. Mol. Gen. Genet. 192:140-148.
144. Yun, T., and Vapnek, D. 1977. Virology 77:376-385.

BACTERIOPHAGE P2

Updated by E. Haggård-Ljungquist, July 1989
Dept. of Microbial Genetics, Karolinska Institutet, Box 60400,
S-10401 Stockholm, Sweden

P2 is a temperate bacteriophage that grows on several enterobacteria. For reviews of phage P2, see ref.1 and 2. The virus particle has an icosahedral head and a contractile tail(3,4), and the particle contains a double stranded DNA molecule of 33 kilobases with 5′protruding cohesive ends (5,6,7,8,9,10). Its DNA circularizes *in vivo* and replicates (11) unidirectionally (12) as a circular monomer (12,13), which is then linearized and packaged (14,15). The replication requires the *E. coli rep* function (16). The P2 late promoters are not recognized by the *E. coli* RNA polymerase, and the P2 *ogr* gene product is a positive regulatory factor required for late transcription (17,18). In lysogenization the circular DNA is integrated at specific sites on the bacterial chromosome by site specific recombination, mediated by the P2*int* gene product (19,20,21,22,23,24,25,26,27). The prefered integration site of *E.coli* C, i.e. *loc*I, shows a 27nt long sequence homologous to the phage *att* site (28). P2 has an intricate control of the switch between the lysogenic and lytic cycles. There are two divergent promoters, the early promoter *P*e and the repressor promoter *P*c which are located so that their transcripts overlap for about 30 nucleotides (29). Transcription from these promoters is mutually exclusive and negatively controlled by the *C* and *cox* gene products that are transcribed from *P*c and *P*e, respectively (30). Among other peculiarities of P2 are the very low frequency of genetic recombination (25,27,31,32), the non-inducibility by UV (1), the lack of complementation of the early essential gene *A* (33), and a control of integration/excision such that derepression alone causes only minimal phage production (26,27,34,35). For transcription data and strand orientation see 18,36,37,38,39,40.

P2 can also function as a helper for the unrelated satellite phage P4 (2). The P2/P4 interactions involves mutual transactivation of two types: transactivation (mediated by the P2 *ogr* gene produt and the P4 *delta* gene product) (41,42), and derepression (mediated by the P2 *cox* gene product and the P4 *epsilon* gene product) (43,44). The P2 *cox* gene product is an activator of the P4 late promoter *P*ll (45).

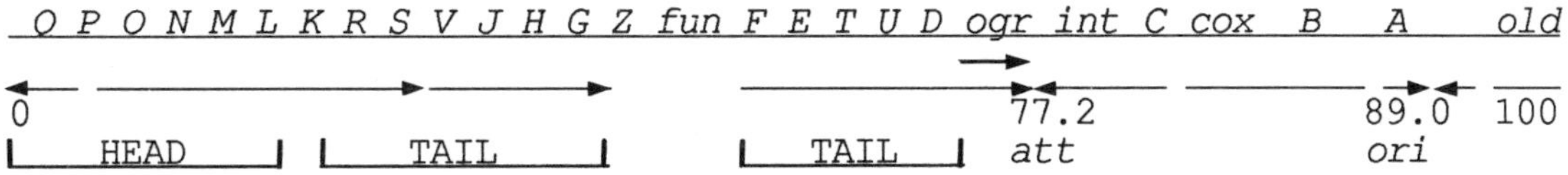

Q P O N M L K R S V J H G Z fun F E T U D ogr int C cox B A old

0 — 77.2 *att* — 89.0 *ori* — 100

HEAD | TAIL | TAIL

GENES, FUNCTIONAL SITES AND STRUCTURAL ABBERATIONS:

A and *B* are early (preduplication) function genes (33,46,47,48); mutations in A are not complementable (33).
C is the immunity repressor gene (29,49,50).
D through *V* are late genes (15,51,52,53,54) needed to make the protein shell of the virus or (gene K) to lyse the cell.
Z mutants (include former *vir20*) give clear plaques, the lysogens being inactivated by a still unidentified low molecular weight bacterial product (55,56).
att, site of exchange during site specific recombination during integration, excision or between phage genomes (28).
cox (or *cox*II, redefined in 43): gene required in addition to *int* for prophage excision (34,57,58). A repressor of the repressor promoter *P*c (5).
*del*1, deletion of gene *old* (32,59).
*del*2, *del*5, deletions of *fun* (32,60,61).
*del*3, deletion of *int*; affects also *C* expression (62,63).
*del*6, deletion of *int*: *C* unaffected (28,34,63).
*del*15 deletion of *ogr* and *int* (18).
*del*19, deletion originating in P2/P4 hybrid *Hy19* (64).
disA,B,C, regions of non-homology vis-a-vis P2 (65) in phage P2 *Hy dis* (66). *disC* extends to the right end of the chromosome and is longer than the corresponding P2 wild type segment.
fun, gene that makes P2 lysogens unusually sensitive to fluoro-uracil (67,68).
int, gene necessary for prophage integration and excision (24,25,27,58).
lg, large plaques (69); probably a mutation in gene *E* or *F* (70).
*nip*1, facilitates phage production after derepression (71); located in the *int* gene (28,34).
ogr, regulatory gene for late transcription (17,18,72).
old, gene that blocks survival of infected *recB,C* bacteria and causes interference with lambda phage (40,73-78).
ori, origine of replication (12,46).
*rlb*1, allows P2 multiplication in *dnaB E. coli* mutants (79). Located in the *B* gene (29).
*sig*5, insertion that allows P2 to integratively suppress the T46 (*dnaA*) *E. coli* mutant (80,81). It contains an *IS*2 insertion in *orfX*, located between genes *cox* and *C* (30).
*sly*1 (29), formerly *cox*I (57), mutation depressing spontaneous phage production in lysogens. Located in the *C* gene (82).
sos, mutations in the early gene region that prevent activation of late genes by satellite phage P4 (83).
*vir*3, *vir*22, *vir*79, *vir*56b, *vir*94: deletion of various lengths, affecting lysogeny and regulation of early functions (32,34,59,61,63,84).
*vir*37, tandem duplication affecting replication controls (85,86,87).

0.0 left end
0.8 *Bam*HI

Q (1.0–3.8)

1.5 *Pss*I
1.7 *Nru*I
1.8 *Eco*RV
2.9 *Nco*I

3.8 *Bst*XI
4.3 *Aat*II

P (4.8–9.2)

4.8 *Eco*RV
5.0 *Avr*II
5.0 *Bgl*I

5.3 *Dra*III
5.5 *Sal*I
*Aac*I
5.8 *Hind*II
6.2 *Aat*II
6.3 *Mlu*I
6.7 *Bam*HI
7.1 *Bst*EII
7.1 *Hind*II
7.1 *Hind*II
7.6 *Eco*RV
7.8 *Pvu*II

O (9.6–11.3)

10.4 *Nru*I
10.5 *Acc*I
10.8 *Eco*RI
11.2 *Bst*EII

N (11.5–15.9)

12.2 *Bst*EII
12.6 *Kpn*I
13.5 *Pst*I
14.2 *Hind*II
14.3 *Aat*II
14.4 *Bst*EII
14.4 *Mlu*I

16.0 *Apa*LI
17.4 *Hpa*I
*Hind*II
17.7 *Eco*RV
18.0 *Eco*RV
18.1 *Tth*I
22.0 *Bst*EII
23.7 *Sna*BI
24.4 *Dra*III
24.4 *Asu*II
24.7 *Sph*I
25.0 *Eco*RV
26.7 *Bst*EII
29.3 *Bal*I

V (30.5–32.5)

30.9 *Apa*LI

32.5 *Apa*LI
32.8 *Eco*RV
32.8 *Aat*II

33.4 *Hpa*I
*Hind*II
35.2 *Afl*II
36.4 *Nru*I
36.8 *Sal*I
*Acc*I
*Hind*II
37.7 *Sma*I
*Ava*I
38.2 *Aat*II
38.4 *Hind*II
38.7 *Stu*I
41.0 *Acc*I
41.8 *Asu*II
43.2 *Hpa*I
*Hind*II
43.3 *Pfl*MI
45.1 *Apa*LI
45.3 *Bam*HI
45.5 left end of del2

fun (46.5–51.3)

47.5 *Eco*RI
47.7 *Sph*I
48.7 *Hpa*I
*Hind*II
49.4 *Eco*RV
49.5 *Cla*I
49.6 *Stu*I
49.7 *Sph*I
50.0 *Afl*II
50.5 *Afl*II
51.2 right end of del2
51.2 *Bgl*II

51.3 *Nde*I
51.4 *Eco*RV
51.4 *Nde*I

F_{I} (52.7–56.1)

53.1 *Bst*EII
53.5 *Bst*EII
53.7 *Bst*XI
54.1 *Bst*EI
55.3 *Nru*I
56.0 *Bst*EII
56.0 *Hpa*I
*Hind*II

F_{II} (56.2–57.5)

57.0 *Acc*I
57.0 *Aat*II

57.6 *Hpa*I
*Hind*II
58.6 *Apa*LI
59.2 *Hind*II
59.4 *Eco*RV
62.6 *Pst*I
63.7 *Hind*II
63.8 *Apa*LI
65.6 *Mlu*I
65.8 *Kpn*I

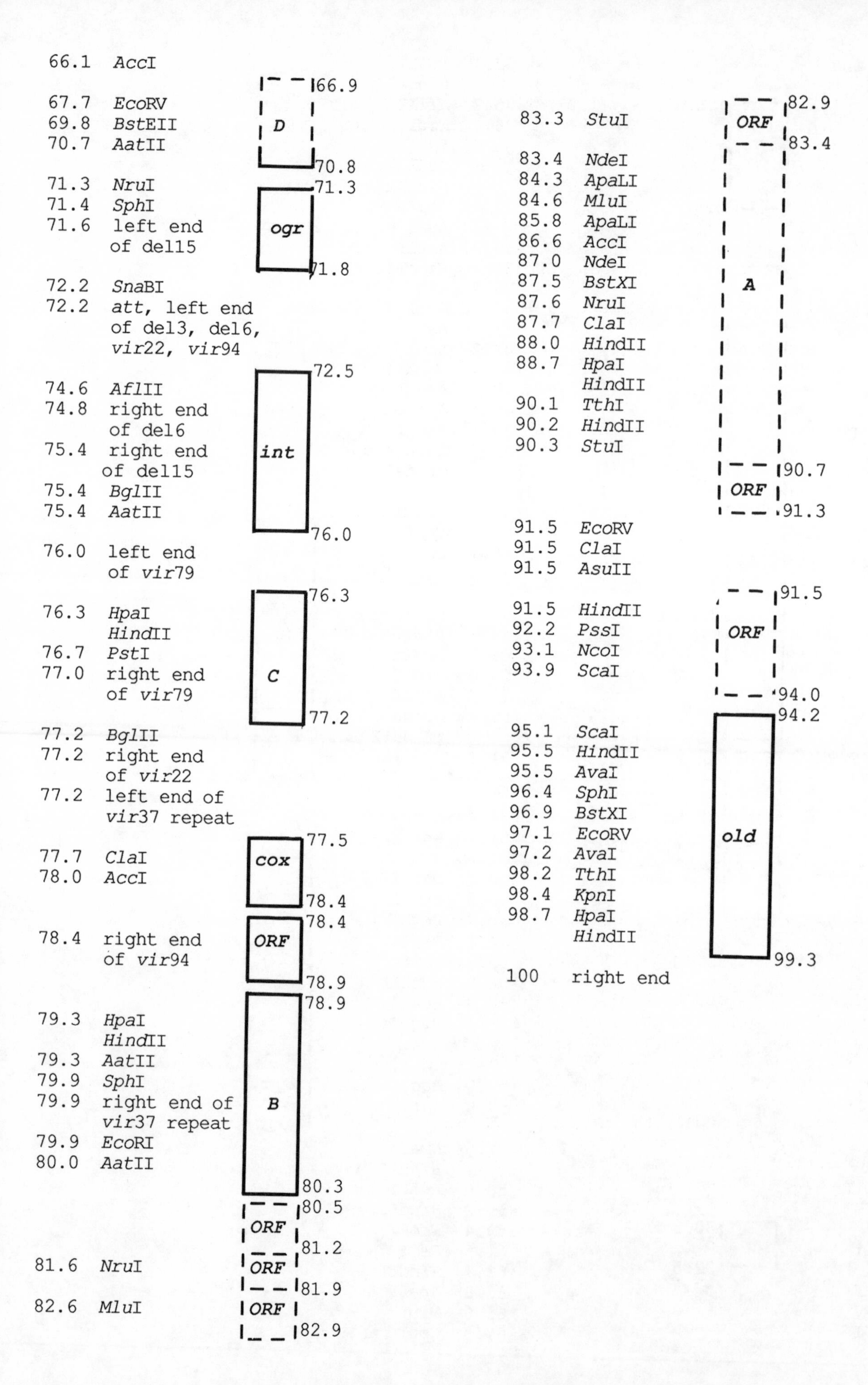

66.1 AccI
67.7 EcoRV
69.8 BstEII
70.7 AatII
71.3 NruI
71.4 SphI
71.6 left end of del15
72.2 SnaBI
72.2 att, left end of del3, del6, vir22, vir94
74.6 AflII
74.8 right end of del6
75.4 right end of del15
75.4 BglII
75.4 AatII
76.0 left end of vir79
76.3 HpaI HindII
76.7 PstI
77.0 right end of vir79
77.2 BglII
77.2 right end of vir22
77.2 left end of vir37 repeat
77.7 ClaI
78.0 AccI
78.4 right end of vir94
79.3 HpaI HindII
79.3 AatII
79.9 SphI
79.9 right end of vir37 repeat
79.9 EcoRI
80.0 AatII
81.6 NruI
82.6 MluI
D
66.9
70.8
ogr
71.3
71.8
int
72.5
76.0
C
76.3
77.2
cox
77.5
78.4
ORF
78.4
78.9
B
78.9
80.3
ORF
80.5
81.2
ORF
81.9
ORF
82.9
83.3 StuI
83.4 NdeI
84.3 ApaLI
84.6 MluI
85.8 ApaLI
86.6 AccI
87.0 NdeI
87.5 BstXI
87.6 NruI
87.7 ClaI
88.0 HindII
88.7 HpaI HindII
90.1 TthI
90.2 HindII
90.3 StuI
91.5 EcoRV
91.5 ClaI
91.5 AsuII
91.5 HindII
92.2 PssI
93.1 NcoI
93.9 ScaI
95.1 ScaI
95.5 HindII
95.5 AvaI
96.4 SphI
96.9 BstXI
97.1 EcoRV
97.2 AvaI
98.2 TthI
98.4 KpnI
98.7 HpaI HindII
100 right end
ORF
82.9
83.4
A
ORF
90.7
91.3
ORF
91.5
94.0
old
94.2
99.3

REFERENCES:

1. Bertani, L.E. and Bertani, G. (1971) Adv. in Genet. 16:199
2. Bertani, L.E. and Six, E.W. (1988) In "The bacteriophages" R. Calendar (ed.), Vol. 2, p. 73
3. Anderson, T.F. (1960) Proceedings, European Conference on Electron Microscopy, Delft, 2:1008 2.
4. Walker, D. (unpublished)
5. Mandel, M. (1967) Mol. Gen. Genet. 99:88
6. Inman, R.B. and Bertani, G. (1969) J. Mol. Biol. 44:533
7. Murray, K., Isaksson-Forsen, A.G., Challberg, M. and Englund, P.T. (1977) J. Mol. Biol. 112:471
8. Padmanabhan, R., Wu,R. and Calendar, R. (1974) J. Biol. Chem. 249:6197
9. Murray, K. and Murray, N.E. (1973) Nature New Biology 243:134
10. Mandel, M. and Berg, A. (1968) Proc. Nat. acad. Sci. USA 60:265
11. Kurosawa, Y. and Okazaki, R. (1975) J. Mol. Biol. 94:229
12. Schnös, M. and Inman, R.B. (1971) J. Mol. Biol. 55:31
13. Lindqvist, B.H. (1971) Mol. Gen. Genet. 110:178
14. Pruss, G.J., Wang, J.C. and Calendar, R. (1975) J. Mol. Biol. 98:465
15. Pruss, G.J. and Calendar, R. (1978) Virology 86:454
16. Calendar, R., Lindqvist, B.H., Sironi, G. and Clark, A.J. (1970) Virology 40:72
17. Sunshine, M.G. and Sauer, B. (1975) Proc. Nat. Acad. Sci USA 72:2770
18. Birkeland, N.K., Christie, G.E. and Lindqvist, B.H. (1988) Gene 73:327
19. Calendar, R., Six, E.W. and Kahn, F. (1977) In: DNA insertion Elements, Plasmids and Episomes (Bukhari, Shapiro, Adhya, Eds.) Cold Spring Harbor Laboratory, Cold Spring Harbor, New York. p. 395
20. Kelly, B. (1963) Virology 19:32
21. Calendar, R. and Lindahl, G. (1969) Virology 39:867
22. Six, E. (1966) Virology 29:106
23. Sunshine, M.G. and Kelly, B.L. (1967) Virology 32:644
24. Choe, B.K. (1969) Molec. Gen. Genet. 105:275
25. Lindahl, G. (1969) Virology 39:861
26. Bertani, L.E. (1970) Proc. Nat. Acad. Sci. USA 65:331
27. Bertani. L.E., Ljungquist, E. and Bertani, G. (1981) In: Microbiology 1981, American Soc. Microbiol., p. 61
28. Yu, A., Bertani, L.E. and Haggård-Ljungquist, E. (1989) Gene 80:1
29. Haggård-Ljungquist, E., Kockum, K. and Bertani, L.E. (1987) Mol. Gen. Genet. 208:52
30. Saha, S., Haggård-Ljungquist, E. and Nordström, K. (1987) EMBO J. 6:3191
31. Hudnik-Plevnik, T. and Bertani, G. (1980) Mol. Gen. Genet. 178:131
32. Bertani, G. (1975) Mol. Gen. Genet. 136:107
33. Lindahl, G. (1970) Virology 42:522
34. Bertani, L.E. (1980) Mol. Gen. Genet. 178:91
35. Bertani, L.E. (1971) Virology 46:426
36. Lindqvist, B.H. and Boevre, K. (1972) Virology 49:690
37. Geisselsoder, J., Mandel, M., Calendar, R. and Chattoraj, D.K. (1973) J. Mol. Biol. 77:405
38. Lindahl, G. (1971) Virology 46:620
39. Funnel, B.E. and Inman, R.B. (1982) J. Mol. Biol. 154:85
40. Haggård-Ljungquist, E., Barreiro, V., Calendar, R., Kurnit, D.M. and Cheng, H. Gene, in press
41. Six, E.W. (1975) Virology 67:249
42. Souza, L., Calendar, R., Six, E.W. and Lindqvist, B.H. (1977) Virology 81:81
43. Six, E.W. and Lindqvist, B.H. (1978) Virology 87:217
44. Geisselsoder, J., Youderian, P., Deho, G., Chidambaram, M., Goldstein, R.N. and Ljungquist, E. (1981) J. Mol. Biol. 148:1

45. Saha, S., Haggård-Ljungquist, E. and Nordström, K. (1989) Proc. Nat. Acad. Sci USA 86:3973
46. Chattoraj, D.K. (1978) Proc. Nat. Acad. Sci USA 75:1685
47. Geisselsoder, J. (1976) J. Mol. Biol. 100:13
48. Lengyel, J.A. and Calendar, R. (1974) Virology 53:305
49. Bertani, L.E. (1968) Virology 36:87
50. Westöö, A. and Ljungquist, E. (1980) Mol. Gen. Genet. (1980) 178:101
51. Lindahl, G. (1974) Mol. Gen. Genet. 128:249
51. Lengyel, J.A., Goldstein, R., Marsh, M. and Calendar, R. (1974) Virology 62:161
53. Lengyel, J.A., Goldstein, R.N., Marsh, M., Sunshine, M.G. and Calendar, R. (1973) Virology 53:1
54. Christie, G.E. and Calendar, R. (1983) J. Mol. Biol. 167:773
55. Bertani, L.E. (1976) Virology 71:85
56. Bertani, L.E. (1978) Mol. Gen. Genet. 166:85
57. Lindahl, G. and Sunshine, M. (1972) Virology 49:180
58. Ljungquist, E. and Bertani, L.E. (1983) Mol. Gen. Genet. 192:87
59. Chattoraj, D.K. and Inman, R.B. (1972) J. Mol. Biol. 66:423
60. Chattoraj, D.K., Oberoi, Y.K. and Bertani, G. (1977) Virology 81:460
61. Chattoraj, D.K., Younghusband, H.B. and Inman, R.B. (1975) Mol. Gen. Genet. 136:139
62. Hyde, J.M. and Bertani, G. (1975) J. Gen. Virol. 28:415
63. Chattoraj, D.K. and Bertani, G. (1980) Mol. Gen. Genet. 178:85
64. Lindqvist, B.H. (1981) Gene 14:243
65. Chattoraj, D.K. and Inman, R.B. (1973) Virology 55:174
66. Cohen, D. (1959) Virology 7:112
67. Bertani, L.E. (1964) Biochim. Biophys. Acta 87:631
68. Bertani, L.E. and Levy, J.A. (1964) Virology 22:634
69. Bertani, G., Choe, B.K. and Lindahl, G. (1969) J. Gen. Virol. 5:97
70. Bertani, G., Ljungquist, E., Jagusztyn-Krynicka, K. and Jupp, S. (1978) J. Gen. Virol. 38:251
71. Calendar, R., Lindahl, G., Marsh, M. and Sunshine, M. (1972) Virology 47:68
72. Christie, G.E., Haggård-Ljungquist, E., Feiwell, R. and Calendar.R. (1986) Proc. Nat. acad. Sci. USA 831:3238
73. Sironi, G. (1969) Virology 37:163
74. Sironi, G., Bialy, H., Lozeron, H.A. and Calendar, R. (1971) Virology 46:387
75. Bregegere, F. (1974) J. Mol. Biol. 90:459
76. Bregegere, F. (1976) J. Mol. Biol. 104:411
77. Bregegere, F. (1978) J. Mol. Biol. 122:113
78. Ghisotti, D.,Zangrossi, S. and Sironi, G. (1979) Mol. Gen. Genet. 169:229
79. Sunshine, M., Usher, D. and Calendar, R. (1975) J. Virol. 16:284
80. Lindahl, G., Hirota, Y. and Jacob, F. (1971) Proc. Nat. Acad. Sci USA 68:2407
81. Kuempel, P.L., Duerr, S.A. and Maglothin, P.D. (1978) J. Bacteriol. 134:902
82. Ljungquist, E., Kockum, K. and Bertani, L.E. (1984) Proc. Nat. Acad. Sci USA 81:3988
83. Barclay, S.L. and Dowe, W.F. (1978) Virology 91:321
84. Bertani, L.E. (1957) Virology 4:53
85. Bertani, L.E. and Bertani, G. (1974) Proc. Nat. Acad. Sci. USA 71:315
86. Chattoraj, D.K. and Inman, R.B. (1974) Proc. Nat Acad. Sci. USA 71:311
87. Bertani, G. and Chattoraj, D.K. (1980) Nucleic Acid Res. 8:1339

RESTRICTION ENZYME USED (number of sites in P2): *Aat*II(10), *Acc*I(8), *Afl*II(4), *Apa*LI(8), *Asu*II(3), *Ava*I(3), *Bal*I(1), *Bam*HI(3), *Bgl*II(3), *Bst*EII(11), *Bst*XI(4), *Cla*I(4), *Dra*III(7), *Eco*RI(3), *Eco*RV(13), *Hind*II(22), *Hpa*I(10), *Kpn*I(3), *Mlu*I(6), *Nco*I(2), *Nde*I(4), *Nru*I(7), *Pfl*MI(1), *Pss*I(2), *Pst*I(3), *Pvu*II(1), *Sal*I(2), *Sca*I(2), *Sma*I(1), *Sna*BI(2), *Sph*I(6), *Stu*I(4), *Tth*I(3).

Numbers are distances from left end of map in % P2 DNA lenght (33kb). Gene extensions drawn in full lines are based on published sequences, and dashed lines are based on unpublished sequences by E. H-L and Gail Christie (F_I and F_{II}).

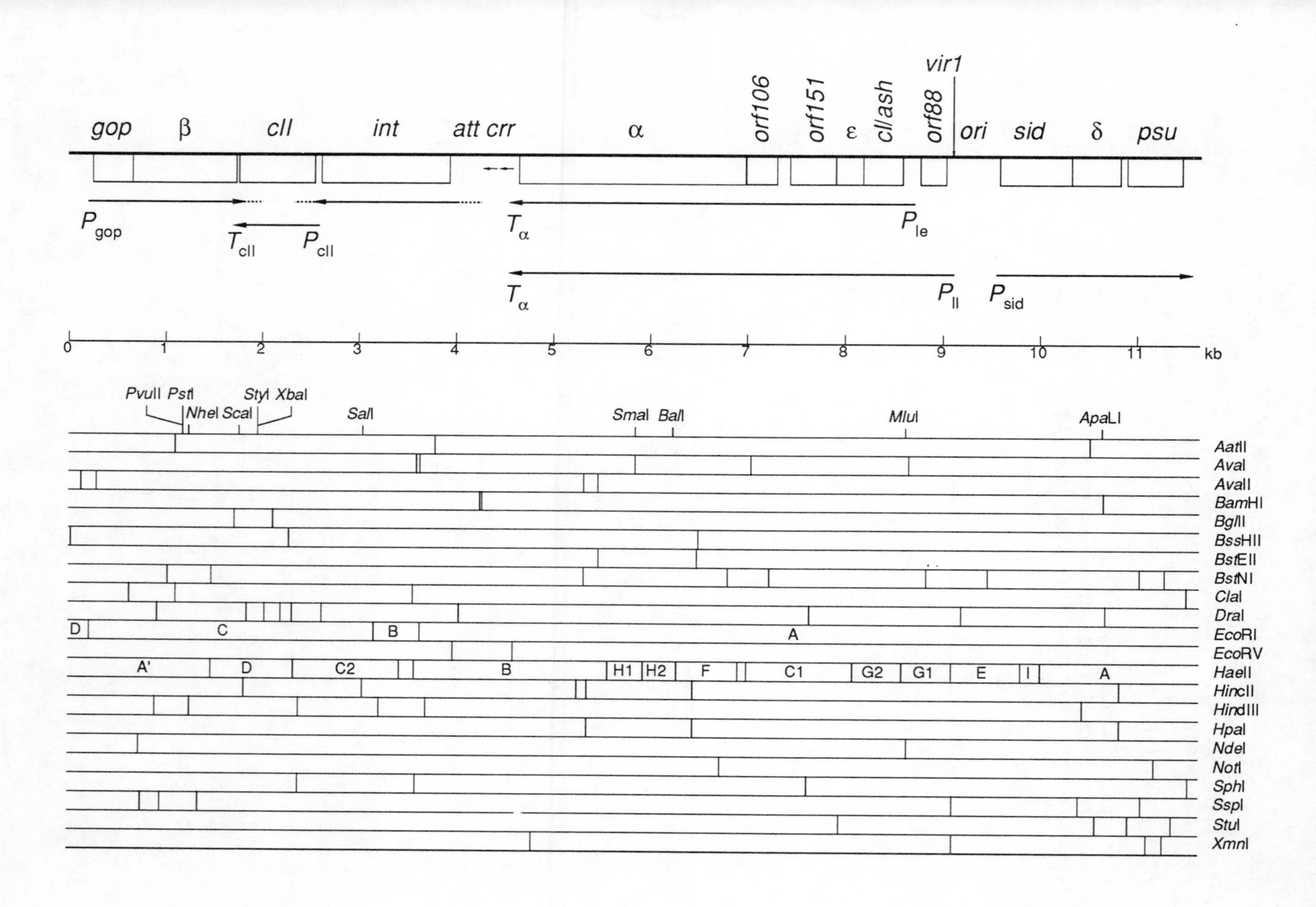
gop
β
cII
int
att crr
α
orf106
orf151
ε
cII/ash
orf88
vir1
ori
sid
δ
psu
P_{gop}
T_{cII}
P_{cII}
T_{α}
P_{le}
T_{α}
P_{II}
P_{sid}
0
1
2
3
4
5
6
7
8
9
10
11
kb
PvuII PstI
NheI ScaI
StyI XbaI
SalI
SmaI BalI
MluI
ApaLI
AatII
AvaI
AvaII
BamHI
BglII
BssHII
BstEII
BstNI
ClaI
DraI
EcoRI
EcoRV
HaeII
HincII
HindIII
HpaI
NdeI
NotI
SphI
SspI
StuI
XmnI
D
C
B
A
A'
D
C2
B
H1
H2
F
C1
G2
G1
E
I
A

BACTERIOPHAGE P4

August, 1989

Conrad Halling and Richard Calendar, Department of Molecular and Cell Biology,
University of California at Berkeley, Berkeley, CA 94720. E-mail: calendar@garnet.berkeley.edu

P4 is a satellite bacteriophage that requires the late genes of a helper of the P2 family to carry out lytic multiplication. The life-cycles of P4 and the P2-like phages were recently reviewed in Ref. 4. The DNA sequence of the entire P4 genome (11,627 bp) has been determined (7, 11, 14, 16, 25, 26, 28, 31, 37). A physical map of P4 is presented in Figure 1. The complete sequence with corrections and annotations is available as a text file on Macintosh-format disks or by electronic mail. The complete sequence will be submitted to GenBank.

P4 genes and functional sites

Gene/site	Description	References
cos	19-base 5′ cohesive ends of linear P4 DNA and flanking DNA (< 100 bp) towards *psu*, required for packaging P4 DNA	28, 37
P_{gop}	promoter for transcription of *gop* and β; mRNA start mapped	14
gop	blocks multiplication of P4 on *E. coli argU* (P2); product identified (M_r 14,874)	5, 8, 13, 14
β	negatively controls activity of *gop*; the single mutation identified in the gene (β*103*) confers a lethal Ts phenotype, which is suppressed by *gop* mutations; product identified (M_r 39,864)	3, 8, 14, 15
T_{cII}	putative Rho-independent terminator for transcription of *cII*; a mutation in the stem (ts*37*) confers a lethal Ts phenotype, probably due to interference of unterminated transcription from P_{cII} with transcription of β	14
cII	function unknown; mutations confer a clear plaque phenotype; product identified (M_r 30,314)	5, 14
P_{cII}	promoter for transcription of *cII*; mRNA start mapped	14
int	integrase; product not identified (predicted M_r 48,000 to 51,000, depending on location of start codon)	5, 30, 31

P4 genes and functional sites (cont.)

Gene/site	Description	References
P_{int}	promoter for transcription of *int*, repressed by *int* protein; mRNA start not mapped	31
att	attachment site for integration	5, 30, 31
crr	*cis*-acting site required for P4 DNA replication	11, 22
T_{α}	putative Rho-independent terminator for transcription through gene α	11
α	primase for P4 DNA replication; product identified (M_r 84,823)	1-3, 11, 15, 21
orf106	may be required for P4 DNA replication; product identified (M_r 11,788)	6, 11
orf151	function unknown; product not identified (predicted M_r 17,706)	26
ε	derepression of helper prophage; product detected (M_r 10,952)	10, 12, 26
cI/ash	function unknown; mutations confer a clear plaque phenotype and allow P4 to make plaques on *E. coli* (P3); some mutations suppress *vir1*; product not identified (predicted M_r 14,833)	5, 23, 25-27
P_{le}	promoter for leftward early transcription of *cI* through α	9, 26
orf88	may channel P4 toward the lytic response; product, not identified (predicted M_r 9,771), contains helix-turn-helix motif	16, 19, 23, 26
P_{ll}	promoter for leftward late transcription of *orf88* through α; positively regulated by P4 δ gene product	9, 26
vir1	a promoter-up mutation of P_{ll} that confers a clear-plaque phenotype and causes partial insensitivity to immunity	5, 9, 26, 29, 35
ori	origin of P4 DNA replication	22, 26
P_{sid}	late transcription of *sid*, δ, and *psu*; positively regulated by P4 δ and P2 *ogr* gene products	7, 9, 20
sid	directs construction of P4-size capsids; product identified (M_r 27,227)	3, 10, 24, 26, 34
δ/*org*	positive control of P4 and helper late transcription; the *org* mutation suppresses *E. coli rpoA109*; product identified (M_r 19,019)	17-19, 26, 32, 36
psu	suppression of polarity of amber mutations in the P2 late genes product detected (M_r 21,282)	3, 7, 33

REFERENCES

1. Barrett, K. J., A. Blinkova, and G. Arnold. 1983. *J. Virol.* **48:**157-169.
2. Barrett, K. J., W. Gibbs, and R. Calendar. 1972. *Proc. Natl. Acad. Sci. USA* **69:**2986-2990.
3. Barrett, K. J., M. L. Marsh, and R. Calendar. 1976. *J. Mol. Biol.* **106:**683-707.
4. Bertani, L. E., and E. W. Six. 1988. In R. Calendar (ed.), *The Bacteriophages,* Plenum Press, N.Y., pp. 73-143.
5. Calendar, R., E. Ljungquist, G. Dehò, D. C. Usher, R. Goldstein, P. Youderian, G. Sironi, and E. W. Six. *Virology* **113:**20-38.
6. Christian, R., J. Flensburg, M. Krevolin, and R. Calendar, unpublished data.
7. Dale, E. C., G. E. Christie, and R. Calendar. 1986. *J. Mol. Biol.* **192:**793-803.
8. Dehò, G. 1983. *Virology* **126:**267-278.
9. Dehò, G., S. Zangrossi, D. Ghisotti, and G. Sironi. 1988. *J. Virol.* **62:**1697-1704.
10. Diana, C., G. Dehò, J. Geisselsoder, L. Tinelli, and R. Goldstein. 1978. *J. Mol. Biol.* **126:**433-445.
11. Flensburg, J., and R. Calendar. 1987. *J. Mol. Biol.* **195:**439-445.
12. Geisselsoder, J., P. Youderian, G. Dehò, M. Chidambaram, R. Goldstein, and E. Ljungquist. 1981. *J. Mol. Biol.* **148:**1-19.
13. Ghisotti, D., S. Zangrossi, and G. Sironi. 1983. *J. Virol.* **48:**616-626.
14. Ghisotti, D., S. Finkel, C. Halling, G. Dehò, G. Sironi, and R. Calendar. *J. Virol.*, in press.
15. Gibbs, W., R. N. Goldstein, R. Wiener, B. Lindqvist, and R. Calendar. 1973. *Virology* **53:**24-39.
16. Halling, C. 1989. Ph.D. dissertation, University of California, Berkeley, CA.
17. Halling, C., and R. Calendar, manuscript in preparation.
18. Halling, C., M. G. Sunshine, K. B. Lane, E. W. Six, and R. Calendar, manuscript in preparation.
19. Kahn, M., D. Ow, B. Sauer, A. Rabinowitz, and R. Calendar. 1980. *Molec. Gen. Genet.* **177:**399-412.
20. Keener, J., E. C. Dale, S. Kustu, and R. Calendar. 1988. *J. Bacteriol.* **170:**3543-3546.
21. Krevolin, M. D., and R. Calendar. 1985. *J. Mol. Biol.* **182:**509-517.
22. Krevolin, M. D., R. B. Inman, D. Roof, M. Kahn, and R. Calendar. *J. Mol. Biol.* **182:**519-527.
23. Lane, K. B. 1985. M.S. thesis, University of Iowa, Iowa City, IA.
24. Lee, S.-J. 1981. Bachelor of Arts Thesis, Harvard University, Cambridge, MA.
25. Lin, C.-S. 1983. Ph.D. dissertation, University of Iowa, Iowa City, IA.
26. Lin, C.-S. 1984. *Nucleic Acids Res.* **12:**8667-8684.
27. Lin, C.-S., and E. W. Six, manuscript in preparation.
28. Lindqvist, B. H. 1981. *Gene* **14:**231-241.
29. Lindqvist, B. H., and E. W. Six. 1971. *Virology* **43:**1-7.
30. Pierson, L. S., III, and M. L. Kahn. 1984. *Mol. Gen Genet.* **195:**44-51.
31. Pierson, L. S., III, and M. L. Kahn. 1987. *J. Mol. Biol.* **196:**487-496.
32. Sauer, B., R. Calendar, E. Ljungquist, E. Six, and M. G. Sunshine. 1982. *Virology***116:**523-534.
33. Sauer, B., D. Ow, L. Ling, and R. Calendar. 1981. *J. Mol. Biol.* **145:**29-46.
34. Shore, D., G. Dehò, J. Tispis, and R. Goldstein. 1978. *Proc. Natl. Acad. Sci. USA* **75:**400-404.
35. Six, E. W., and C. A. C. Klug. 1973. *Virology* **51:**327-344.
36. Souza, L., R. Calendar, E. W. Six, and B. H. Lindqvist. 1977. *Virology* **81:**81-90.
37. Ziermann, R., and R. Calendar, unpublished data.

GENETIC AND RESTRICTION MAPS OF *BACILLUS* BACTERIOPHAGE ⌀29

November, 1987

Václav Pačes and Čestmír Vlček
Institute of Molecular Genetics,
Czechoslovak Academy of Sciences,
166 37 Prague,
Czechoslovakia

Bacteriophage ⌀29 lytically infects cells of *Bacilli* (for review see 10). Its genome consists of linear duplex DNA 19 285 bp long. The complete nucleotide sequence of the genome was determined (5-9,16,17,21,25-27).

Eighteen genes were identified in the genome by genetic and biochemical methods (1,3,4,11,15,22) (Table I). They are transcribed in three groups (Fig.1). Major (left) early region contains genes 1 to 6, minor (right) early region contains gene 17. The two early regions are transcribed from the l strand in the right to left orientation. Genes 7, 8, 8.5 and 9 to 16 are late genes transcribed from the central part of the r DNA strand in the left to right orientation. Early genes 2, 3, 5 and 6 code for proteins involved in phage DNA replication (Table I). Genes 1 and 17 are probably also involved in DNA replication; however, their function in this process is not understood. Phage genome is replicated in the form of linear DNA (for review see 23,24). Gene 3 product is covalently attached to the 5' end of the DNA strands and it serves as a primer for replication. Gene 2 product is the phage specific DNA polymerase.

The switch from early to late transcription is regulated by the gene 4 product. Several early promoters and one late promoter were well characterized (2,13,14). A bidirectional terminator of transcription was found between the right early region and the late region (9,17).

Several closely related phages are studied, namely phage PZA and phage ⌀15 (12,18-21). Complete nucleotide sequence of the phage PZA genome has been determined (19-21). Genomes of ⌀29 and PZA are identically organized with the same set of genes. Comparison of the two nucleotide sequences allowed to determine tolerated variations in these genomes and their evolutionary history (16,18).

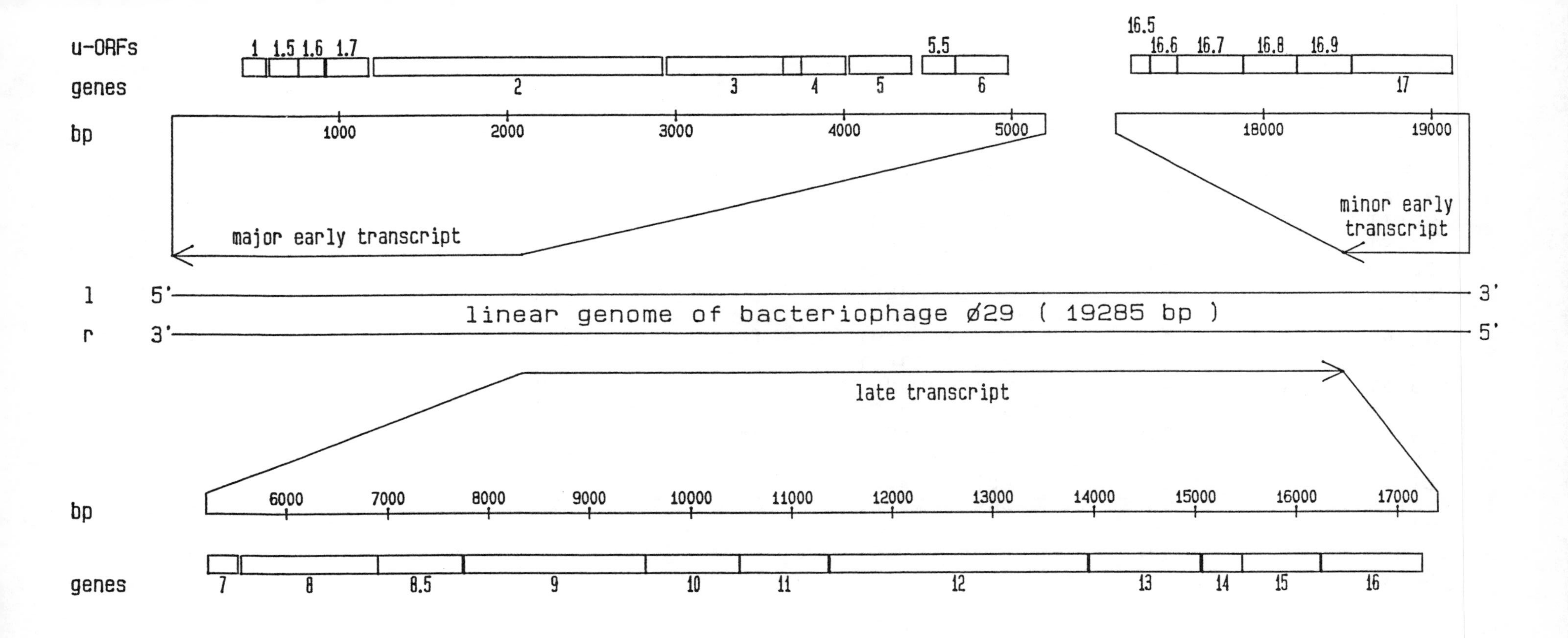

Fig.1 Genetic and transcriptional maps of bacteriophage ø29. u-ORFs, unidentified open reading frames coded by the 5'-3' strand. No ORF was assigned yet to gene 1. Genes 2-6 and 17 are coded also by the 5'-3' strand. Genes 7-16 are coded by the 3'-5' strand.

Table I ø29 Genes, proteins and functions

Gene or open reading frame	Transcript	Protein or phenotype	Nucleotides of the l DNA strand	Base pairs	Amino acids	Calculated mol.wt. of the gene product
1		?	510- 367	144	47	5096
1.5		DNA synthesis (?)	703- 527	177	58	6508
1.6		?	870- 700	171	56	6525
1.7		?	1117- 857	261	86	9219
2	Major early	DNA polymerase	2863- 1145	1719	572	63631
3		Terminal protein	3686- 2886	801	266	29536
4		Late transcription	3952- 3575	378	125	14097
5		DNA synthesis	4345- 3971	375	124	13656
5.5		?	4607- 4410	198	65	6961
6		DNA synthesis	4921- 4607	315	104	12047
7		Head morphogenesis	5222- 5518	297	98	11406
8		Major head protein	5549- 6895	1347	448	49850
8.5		Head fibre protein	6895- 7737	843	280	31372
9		Tail protein	7751- 9550	1800	599	67744
10	Late	Upper collar protein	9555-10484	930	309	35117
11		Lower collar protein	10477-11358	882	293	34491
12		Appandage precursor	11371-13935	2565	854	94960
13		Morphogenesis	13947-15044	1098	365	40957
14		Lysis	15063-15458	396	131	14161
15		Morphogenesis+lysis	15460-16236	777	258	28969
16		Encapsidation, ATPase	16247-17245	999	332	36756
16.5		?	17389-17276	114	37	4267
16.6		?	17550-17386	166	54	5922
16.7	Minor early	?	17937-17547	393	130	15121
16.8		?	18258-17941	318	105	11928
16.9		?	18581-18255	327	108	12038
17		DNA synthesis (?)	19083-18583	501	166	18897

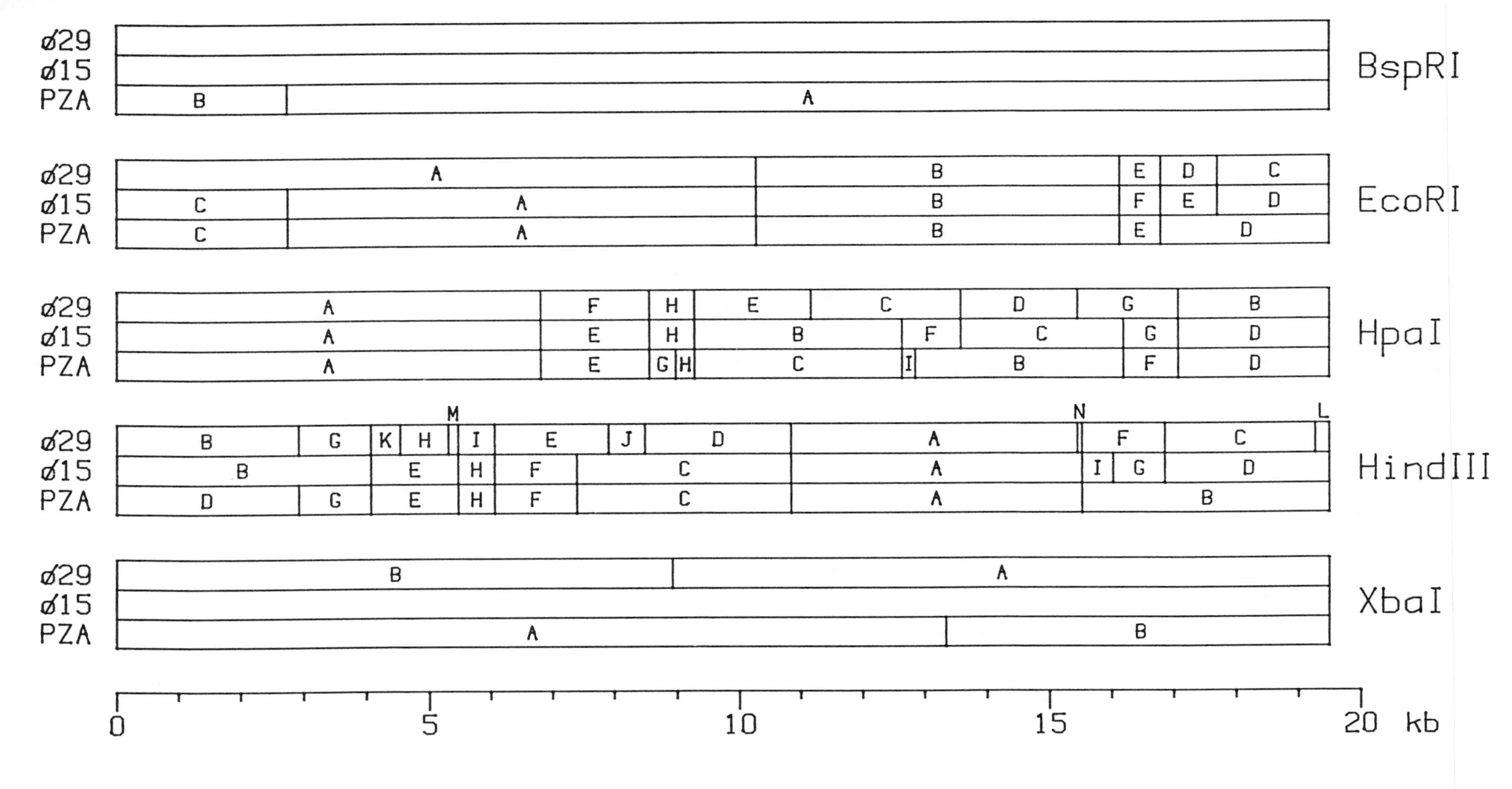

Fig.2 Restriction maps of bacteriophages ø29, ø15 and PZA. Restriction fragments A,B,C etc. are designated according to decreasing size.

REFERENCES

1. Anderson,D.L. and Reilly,B.E. 1974. J. Virol. 13:211-221
2. Barthelemy,I., Salas,M. and Mellado,R.P. 1986. J. Virol. 60:874-879
3. Camacho,A., Moreno,F., Carrascosa,J.L., Viñuela,E. and Salas,M. 1974. Eur.J.Biochem. 47:199-205
4. Carrascosa,J.L., Camacho,A., Moreno,F., Jimenez,F., Mellado,R.P., Viñuela,E. and Salas,M. 1976. Eur.J.Biochem. 66:229-241
5. Escarmis,C. and Salas,M. 1981. Proc. Natl. Acad. Sci. USA 78:1446-1450
6. Escarmis,C. and Salas,M. 1982. Nucleic Acids Res. 10:5785-5798
7. Garcia,G.A., Mendez,E. and Salas,M. 1984. Gene 30:87-98
8. Garvey,K.J., Saedi.M.S. and Ito,J. 1985. Gene 40:311-316
9. Garvey,K.J., Yoshikawa,H. and Ito,J. 1985. Gene 40:301-309
10. Geiduschek,E.P. and Ito,J. 1982. The Molecular Biology of the Bacilli, Vol.1. Academic Press, New York. 203-218
11. Hagen,E.W., Reilly,B.E., Tosi,M.E. and Anderson,D.L. 1976. J. Virol. 19:501-517
12. Hostomsky,Z., Paces,V. and Zadrazil,S. 1985. FEBS Letters 188:123-126
13. Mellado,R.P., Barthelemy,I. and Salas,M. 1986. Nucleic Acids Res. 14:4731-4342
14. Mellado,R.P., Barthelemy,I. and Salas,M. 1986. J. Mol. Biol. 191:191-198
15. Mellado,R.P., Moreno,F., Viñuela,E., Salas,M., Reilly,B.E. and Anderson,D.L. 1976. J. Virol. 19:495-500
16. Pačes,V., Hostomský,Z., Fučík,V., Pivec,L. and Zadražil,S. 1984. Folia Biologica (Prague) 30:52-64
17. Pačes,V., Hostomský,Z., Vlček,C., Urbánek,P. and Zadražil,S. 1985. Gene Manipulation and Expression. Croom Helm, London. 209-224
18. Pačes,V., Vlček,C., Šmarda,J., Zadražil,S. and Fučík,V. 1986. Gene 46:215-225
19. Pačes,V., Vlček,C. and Urbánek,P. 1986. Gene 44:107-114
20. Pačes,V., Vlček,C., Urbánek,P. and Hostomský,Z. 1985. Gene 38:45-56
21. Pačes,V., Vlček,C., Urbánek,P. and Hostomský,Z. 1986. Gene 44:115-120
22. Reilly,B.E., Nelson,R.A. and Anderson,D.L. 1977. J. Virol. 29:367-377
23. Salas,M. 1987. Current Topics in Microbiology and Immunology. In press.
24. Salas,M., Blanco,L., Prieto,I., Garcia,J.A., Mellado,R.P., Lazaro,J.M. and Hermoso,J.M. 1984. Proteins Involved in DNA Replication. Plenum Press, New York. 35-44
25. Vlček,C. and Pačes,V. 1986. Gene 46:215-225
26. Yoshikawa,H., Friedmann,T. and Ito,J. 1981. Proc. Natl. Acad. Sci. USA 78:1336-1340
27. Yoshikawa,H. and Ito,J. 1982. Gene 17:323-335

GENETIC AND RESTRICTION MAP OF BACTERIOPHAGE ØX174

Peter Weisbeek
Dept. of Molecular Cell Biology and
Institute of Molecular Biology
University of Utrecht
The Netherlands

October, 1986

The icosahedral bacteriophage ØX174 has a single-stranded circular DNA molecule of 5386 nucleotides (21). It contains genetic information for at least 10 different proteins. Five of these genes share their nucleotide sequence with another gene (overlapping genes). Two of these (A and A*) use the same sequence in the same reading frame; the A* protein starts from an in-frame internal initiation AUG codon. The other overlapping genes (B + A/A* and D + E) read the common sequence in a different frame.
The genome contains several non-coding sequences (intercistronic regions) of varying size. For two of these intercistronic regions it is demonstrated that they carry genetic information. The region between the genes F and G is essential for the initiation of the synthesis of the complementary DNA strand; it contains the recognition sequence for the prepriming n' protein (29). The sequence between the genes H and A contains the major transcription terminator (21) and information for *in vivo* DNA synthesis (30). The region between the genes J and F appears to carry no essential information (apart from the ribosome binding site for gene F) (27,28).
The phage DNA contains three major promoters: P_A, before gene A; P_B, in front of the B gene and P_D, before gene D. Apart from the central terminator in the intercistronic region between the genes H and A, there are minor (possibly rho-dependent) terminators in the region coding for the late (coat) proteins (31,32). These transcription signals permit the phage to regulate the expression of its genes in great detail. The recent advance in sequencing and in the synthesis of short nucleotide sequences has given a new dimension to genetic analysis. By changing, deleting or adding preselected nucleotides a much more direct analysis of the genome has become possible as is demonstrated very elegantly in the DNA replication studies of Baas *et al.* (25).

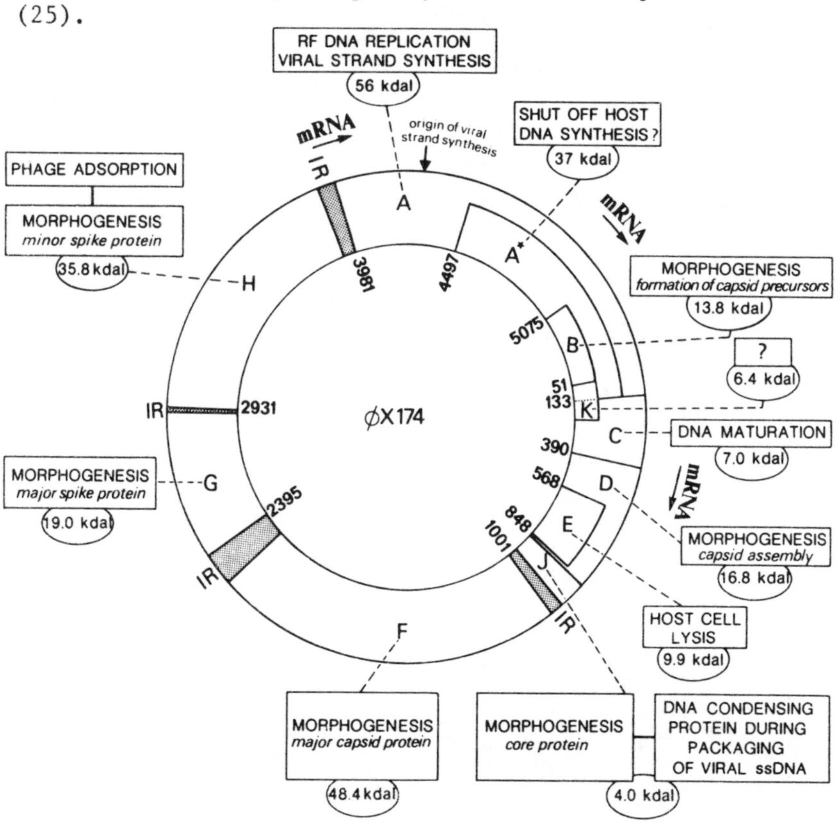

Gene	Protein		Mutant	Mode of DNA synthesis blocked	Reference
	Mr	Function			
A	56,000	DNA replication	am86, am50	RF → RF DNA	
			am8, am30, am33		
			ts128		1
			amN14, amR7		
			amH532Y,ts7		
			amS7		2,3
			to5, tsR3-1		4,5
A* (and A)	35,000	DNA replication	am18, lm35		1
			amR8-1		4
			op101		6
			amH90, amS23		
			amS29, amH29		
			amS1, ts56		2,3
			Am62, opE15	RF → ss DNA	7,8
K	6,000	unknown	amKsB1		37
B	14,000	Morphogenesis	am14, am16	RF → ss DNA	
			ts9, ts116, och5		1
			amH210, anH34		
			ts173		2,3
			to8, ts6, tsR5-2		4,8
C	10,000	Phage maturation	och6	RF → ss DNA	9
			ts35		3
D	16,800	Phage assembly	am10, am42	RF → ss DNA	1
			amH56, amH81		3
			tsZ7-1		5
E	10,000	Lysis	am3, am27	None	1
			amN11		3
			am61		7
J	4,200	Phage coat	sK1	Unknown	10,15
F	46,400	Phage coat	op6, op9, tsh6	RF → ss DNA	
			ts41D		1
			am87, am88		
			am89		1,9
			amH57, amH236		
			amH244, amH299		
			amH375, amH600		3
			ts27, tsγ, tsH1		
			tsZ4-1		5
			cs70 (in am3cs70)		15
			sK2		15
			sB1		16
G	19,050	Phage coat	am9, am32	RF → ss DNA	
			ts79		1
			amH116, amH379		3
			tsZ11, tsZ12, tsZ13		8
			em25		11
			G∇/FSZH1 (deletion)		12
			h8, h10, h11		13
			Gms143 RC2		34
			Gms95 RC8		34
H	35,800	Phage coat	am23, am80	None	
			am90, ts4		1
			amN1, amH257		
			amH497		3
			hr, HaHb		14

SEQUENCED MUTANTS

Gene	Mutant	Nucleotide number of mutation	Base change	Reference
A	am86	4116	C → T	17
	am33	4380	C → T	17
	tsR3-1	4387	G → A	18
	op101	5078	G → A	6
	am18	23	C → T	19
	am35	23	C → T	19
	am62	117	G → A	19
K	amKsB1	70	T → A	37
B	am16	5276	G → T	20
	ts116	23	C → T	
		25	G → C	19
C	och6	196	C → T	21
E	am3	587	G → A	22
	am61	587	G → A	19
	am27	587	G → A	22
	amN11	587	G → A	22
J	sK1	874	C → T	15
F	sK2	1864	G → A	15
F	sB1	1641	A → T	16
G	amG2401	2401	C → T	24

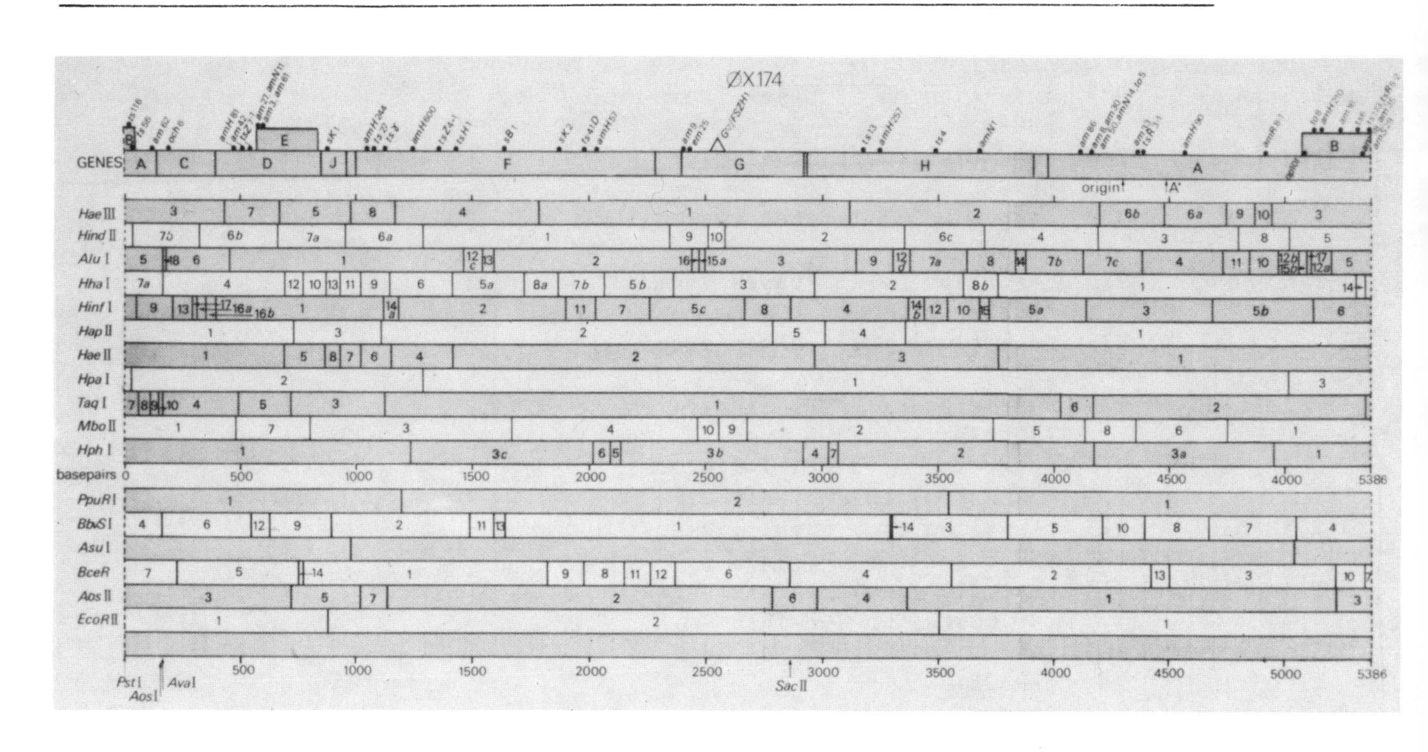

TYPES OF MUTANTS

am to em	: amber (UAG) nonsense codon	h	: host range (extended)
och	: ochre (UAA) nonsense codon	sK	: sensitive to restriction by *E. coli* K
op	: opal (UGA) nonsense codon	sB	: sensitive to restriction by *E. coli* B
ts	: thermosensitive mutant		

Mutant	Position		Base change	Reference
ori-6	4304	in recognition sequence for initiator A protein	T → C	25
ori-10	4308	idem	A → T	33
ori-12	4310	idem	T → C	25
ori-13	4311	idem	A → T	33
ori-14.1	4312	idem	A → T	33
ori-14.2	4312	idem	A → C	33
ori-14.3	4312	idem	A → G	33
ori-15.1	4313	idem	T → C	33
ori-15.2	4313	idem	T → A	33
ori-17	4315	idem	A → C	33
ori-18	4316	idem	C → T	33
ori-19.1	4317	idem	A → G	33
ori-19.2	4317	idem	A → C	33
ori-20	4318	idem	C → T	33
ori-21	4319	idem	T → A	33
ori-22	4320	idem	A → C	33
G/H de2925	2925	ribosome binding sequence of gene H	T deleted	24,26
J-F ins1	978	insertion of 94 bases from pBR322		27,28
J-F ins2	978	insertion of 44 bases from pBR322		27,28
J-F ins3	978	insertion of 55 bases from pBR322		27,28
J-F ins4	978	insertion of 93 bases from pBR322		27,28
J-F ins5	978	insertion of 64 bases from pBR322		27,28
J-F ins6	978	insertion of 115 bases from pBR322		27,28
J-F ins6003	978	insertion of 181 bases from pBR322		35
J-F ins6007	978	insertion of 344 bases from pBR322		35
J-F ins6430	978	insertion of BamHI linker (10 bases) in J-F ins6		36
J-F ins6420	978	insertion of 2 BamHI linkers (20 bases) in J-F ins6		36
J-F ins6461	978	insertion of 3 BamHI linkers (30 bases) in J-F ins6		36
J-F ins6463	978	insertion of 4 BamHI linkers (40 bases) in J-F ins6		36
J-F ins7	978	insertion of 48 bases from pBR322		27,28
J-F ins8	978	insertion of 45 bases from pBR322		27,28
J-F ins9	978	insertion of 89 bases from pBR322		27,28
J-F ins10	978	insertion of 55 bases from pBR322		27,28
J-F ins11	978	insertion of 163 bases from pBR322		27,28
J-F del1	978	deletion of 7 bases		27,28
J-F del2	978	deletion of 4 bases		27,28
J-F del3	978	deletion of 27 bases		27,28

REFERENCES

1. Benbow, R.M., et al. 1971. J. Virol. 7:549-558
2. Linney, E., M. Hayashi. 1974. Nature 249:345-348
3. Hayashi, M., M. Hayashi. 1974. J. Virol. 14:1142-1151
4. Borrias, W.E., et al. 1976. Nature 261:245-248
5. Weisbeek, P.J. and G.A. van Arkel. 1978. The single-stranded DNA phages. Cold Spring Harbor Laboratory. 31-49
6. Heidekamp, F. and H.S. Jansz, personal commun.
7. Weisbeek, P.J., et al. 1977. Proc. Natl.Acad. Sci. USA 74:2504-2508
8. This Laboratory
9. Funk, F. and R.L. Sinsheimer. 1970. J. Virol. 6:12-19
10. Freymeyer, D.K., at al. 1977. Biochemistry 16:4551-4554
11. Borrias, W.E., et al. 1969. Molec. Gen. Genetics 105:152-163
12. Humayan, M.Z. and R.W. Chambers. 1979. Nature 278:524-529
13. Weisbeek, P.J., et al. 1973. Virology 52:408-416
14. Sinsheimer, R.L. 1968. Progress in Nucl. Ac. Res. 8:115-169
15. Chen Kan, N., et al. 1979. J. Mol. Biol. 130:191-209
16. Lautenberger, J.A., et al. 1978. Proc. Natl. Acad. Sci. USA 75:2271-2275
17. Sanger, F., et al. 1977. Nature 265:687-695
18. Van Mansfeld, A.D.M., personal commun.
19. Smith, M., et al. 1977. Nature 265:702-705
20. Brown, N.L. and M. Smith. 1977. J. Mol. Biol. 116:1-30
21. Sanger, F., et al. 1978. J. Mol. Biol. 125:225-246
22. Barrel, B.G., et al. 1976. Nature 264:34-41
23. Roberts, R.J. 1980. Nucl. Ac. Res. 8:63-80
24. Bhanot, O.S. et al. 1979. J. Biological Chemistry 254:12684-12693
25. Baas, P.D., et al. 1981. J. Mol. Biol. 152:615-639
26. Gillam, S., et al. 1980. Gene 12:129-137
27. Müller, U.R. and R.D. Wells. 1980. J. Mol. Biol. 141:1-24
28. Müller, U.R. and R.D. Wells. 1980. J. Mol. Biol. 141:25-41
29. Shlomar, J. and A. Kornberg 1980. Proc. Natl. Acad. Sci. USA 77:799-803
30. Van der Avoort, H.G.A.M., et al. 1982. J. Virol. 42:1-11
31. Hayashi, M.N. and Hayashi, M. 1981. J. Virol. 37:506-510
32. Patrushev, L.I., et al. 1981. Molec. Gen. Genet. 182:471-476
33. Baas, P.D. 1985. Biochim. Biophys. Acta. 825:111-139
34. Chambers, R.W., et al. 1982. Nucl. Ac. Res. 10:6465-6473
35. Russel, P.W. and U.R. Müller. 1984. J. Virol. 52:822-827
36. Müller, U.R. and M.A. Turnage. 1986. J. Mol. Biol. 189:285-292
37. Gillam, S. et al. 1985. J. Virol. 53:708-709

Simian Virus 40 Genetic Map

James Remenick and John Brady
Laboratory of Molecular Virology
National Cancer Institute
Bethesda, Maryland 20892
August, 1989

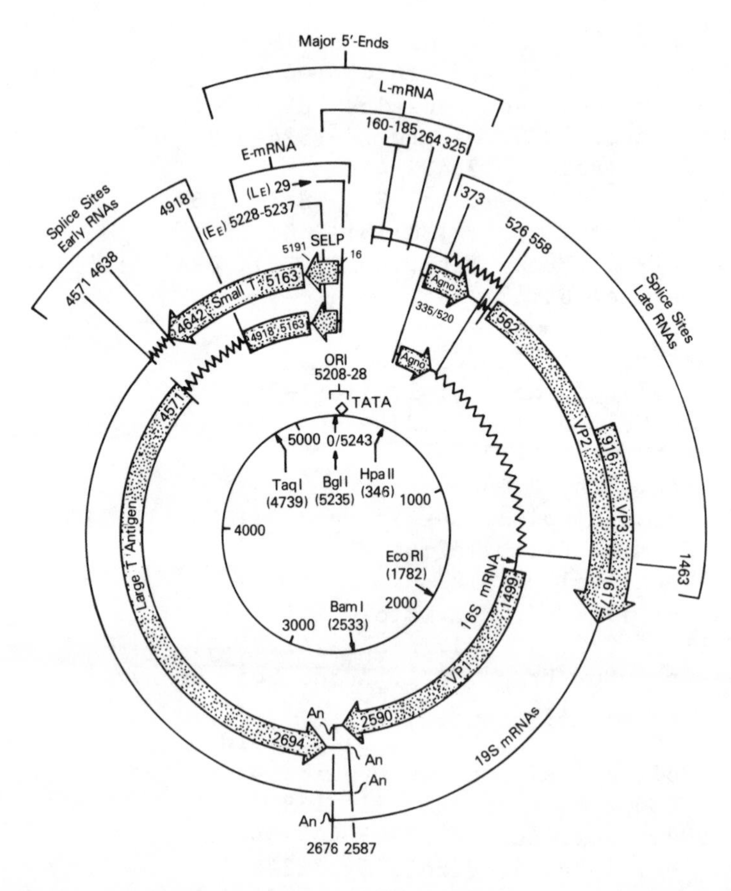

For this edition of **GENETIC MAPS**, we have focused on the explosion of information regarding the regulation of SV40 transcription. Presented on the following page, is a detailed map of the control region of SV40 with the recognition elements and DNase protected regions of the more well characterized transcription factors indicated. Although the figure is comprehensive, a number of the less well defined factors are not included. The high mobility group proteins, HMG-1 and HMG-2 (78), the EBP-1 protein identified in HeLa cells (11), the protein or proteins which bind to the CArG box region of the alpha actin gene (55) and the murine Hox-1.3 homeodomain protein (58), have all been shown to interact with the enhancer region. The SV40 nucleotide sequence presented can be found in GenBank, the genetic sequence data bank (release 45.0, September, 1986, The Los Alamos Laboratory). A map of viral protein coding sequences, start and stop sites, unique restriction sites and other important structural elements is also presented. The positions were taken from the GenBank features list for SV40. To simplify the map, the mRNA initiation and termination sites are not indicated. SV40 virus DNA contains a total of 5243-bp. Its sequence numbering origin is the central G base of a 27 nucleotide palindrome at or near the origin of replication. The first three nucleotides are the last three bases of the unique Bgl I restriction site. The sequence shown is that of strain 776 (25,64) with the numbering system presented by Buchman *et al.* (77), and corrected for the additional 17 bases starting at nucleotide 165. The polarity of the sequence numbering is that of the late mRNAs. The title page map was adapted from Brady and Salzman (9) and DePamphilis and Wassarman (15).

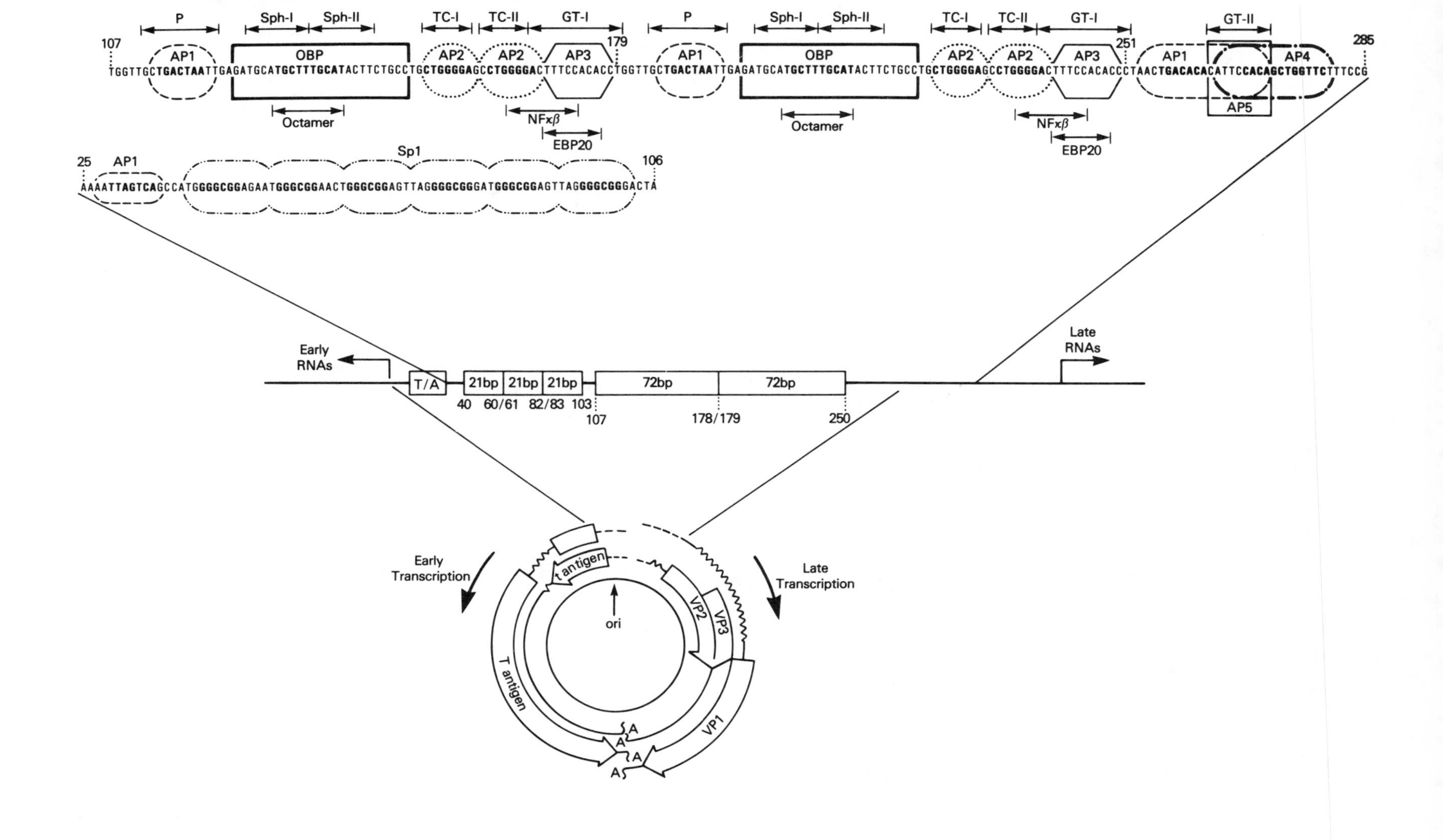

P
Sph-I
Sph-II
TC-I
TC-II
GT-I
P
Sph-I
Sph-II
TC-I
TC-II
GT-I
GT-II
107
179
251
285
AP1
OBP
AP2
AP2
AP3
AP1
OBP
AP2
AP2
AP3
AP1
AP4
AP5
TGGTTGCTGACTAATTGAGATGCATGCTTTGCATACTTCTGCCTGCTGGGGAGCCTGGGGACTTTCCACACCTGGTTGCTGACTAATTGAGATGCATGCTTTGCATACTTCTGCCTGCTGGGGAGCCTGGGGACTTTCCACACCCTAACTGACACACATTCCACAGCTGGTTCTTTCCG
Octamer
NFκβ
EBP20
Octamer
NFκβ
EBP20
25
AP1
Sp1
106
AAAATTAGTCAGCCATGGGGCGGAGAATGGGCGGAACTGGGCGGAGTTAGGGGCGGGATGGGCGGAGTTAGGGGCGGGACTA
Early RNAs
Late RNAs
T/A
21bp
21bp
21bp
72bp
72bp
40
60/61
82/83
103
107
178/179
250
Early Transcription
Late Transcription
t antigen
ori
VP2
VP3
VP1
T antigen
A
A
A
A

SV40 Transcription Factors

AP1
27-36
113-122
185-194
253-271

A 47 kD HeLa cell protein, induced by phorbol esters, that binds to the P motif (1,46). Mutations of the P motif significantly reduce enhancer activity and purified protein activates SV40 early and late transcription in vitro (46,51). AP1 (c-Jun) is a homologue of the fibroblast derived transcription factor, PAE1, and together with c-Fos, c-Myc and GCN4 comprise a family of related transcription factors containing leucine zipper domains (5,53,57,80). Consensus sequence: TT/GAGTCA

AP2
152-168
224-240

A phorbol ester inducible HeLa cell protein of 52 kD, also found in MPC-11 cells, which interacts at the TC-I and TC-II motifs and functions in concert with AP3 and Sp1 to regulate early transcription (12,23). Interestingly, only mutations in TC-I show decreased enhancer activity (38,52). Consensus sequence: CCCCAGGC

AP3
169-178
241-250

A 57 kD protein of HeLa cells which interacts with the SV40 enhancer at the GT-I motif that, in association with AP2, produce partial enhancer activity (12,41). Consensus sequence:GGGTGTGGAAAG

AP4
264-280

A 48 kD protein that functions with AP1 in directing late gene expression and is found in HeLa and MPC-11 cells (41,51,82).

AP5
263-271

A protein identified in lymphoid and nonlymphoid cells by DNase protection studies and mutational analysis of the GT-II motif (41). AP5, in conjunction with AP1 and AP4, may be involved in tissue specific regulation of transcription (82).

OBP
125-149
197-221

The octamer motifs, Sph-I and Sph-II, bind the octamer binding group of proteins, and are similar to sequences found in the immunoglobulin heavy and light chain enhancer elements, the U1 and U2 snRNA genes and the human histone H2B promoter (23,27,28,41,60,66, 68,74,81). These proteins demonstrate tissue specificity and interact synergistically to mediate enhancer activity in vivo and in vitro. Octamer binding proteins possess a core homeodomain and are believed to be ubiquitous in mammalian cells (6,14,29,53,75). Consensus sequence: ATTTGCAT

Sp1
40-103

A HeLa cell protein of 95 kD and 105 kD, that contains a zinc finger domain and can direct accurate initiation of early and late transcription (10,19,20,53,79,81). Sp1 is post-translationally modified by the addition of multiple O-linked N-acetylglucosamine monosaccharide residues (39). Of the six Sp1 binding elements, the outer two sites on each side are the most important for directing early and late transcription, respectively. The inner pair bind one Sp1 protein which can stimulate transcription in either direction (34). Consensus sequence: GGGCGG

EBP20

A rat liver protein, which competes for binding with AP3 to the GT-I motif (40). Consensus sequence: TGTGGA/TA/TA/TG or CCAAT

NFKB

A lymphoid specific transcription factor, originally identified by its binding affinity to the immunoglobulin kappa light chain gene enhancer. Its consensus sequence overlaps the AP2 (TC-II motif) and AP3 (GT-I motif) binding sites (47,59,67). Consensus sequence: GGGGACTTTCC

mHox-1.3

A 35-43 kD murine homeodomain protein whose consensus sequence (5'-CPyPyNATTAT/GPy-3') apparently overlaps the AP1 and OBP binding sites (58).

EBP-1

A 57-60 kD HeLa cell protein that binds to the sequence 5'-GTGGAAAGTCCCCAGGC-3', which overlaps the GT-I and TC-II motifs. EBP-I competes the binding of NFKB to the SV40 enhancer (11).

CArG Box Factor

CBF is a muscle specific protein found in human and mouse cells which binds to the sequence $CC(A+T)_6GG$. DNA fragments which contain the CArG Box sequence compete in vivo for factors essential for SV40 expression in muscle cells (55).

REGULATORY SEQUENCES

DNA Replication

ori
5192-31
Site of interaction of host cell DNA polymerase alpha-primase complex with the SV40 genome to initiate viral DNA replication (13,24,35,44,56,70-73,76).

T-Antigen Binding:
Site I *5184-5209*
Site II *5231-13*
Site III *56-61*
T-antigen binding sites which, when bound, trigger the onset of viral DNA replication, the autoregulation of early transcription and the expression of SV40 late genes (50,69,76).

Transcription

T/A Box
20-31
Site equivalent to the TATA box of other promoters which is required for accurate and efficient transcription. Located approximately 25 to 30 basepairs upstream of the initiation site of early transcription (2,3,25,26,31,35,48,61,65).

21bp repeats
40-60
62-82
83-103
Highly guanosine/cytosine-rich region containing six Sp1 binding elements essential for maximal activity of the enhancer (8,21,22,30,35).

72bp repeats
107-178
179-250
Tandemly repeated sequences containing the majority of elements required for early transcription that together constitute the classic enhancer motif (7,26,30,36,37,54,64).

Protein Coding Sequences

Protein	RNA	Coding Sequence
T-antigen	early 18S mRNA	5163-4918 4571-2694
t-antigen	early 19S mRNA	5163-4642
SELP	early mRNA	16-5191
VP1	late 16S mRNA	1499-2590
VP2	late 19S mRNA	562-1617
VP3	late 19S mRNA	916-1617
Agnoprotein	late mRNA	335-520

RNA Processing

	position	reference
Splice Junctions		
early 18S mRNA	4918-4571	(4,49,63)
early 19S mRNA	4638-4571	(4,49,63)
late 16S mRNA	526- 1463	(32,33,45)
late 19S mRNA	526- 558	(32,33,43,45)
	373- 558	
mRNA Processing (AAUAAA)		
early 18S/19S	2632-2637	(25,42,64,70)
late 16S/19S	2657-2662	(25,42,64,70)
poly A addition		
early 18S/19S	2587	(4,18,63)
late 16S/19S	2676	(16-18,26,43,62)

Abreviations used in SV40 Sequence

ORGRPL	Origin of replication
RPT	One of a series of tandem repeated sequences found near the origin of replication
T-antigen Binding	Potential or actual site of T-antigen binding near the replication origin
CDS Start	The initiator (ATG) codon for one of the early or late proteins
CDS End	Termination codon of viral protein translation
Dashed Line (------)	Designates a unique recognition site for the indicated restriction enzyme

The Nucleotide Sequence of SV40

```
        10         20         30         40         50         60         70         80         90        100
GCCTCGGCCT CTGCATAAAT AAAAAAAATT AGTCAGCCAT GGGGCGGAGA ATGGGCGGAA CTGGGCGGAG TTAGGGGCGG GATGGGCGGA GTTAGGGGCG
CGGAGCCGGA GACGTATTTA TTTTTTTTAA TCAGTCGGTA CCCCGCCTCT TACCCGCCTT GACCCGCCTC AATCCCCGCC CTACCCGCCT CAATCCCCGC
........ORGRPL CORE.............||--->                        ORGRPL AUX         <---|
SELP CDS Start<-|                   |---> 21 bp RPT  <---|  |---> 21 bp RPT  <---||---> 21 bp RPT  <-
T-ag 11  <---|           |--->    T-ag Binding site III   <---|

       110        120        130        140        150        160        170        180        190        200
GGACTATGGT TGCTGACTAA TTGAGATGCA TGCTTTGCAT ACTTCTGCCT GCTGGGGAGC CTGGGGACTT TCCACACCTG GTTGCTGACT AATTGAGATG
CCTGATACCA ACGACTGATT AACTCTACGT ACGAAACGTA TGAAGACGGA CGACCCCTCG GACCCCTGAA AGGTGTGGAC CAACGACTGA TTAACTCTAC
--|  |--->                     72 bp RPT                                  <---||--->        72 bp RPT

       210        220        230        240        250        260        270        280        290     ------

CATGCTTTGC ATACTTCTGC CTGCTGGGGA GCCTGGGGAC TTTCCACACC CTAACTGACA CACATTCCAC AGCTGGTTCT TTCCGCCTCA GAAGGTACCT
GTACGAAACG TATGAAGACG GACGACCCCT CGGACCCCTG AAAGGTGTGG GATTGACTGT GTGTAAGGTG TCGACCAAGA AAGGCGGAGT CTTCCATGGA
                                                <---|                                                   Kpn I

                                 |---> Agnoprotein CDS Start
       310        320        330  | 340      ----         360        370        380        390        400
AACCAAGTTC CTCTTTCAGA GGTTATTTCA GGCCATGGTG CTGCGCCGGC TGTCACGCCA GGCCTCCGTT AAGGTTCGTA GGTCATGGAC TGAAAGTAAA
TTGGTTCAAG GAGAAAGTCT CCAATAAAGT CCGGTACCAC GACGCGGCCG ACAGTGCGGT CCGGAGGCAA TTCCAAGCAT CCAGTACCTG ACTTTCATTT
                                                Hpa II

       410        420        430        440        450        460        470        480        490        500
AAAACAGCTC AACGCCTTTT TGTGTTTGTT TTAGAGCTTT TGCTGCAATT TTGTGAAGGG GAAGATACTG TTGACGGGAA ACGCAAAAAA CCAGAAAGGT
TTTTGTCGAG TTGCGGAAAA ACACAAACAA AATCTCGAAA ACGACGTTAA AACACTTCCC CTTCTATGAC AACTGCCCTT TGCGTTTTTT GGTCTTTCCA

  Agnoprotein
   CDS End <---|                                                   |---> VP2 CDS Start
       510        520        530        540        550        560 |      570        580   --------        600
TAACTGAAAA ACCAGAAAGT TAACTGGTAA GTTTAGTCTT TTTGTCTTTT ATTTCAGGTC CATGGGTGCT GCTTTAACAC TGTTGGGGGA CCTAATTGCT
ATTGACTTTT TGGTCTTTCA ATTGACCATT CAAATCAGAA AAACAGAAAA TAAAGTCCAG GTACCCACGA CGAAATTGTG ACAACCCCCT GGATTAACGA
                                                                                                PpuM I

       610        620        630        640        650        660        670        680        690        700
ACTGTGTCTG AAGCTGCTGC TGCTACTGGA TTTTCAGTAG CTGAAATTGC TGCTGGAGAG GCCGCTGCTG CAATTGAAGT GCAACTTGCA TCTGTTGCTA
TGACACAGAC TTCGACGACG ACGATGACCT AAAAGTCATC GACTTTAACG ACGACCTCTC CGGCGACGAC GTTAACTTCA CGTTGAACGT AGACAACGAT

       710        720        730        740        750        760     ------         780        790        800
CTGTTGAAGG CCTAACCACC TCTGAGGCAA TTGCTGCTAT AGGCCTCACT CCACAGGCCT ATGCTGTGAT ATCTGGGGCT CCTGCTGCTA TAGCTGGATT
GACAACTTCC GGATTGGTGG AGACTCCGTT AACGACGATA TCCGGAGTGA GGTGTCCGGA TACGACACTA TAGACCCCGA GGACGACGAT ATCGACCTAA
                                                                          EcoR V

       810        820        830   ------840        850        860        870        880        890        900
TGCAGCTTTA CTGCAAACTG TGACTGGTGT GAGCGCTGTT GCTCAAGTGG GGTATAGATT TTTTAGTGAC TGGGATCACA AAGTTTCTAC TGTTGGTTTA
ACGTCGAAAT GACGTTTGAC ACTGACCACA CTCGCGACAA CGAGTTCACC CCATATCTAA AAAATCACTG ACCCTAGTGT TTCAAAGATG ACAACCAAAT
                                   Hae II

                |---> VP3 CDS Start
       910  |     920        930        940        950        960        970        980        990       1000
TATCAACAAC CAGGAATGGC TGTAGATTTG TATAGGCCAG ATGATTACTA TGATATTTTA TTTCCTGGAG TACAAACCTT TGTTCACAGT GTTCAGTATC
ATAGTTGTTG GTCCTTACCG ACATCTAAAC ATATCCGGTC TACTAATGAT ACTATAAAAT AAAGGACCTC ATGTTTGGAA ACAAGTGTCA CAAGTCATAG

      1010       1020       1030       1040       1050       1060       1070       1080       1090       1100
TTGACCCCAG ACATTGGGGT CCAACACTTT TTAATGCCAT TTCTCAAGCT TTTTGGCGTG TAATACAAAA TGACATTCCT AGGCTCACCT CACAGGAGCT
AACTGGGGTC TGTAACCCCA GGTTGTGAAA AATTACGGTA AAGAGTTCGA AAAACCGCAC ATTATGTTTT ACTGTAAGGA TCCGAGTGGA GTGTCCTCGA

      1110       1120       1130       1140       1150       1160       1170       1180       1190       1200
TGAAAGAAGA ACCCAAAGAT ATTTAAGGGA CAGTTTGGCA AGGTTTTTAG AGGAAACTAC TTGGACAGTA ATTAATGCTC CTGTTAATTG GTATAACTCT
ACTTTCTTCT TGGGTTTCTA TAAATTCCCT GTCAAACCGT TCCAAAAATC TCCTTTGATG AACCTGTCAT TAATTACGAG GACAATTAAC CATATTGAGA

      1210       1220       1230       1240       1250       1260       1270       1280       1290       1300
TTACAAGATT ACTACTCTAC TTTGTCTCCC ATTAGGCCTA CAATGGTGAG ACAAGTAGCC AACAGGGAAG GGTTGCAAAT ATCATTTGGG CACACCTATG
AATGTTCTAA TGATGAGATG AAACAGAGGG TAATCCGGAT GTTACCACTC TGTTCATCGG TTGTCCCTTC CCAACGTTTA TAGTAAACCC GTGTGGATAC

      1310       1320       1330       1340       1350       1360       1370       1380       1390       1400
ATAATATTGA TGAAGCAGAC AGTATTCAGC AAGTAACTGA GGAGGTGGAA GCTCAAAGCC AAAGTCCTAA TGTGCAGTCA GGTGAATTTA TTGAAAAATT
TATTATAACT ACTTCGTCTG TCATAAGTCG TTCATTGACT CCTCCACCTT CGAGTTTCGG TTTCAGGATT ACACGTCAGT CCACTTAAAT AACTTTTTAA

      1410       1420       1430       1440       1450       1460       1470       1480       1490       1500
TGAGGCTCCT GGTGGTGCAA ATCAAAGAAC TGCTCCTCAG TGGATGTTGC CTTTACTTCT AGGCCTGTAC GGAAGTGTAA CTTCTGCTCT AAAAGCTTAT
ACTCCGAGGA CCACCACGTT TAGTTTCTTG ACGAGGAGTC ACCTACAACG GAAATGAAGA TCCGGACATG CCTTCACATT GAAGACGAGA TTTTCGAATA

--> VP1 CDS Start
      1510       1520       1530       1540       1550       1560       1570       1580       1590       1600
GAAGATGGCC CCAACAAAAA GAAAAGGAAG TTGTCCAGGG GCAGCTCCCA AAAAACCAAA GGAACCAGTG CAAGTGCCAA AGCTCGTCAT AAAAGGAGGA
CTTCTACCGG GGTTGTTTTT CTTTTCCTTC AACAGGTCCC CGTCGAGGGT TTTTTGGTTT CCTTGGTCAC GTTCACGGTT TCGAGCAGTA TTTTCCTCCT

VP2, VP3
 CDS End <---|
      1610       1620    --- ---    1640       1650       1660       1670       1680       1690       1700
ATAGAAGTTC TAGGAGTTAA AACTGGAGTA GACAGCTTCA CTGAGGTGGA GTGCTTTTTA AATCCTCAAA TGGGCAATCC TGATGAACAT CAAAAAGGCT
TATCTTCAAG ATCCTCAATT TTGACCTCAT CTGTCGAAGT GACTCCACCT CACGAAAAAT TTAGGAGTTT ACCCGTTAGG ACTACTTGTA GTTTTTCCGA
                                  Acc I

      1710       1720       1730       1740       1750       1760       1770       1780   ------       1800
TAAGTAAAAG CTTAGCAGCT GAAAAACAGT TTACAGATGA CTCTCCAGAC AAAGAACAAC TGCCTTGCTA CAGTGTGGCT AGAATTCCTT TGCCTAATTT
ATTCATTTTC GAATCGTCGA CTTTTTGTCA AATGTCTACT GAGAGGTCTG TTTCTTGTTG ACGGAACGAT GTCACACCGA TCTTAAGGAA ACGGATTAAA
                                                                                 EcoR I

      1810       1820       1830       1840       1850       1860       1870       1880       1890       1900
AAATGAGGAC TTAACCTGTG GAAATATTTT GATGTGGGAA GCTGTTACTG TTAAAACTGA GGTTATTGGG GTAACTGCTA TGTTAAACTT GCATTCAGGG
TTTACTCCTG AATTGGACAC CTTTATAAAA CTACACCCTT CGACAATGAC AATTTTGACT CCAATAACCC CATTGACGAT ACAATTTGAA CGTAAGTCCC

      1910       1920       1930       1940       1950       1960       1970       1980       1990       2000
ACACAAAAAA CTCATGAAAA TGGTGCTGGA AAACCCATTC AAGGGTCAAA TTTTCATTTT TTTGCTGTTG GTGGGGAACC TTTGGAGCTG CAGGGTGTGT
TGTGTTTTTT GAGTACTTTT ACCACGACCT TTTGGGTAAG TTCCCAGTTT AAAAGTAAAA AAACGACAAC CACCCCTTGG AAACCTCGAC GTCCCACACA

      2010       2020       2030       2040       2050       2060       2070       2080       2090       2100
TAGCAAACTA CAGGACCAAA TATCCTGCTC AAACTGTAAC CCCAAAAAAT GCTACAGTTG ACAGTCAGCA GATGAACACT GACCACAAGG CTGTTTTGGA
ATCGTTTGAT GTCCTGGTTT ATAGGACGAG TTTGACATTG GGGTTTTTTA CGATGTCAAC TGTCAGTCGT CTACTTGTGA CTGGTGTTCC GACAAAACCT

      2110       2120       2130       2140       2150       2160       2170       2180       2190       2200
TAAGGATAAT GCTTATCCAG TGGAGTGCTG GGTTCCTGAT CCAAGTAAAA ATGAAAACAC TAGATATTTT GGAACCTACA CAGGTGGGGA AAATGTGCCT
ATTCCTATTA CGAATAGGTC ACCTCACGAC CCAAGGACTA GGTTCATTTT TACTTTTGTG ATCTATAAAA CCTTGGATGT GTCCACCCCT TTTACACGGA

      2210       2220       2230       2240       2250    --- ----       2270       2280       2290       2300
CCTGTTTTGC ACATTACTAA CACAGCAACC ACAGTGCTTC TTGATGAGCA GGGTGTTGGG CCCTTGTGCA AAGCTGACAG CTTGTATGTT TCTGCTGTTG
GGACAAAACG TGTAATGATT GTGTCGTTGG TGTCACGAAG AACTACTCGT CCCACAACCC GGGAACACGT TTCGACTGTC GAACATACAA AGACGACAAC
                                                                 Apa I

      2310       2320       2330       2340       2350       2360       2370       2380       2390       2400
ACATTTGTGG GCTGTTTACC AACACTTCTG GAACACAGCA GTGGAAGGGA CTTCCCAGAT ATTTTAAAAT TACCCTTAGA AAGCGGTCTG TGAAAAACCC
TGTAAACACC CGACAAATGG TTGTGAAGAC CTTGTGTCGT CACCTTCCCT GAAGGGTCTA TAAAATTTTA ATGGGAATCT TTCGCCAGAC ACTTTTTGGG

      2410       2420       2430       2440       2450       2460       2470       2480       2490       2500
CTACCCAATT TCCTTTTTGT TAAGTGACCT AATTAACAGG AGGACACAGA GGGTGGATGG GCAGCCTATG ATTGGAATGT CCTCTCAAGT AGAGGAGGTT
GATGGGTTAA AGGAAAAACA ATTCACTGGA TTAATTGTCC TCCTGTGTCT CCCACCTACC CGTCGGATAC TAACCTTACA GGAGAGTTCA TCTCCTCCAA

                                                                                 VP1 CDS End <---|
      2510       2520       2530     ------         2550       2560       2570       2580       2590       26
AGGGTTTATG AGGACACAGA GGAGCTTCCT GGGGATCCAG ACATGATAAG ATACATTGAT GAGTTTGGAC AAACCACAAC TAGAATGCAG TGAAAAAA
TCCCAAATAC TCCTGTGTCT CCTCGAAGGA CCCCTAGGTC TGTACTATTC TATGTAACTA CTCAAACCTG TTTGGTGTTG ATCTTACGTC ACTTTTTT
                                 BamH I

      2610       2620       2630       2640       2650       2660       2670       2680       2690       27
GCTTTATTTG TGAAATTTGT GATGCTATTG CTTTATTTGT AACCATTATA AGCTGCAATA AACAAGTTAA CAACAACAAT TGCATTCATT TTATGTTT
CGAAATAAAC ACTTTAAACA CTACGATAAC GAAATAAACA TTGGTAATAT TCGACGTTAT TTGTTCAATT GTTGTTGTTA ACGTAAGTAA AATACAAA
                                                                                                    |--->
      2710       2720       2730       2740       2750       2760          - ----- 2780       2790       28
GGTTCAGGGG GAGGTGTGGG AGGTTTTTTA AAGCAAGTAA AACCTCTACA AATGTGGTAT GGCTGATTAT GATCATGAAC AGACTGTGAG GACTGAGG
CCAAGTCCCC CTCCACACCC TCCAAAAAAT TTCGTTCATT TTGGAGATGT TTACACCATA CCGACTAATA CTAGTACTTG TCTGACACTC CTGACTCC
Large T EXON 2 CDS End                                                       Bcl I

      2810       2820       2830       2840       2850       2860       2870       2880       2890       29
CCTGAAATGA GCCTTGGGAC TGTGAATCAA TGCCTGTTTC ATGCCCTGAG TCTTCCATGT TCTTCTCCCC ACCATCTTCA TTTTTATCAG CATTTTCC
GGACTTTACT CGGAACCCTG ACACTTAGTT ACGGACAAAG TACGGGACTC AGAAGGTACA AGAAGAGGGG TGGTAGAAGT AAAAATAGTC GTAAAAGG

      2910       2920       2930       2940       2950       2960       2970       2980       2990       30
GCTGTCTTCA TCATCATCAT CACTGTTTCT TAGCCAATCT AAAACTCCAA TTCCCATAGC CACATTAAAC TTCATTTTTT GATACACTGA CAAACTAA
CGACAGAAGT AGTAGTAGTA GTGACAAAGA ATCGGTTAGA TTTTGAGGTT AAGGGTATCG GTGTAATTTG AAGTAAAAAA CTATGTGACT GTTTGATT

      3010       3020       3030       3040       3050       3060       3070       3080       3090       31
TCTTTGTCCA ATCTCTCTTT CCACTCCACA ATTCTGCTCT GAATACTTTG AGCAAACTCA GCCACAGGTC TGTACCAAAT TAACATAAGA AGCAAAGC
AGAAACAGGT TAGAGAGAAA GGTGAGGTGT TAAGACGAGA CTTATGAAAC TCGTTTGAGT CGGTGTCCAG ACATGGTTTA ATTGTATTCT TCGTTTCG

      3110       3120       3130       3140       3150       3160       3170       3180       3190       32
TGCCACTTTG AATTATTCTC TTTTCTAACA AAAACTCACT GCGTTCCAGG CAATGCTTTA AATAATCTTT GGGCCTAAAA TCTATTTGTT TTACAAAT
ACGGTGAAAC TTAATAAGAG AAAAGATTGT TTTTGAGTGA CGCAAGGTCC GTTACGAAAT TTATTAGAAA CCCGGATTTT AGATAAACAA AATGTTTA

      3210       3220       3230       3240       3250       3260       3270       3280       3290       33
GGCCTGCAGT GTTTTAGGCA CACTGTACTC ATTCATGGTG ACTATTCCAG GGGGAAATAT TTGAGTTCTT TTATTTAGGT GTTTCTTTTC TAAGTTTA
CCGGACGTCA CAAAATCCGT GTGACATGAG TAAGTACCAC TGATAAGGTC CCCCTTTATA AACTCAAGAA AATAAATCCA CAAAGAAAAG ATTCAAAT

      3310       3320       3330       3340       3350       3360       3370       3380       3390       34
TTAACACTGC CATCCAAATA ATCCCTTAAA TTGTCCAGGT TATTAATTCC CTGACCTGAA GGCAAATCTC TGGACTCCCC TCCAGTGCCC TTTACATC
AATTGTGACG GTAGGTTTAT TAGGGAATTT AACAGGTCCA ATAATTAAGG GACTGGACTT CCGTTTAGAG ACCTGAGGGG AGGTCACGGG AAATGTAG

      3410       3420       3430       3440       3450       3460       3470       3480       3490       35
CAAAAACATC TAAAAACTGG TCAATAGCTA CTCCTAGCTC AAAGTTCAGC CTGTCCAAGG GCAAATTAAC ATTTAAAGCT TTCCCCCCAC ATAATTCA
GTTTTTGATG ATTTTTGACC AGTTATCGAT GAGGATCGAG TTTCAAGTCG GACAGGTTCC CGTTTAATTG TAAATTTCGA AAGGGGGGTG TATTAAGT

      3510       3520       3530       3540       3550       3560       3570       3580       3590       36
CAAAGCAGCT GCTAATGTAG TTTTACCACT ATCAATTGGT CCTTTAAACA GCCAGTATCT TTTTTTAGGA ATGTTGTACA CCATGCATTT TAAAAAGT
GTTTCGTCGA CGATTACATC AAAATGGTGA TAGTTAACCA GGAAATTTGT CGGTCATAGA AAAAAATCCT TACAACATGT GGTACGTAAA ATTTTTCA

      3610       3620       3630       3640       3650       3660       3670       3680       3690       37
TACACCACTG AATCCATTTT GGGCAACAAA CAGTGTAGCC AAGCAACTCC AGCCATCCAT TCTTCTATGT CAGCAGAGCC TGTAGAACCA AACATTAT
ATGTGGTGAC TTAGGTAAAA CCCGTTGTTT GTCACATCGG TTCGTTGAGG TCGGTAGGTA AGAAGATACA GTCGTCTCGG ACATCTTGGT TTGTAATA

      3710       3720       3730       3740       3750       3760       3770       3780       3790       38
CCATCCTATC CAAAAGATCA TTAAATCTGT TTGTTAACAT TTGTTCTCTA GTTAATTGTA GGCTATCAAC CCGCTTTTTA GCTAAAACAG TATCAACA
GGTAGGATAG GTTTTCTAGT AATTTAGACA AACAATTGTA AACAAGAGAT CAATTAACAT CCGATAGTTG GGCGAAAAAT CGATTTTGTC ATAGTTGT

      3810       3820       3830       3840       3850       3860       3870       3880       3890       39
CTGTTGGCAT ATGGTTTTTT GGTTTTTGCT GTCAGCAAAT ATAGCAGCAT TTGCATAATG CTTTTCATGG TACTTATAGT GGCTGGGCTG TTCTTTTT
GACAACCGTA TACCAAAAAA CCAAAAACGA CAGTCGTTTA TATCGTCGTA AACGTATTAC GAAAAGTACC ATGAATATCA CCGACCCGAC AAGAAAAA

      3910       3920       3930       3940       3950       3960       3970       3980       3990       40
ATACATTTTA AACACATTTC AAAACTGTAC TGAAATTCCA AGTACATCCC AAGCAATAAC AACACATCAT CACATTTTGT TTCCATTGCA TACTCTGT
TATGTAAAAT TTGTGTAAAG TTTTGACATG ACTTTAAGGT TCATGTAGGG TTCGTTATTG TTGTGTAGTA GTGTAAAACA AAGGTAACGT ATGAGACA

      4010       4020       4030       4040       4050       4060       4070       4080       4090       41
CAAGCTTCCA GGACACTTGT TTAGTTTCCT CTGCTTCTTC TGGATTAAAA TCATGCTCCT TTAACCCACC TGGCAAACTT TCCTCAATAA CAGAAAAT
GTTCGAAGGT CCTGTGAACA AATCAAAGGA GACGAAGAAG ACCTAATTTT AGTACGAGGA AATTGGGTGG ACCGTTTGAA AGGAGTTATT GTCTTTTA

      4110       4120       4130       4140       4150       4160       4170       4180       4190       42
ATCTCTAGTC AAGGCACTAT ACATCAAATA TTCCTTATTA ACCCCTTTAC AAATTAAAAA GCTAAAGGTA CACAATTTTT GAGCATAGTT ATTAATAG
TAGAGATCAG TTCCGTGATA TGTAGTTTAT AAGGAATAAT TGGGGAAATG TTTAATTTTT CGATTTCCAT GTGTTAAAAA CTCGTATCAA TAATTATC

      4210       4220       4230       4240       4250       4260       4270       4280       4290       43
GACACTCTAT GCCTGTGTGG AGTAAGAAAA AACAGTATGT TATGATTATA ACTGTTATGC CTACTTATAA AGGTTACAGA ATATTTTTCC ATAATTTT
CTGTGAGATA CGGACACACC TCATTCTTTT TTGTCATACA ATACTAATAT TGACAATACG GATGAATATT TCCAATGTCT TATAAAAAGG TATTAAAA

      4310       4320       4330       4340       4350       4360       4370       4380       4390       44
TGTATAGCAG TGCAGCTTTT TCCTTTGTGG TGTAAATAGC AAAGCAAGCA AGAGTTCTAT TACTAAACAC AGCATGACTC AAAAAACTTA GCAATTCT
ACATATCGTC ACGTCGAAAA AGGAAACACC ACATTTATCG TTTCGTTCGT TCTCAAGATA ATGATTTGTG TCGTACTGAG TTTTTTGAAT CGTTAAGA

      4410       4420       4430       4440       4450       4460       4470       4480       4490
AGGAAAGTCC TTGGGGTCTT CTACCTTTCT CTTCTTTTTT GGAGGAGTAG AATGTTGAGA GTCAGCAGTA GCCTCATCAT CACTAGATGG CATTTC
TCCTTTCAGG AACCCCAGAA GATGGAAAGA GAAGAAAAAA CCTCCTCATC TTACAACTCT CAGTCGTCAT CGGAGTAGTA GTGATCTACC GTAAAG

      4510       4520       4530       4540       4550       4560       4570       4580       4590
GAGCAAAACA GGTTTTCCTC ATTAAAGGCA TTCCACCACT GCTCCCATTC ATCAGTTCCA TAGGTTGGAA TCTAAAATAC ACAAACAATT AGAATC
CTCGTTTTGT CCAAAAGGAG TAATTTCCGT AAGGTGGTGA CGAGGGTAAG TAGTCAAGGT ATCCAACCTT AGATTTTATG TGTTTGTTAA TCTTAG
                                                 Large T EXON 2 CDS Start <---||--->      Large T INTRON

      4610       4620       4630       4640       4650       4660       4670       4680       4690
GTTTAACACA TTATACACTT AAAAATTTTA TATTTACCTT AGAGCTTTAA ATCTCTGTAG GTAGTTTGTC CAATTATGTC ACACCACAGA AGTAAC
CAAATTGTGT AATATGTGAA TTTTTAAAAT ATAAATGGAA TCTCGAAATT TAGAGACATC CATCAAACAG GTTAATACAG TGTGGTGTCT TCATTC
                                            |--->  Small t CDS End

      4710       4720       4730      -- --      4750      -- ----------       4780       4790

CTTCACAAAG ATCAAGTCCA AACCACATTC TAAAGCAATC GAAGCAGTAG CAATCAACCC ACACAAGTGG ATCTTTCCTG TATAATTTTC TATTTT
GAAGTGTTTC TAGTTCAGGT TTGGTGTAAG ATTTCGTTAG CTTCGTCATC GTTAGTTGGG TGTGTTCACC TAGAAAGGAC ATATTAAAAG ATAAAA
                                        Taq I                      BstX I

      4810       4820       4830       4840       4850       4860       4870       4880       4890
CTTCATCCTC AGTAAGCACA GCAAGCATAT GCAGTTAGCA GACATTTTCT TTGCACACTC AGGCCATTGT TTGCAGTACA TTGCATCAAC ACCAGC
GAAGTAGGAG TCATTCGTGT CGTTCGTATA CGTCAATCGT CTGTAAAAGA AACGTGTGAG TCCGGTAACA AACGTCATGT AACGTAGTTG TGGTCC

      4910       4920       4930       4940       4950       4960       4970       4980       4990
AAGGAAGAAG CAAATACCTC AGTTGCATCC CAGAAGCCTC CAAAGTCAGG TTGATGAGCA TATTTTACTC CATCTTCCAT TTTCTTGTAC AGAGTA
TTCCTTCTTC GTTTATGGAG TCAACGTAGG GTCTTCGGAG GTTTCAGTCC AACTACTCGT ATAAAATGAG GTAGAAGGTA AAAGAACATG TCTCAT
Lg T INTRON <---||---> Large T EXON I CDS End

      5010       5020       5030       5040       5050       5060       5070       5080       5090
TTTTCTTCAT TTTTTCTTCA TCTCCTCCTT TATCAGGATG AAACTCCTTG CATTTTTTTA AATATGCCTT TCTCATCAGA GGAATATTCC CCCAGC
AAAAGAAGTA AAAAAGAAGT AGAGGAGGAA ATAGTCCTAC TTTGAGGAAC GTAAAAAAAT TTATACGGAA AGAGTAGTCT CCTTATAAGG GGGTCC

      5110       5120       5130       5140       5150       5160       5170       5180       5190
CCTTTCAAGA CCTAGAAGGT CCATTAGCTG CAAAGATTCC TCTCTGTTTA AAACTTTATC CATCTTTGCA AAGCTTTTTG CAAAAGCCTA GGCCTC
GGAAAGTTCT GGATCTTCCA GGTAATCGAC GTTTCTAAGG AGAGACAAAT TTTGAAATAG GTAGAAACGT TTCGAAAAAC GTTTTCGGAT CCGGAG
                          Large T EXON I, Small t CDS Start <---|                        |         |...
                                                                                        | T-ag Binding
                                                                                         |->SELP CDS
                                                                                            End
      5210       5220       5230       5240
AAAGCCTCCT CACTACTTCT GGAATAGCTC AGAGGCCGAG GCG

TTTCGGAGGA GTGATCAAGA CCTTATCGAG TCTCCGGCTC CGC
.............ORGRPL CORE......................
I  <---|                         |-> T-ag Binding Site II
```

References

1. Angel, P., et al. (1987) Cell 49:729-789.
2. Benoist, C. and Chambon, P. (1980) Proc. Natl. Acad. Sci. USA 77:3865-3869.
3. Benoist, C. and Chambon, P. (1981) Nature 290:304-310.
4. Berk, A. and Sharp, P.A. (1978) Proc. Natl. Acad. Sci. USA 75:1274-1278.
5. Bohmann, D., et al. (1987) Science 238:1386-1392.
6. Bohmann, D., et al. (1987) 325:268-272.
7. Brady, J., et al. (1982) Cell 31:625-633.
8. Brady, J., et al. (1984) Mol. Cell. Biol. 4:133-141.
9. Brady, J. and Salzman, N. (1987) The Papovaviruses, Plenum Press.
10. Briggs, M.R., et al. (1986) Science 234:47-52.
11. Clark, L., Pollock, R.M. and Hay, R.T. (1988) Genes Dev. 2:991-1002.
12. Chiu, R., et al. (1987) Nature 329:648-651.
13. Danna, K.J. and Nathans, D. (1972) Proc. Natl. Acad. Sci. USA 69:3097-3101.
14. Davidson, I., et al. (1986) Nature 323:544-548.
15. DePamphilis, M.L. and Wassarman, P.M. (1983) Organization and Replication of Viral DNA, CRC Press, 37-114.
16. Dhar, R., et al. (1977) Proc. Natl. Acad. Sci. USA 74:827-831.
17. Dhar, R., et al. (1975) INSERM 47:25-32.
18. Dhar, R., et al. (1974) Cold Spring Harbor Symp. Quant. Biol. 39:153-160.
19. Dynan, W.S. and Tjian, R. (1983) Cell 35: 79-87.
20. Dynan, W.S., et al. (1985) Proc. Natl. Acad. Sci. USA 82:4915-4919.
21. Dynan, W.S. and Tjian, R. (1985) Nature 316:774-778.
22. Everett, R.D., et al. (1983) Nucleic Acids Res. 11:2447-2464.
22. Dynan, W.S. and Cherritz, S.A. (1989) J. Virol. 63:1420-1427.
23. Falkner, F.G. and Zachau, H.G. (1984) Nature 310:71-74.
24. Fareed, G.C., et al. (1972) J. Virol. 10:484-491.
25. Fiers, W., et al. (1978) Nature 273:113-120.
26. Fitzgerald, M. and Shenk, T. (1980) Ann. N.Y. Acad. Sci. 354:53-59.
27. Fletcher, C., Heintz, N. and Roeder, R.G. (1987) Cell 51:773-781.
28. Fritz, A., et al. (1984) EMBO J. 178:272-285.
29. Fromental, C., et al. (1988) Cell 54:943-953.
30. Fromm, M. and Berg, P. (1982) J. Mol. Appl. Genet. 1:457-481.
31. Gannon, F., et al. (1979) Nature 278:428-434.
32. Ghosh, P.K., et al. (1978) J. Biol. Chem. 253:3643-3647.
33. Ghosh, P.K., et al. (1978) J. Mol. Biol. 126:813-846.
34. Gidoni, D., et al. (1985) Science 230:511-517.
35. Gluzman, Y. (1980) Proc. Natl. Acad. Sci. USA 77:3898-3902.
36. Gruss, P., et al. (1981) Proc. Natl. Acad. Sci. USA 78:943-947.
37. Hansen, U. and Sharp, P. (1983) EMBO J. 2:2293-2303.
38. Imagawa, M., Chiu, R. and Karin, M. (1987) Cell 51:251-260.
39. Jackson, S.P. and Tjian, R. (1988) Cell 55:125-133.
40. Johnson, P.F., et al. (1987) Genes Dev. 1:133-146.
41. Jones, N.C., Rigby, P.W.J. and Ziff, E.B. (1988) Genes Dev. 2:267-281.
42. Khoury, G., et al. (1979) Cell 18:85-92.
43. Lai, C.-J., et al. (1978) Cell 14:971-982.
44. Lai, C.-J. and Nathans, D. (1975) J. Mol. Biol. 97:113-118.
45. Lai, C.-J. and Khoury, G. (1979) Proc. Natl. Acad. Sci. USA 76:71-75.
46. Lee, W., Mitchell, P.J. and Tjian R. (1987) Cell 49:741-752.
47. Leung, K. and Nabel, G.J. (1988) Nature 333:776-778.
48. Mathis, D. and Chambon, P. (1981) Nature 290:310-315.
49. May, E., et al. (1978) Nucleic Acids Res. 5:3083-3099.
50. McKay, R. and DiMaio, D. (1981) Nature 289:810-813.
51. Mermod, N., Williams, T.J. and Tjian, R. (1988) Nature 332:557-561.
52. Mitchell, P.J., Wang, C. and Tjian, R. (1987) Cell 50:847-861.
53. Mitchell, P.J. and Tjian, R. (1989) Science 245:371-378.
54. Moreau, P., et al. (1981) Nucleic Acids Res. 9:6047-6068.
55. Muscat, G.E., Gustafson, T.A. and Kedes, L. (1988) Mol. Cell. Biol. 8:4120-4133.
56. Myers, R. and Tjian, R. (1980) Proc. Natl. Acad. Sci. USA 77:6491-6495.
57. Neuberg, M., et al. (1989) Nature 338:589-590.
58. Odenwald, W.F., et al. (1989) Genes Dev. 3:158-172.
59. Osborn, L., Kunkel, S. and Nabel, G.J. (1989) Proc. Natl. Acad. Sci. USA 86:2336-2340.
60. Parslow, T.G., et al. (1984) Proc. Natl. Acad. Sci. USA 81:2650-2654.
61. Pribnow, D. (1975) Proc. Natl. Acad. Sci. USA 72:784-788.
62. Reddy, V.B., et al. (1978) Nucleic Acids Res. 5:4195-4213.
63. Reddy, V.B., et al. (1979) J. Virol. 30:279-296.
64. Reddy, V.B., et al. (1978) Science 200:494-502.
65. Rio, D. and Tjian, R. (1980) Proc. Natl. Acad. Sci. USA 77:5706-5710.
66. Rosales, R., et al. (1987) EMBO J. 6:3015-3025.
67. Sen, R. and Baltimore, D. (1986) Cell 46:705-716.
68. Sive, H.L. and Roeder, R.G. (1986) Proc. Natl. Acad. Sci. USA 83:6382-6386.
69. Sleigh, M.J., et al. (1978) Cell 14:79-88.
70. Shenk, T. (1978) Cell 13:791-798.
71. Shortle, D. and Nathans, D. (1979) J. Mol. Biol. 131:801-817.
72. Shortle, D. and Nathans, D. (1978) Proc. Natl. Acad. Sci. USA 75:2170-2174.
73. Shortle, D., et al. (1979) Proc. Natl. Acad. Sci. USA 76:6128-6131.
74. Strum, R., et al. (1987) Genes Dev. 1:1147-1160.
75. Tanaka, M., et al. (1988) Genes Dev. 2:1764-1778.
76. Tjian, R. (1978) Cell 13:165-180.
77. Tooze, J. (1980) Mol. Biol. of Tumor Viruses, 2nd Ed., Part 2.
78. Tremethick, D.J. and Molloy, P.L. (1985) FEBS Lett. 242:346-350.
79. Vigneron, M.H., et al. (1984) EMBO J. 2373-2382.
80. Wasylyk, C., Imler, J.L. and Wasylyk, B. (1988) EMBO J. 7:2475-2483.
81. Wingender, E. (1988) 16:1879-1902.
82. Xiao, J.H., et al. (1987) EMBO J. 6:3005-3013.

[1]Department of Viral Oncology
Institute for Virus Research
Kyoto University
Kyoto, 606, Japan

[2]Department of Virology
Royal Postgraduate Medical School
DuCane Road
London W12, England

August, 1989

YOSHIAKI ITO[1] and BEVERLY E. GRIFFIN[2]

Polyoma virus induced tumors in newborn mice, hamsters, and rats and grows well in mouse cells in culture. The virus can also transform cultured rodent cells. The genome of the virus is a double-stranded, supercoiled circular DNA that can be divided into two distinct biological regions, the so-called early and late regions (49). One large plaque strain (A2) of the virus contains 5292 base pairs (80). The virus codes for three early proteins: large-(100K), middle-(55K), and small (22K) T-antigens (reviewed in ref. 47). Large T-antigen is a product of the tsa gene (27,44) and is required for the initiation of viral DNA replication (26) and the control of the transcription of the early RNA (15). The hrt mutants have lesions in both middle and small T-antigens and are host range specific for growth (3,43,45,46,73). The three T-antigens may work cooperatively in a multistep cell transformation process (71). A recombinant DNA clone that is capable of expressing only the amino-terminal 40% of large T-antigen appears to be able to immortalize primary rodent cells in culture (72); established fibroblast cell lines can be transformed by a recombinant DNA clone capable of expressing only middle T-antigen (93). Small T-antigen stimulates the growth of an established fibroblast cell line beyond the saturation of the cells (68). Some portion of the region of the genome where both large and middle T-antigens are encoded in different reading frames can be deleted without harming viability. Some of these deletion mutants (mlt mutants) have altered growth and transformation properties (35,56,76).

The late region of the genome encodes three capsid proteins: VP1, VP2, and VP3. There are two groups of VP1 mutants, represented by ts 1260 and ts 10 (20), which complement each other. They appear to correspond to the B and C groups of SV40 late ts mutants. The BC mutants of SV40 complement the early mutants but do not complement either the B or C mutants. The ts c mutant of polyoma virus (64) is considered to be equivalent to the BC mutant of SV40 (28). The polyoma virus mutant, ts 3, does not complement any other known mutant; it is defective in uncoating and is considered to belong to an independent group (21), equivalent to the SV40 group D mutants. Mutants that grow in embryonal carcinoma and trophoblast cells have their lesions mapped to the late, noncoding region of the genome, near the viral origin of DNA replication (31,50,52,75,87). The noncoding region near the origin of DNA replication contains regulatory signals for replication and transcription including large T-antigen binding sites (32,70) and enhancer sequences (15).

The circular map of the viral genome is divided into 100 units and numbered clockwise, with the single Eco RI restriction enzyme recognition site taken as map unit 0 (Fig. 1). The restriction enzyme Hpa II recognition site at map unit 70.6 probably includes the origin of viral DNA replication. The center of the recognition sequence, 5'-CCGG-3', at this site has been arbitrarily chosen for numbering the nucleotides in the genome, with the G residue at the 5' end designated nucleotide one (80). The numbering then proceeds in a 5' to 3' direction on the strand that has the same polarity as the early RNA. Landmarks in the regulatory region are shown in Fig. 2.

Fig.1 POLYOMAVIRUS (A2 STRAIN)

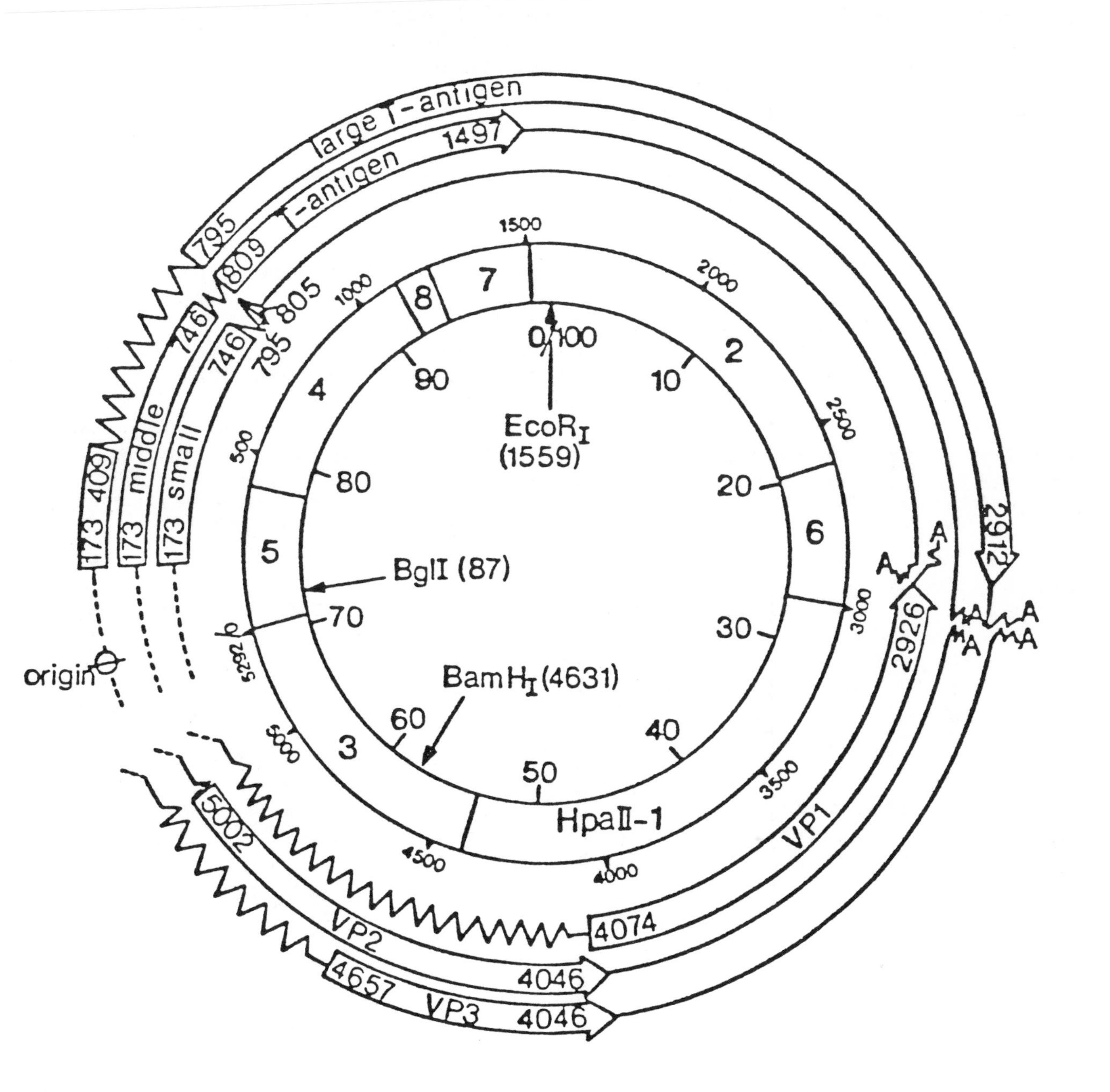

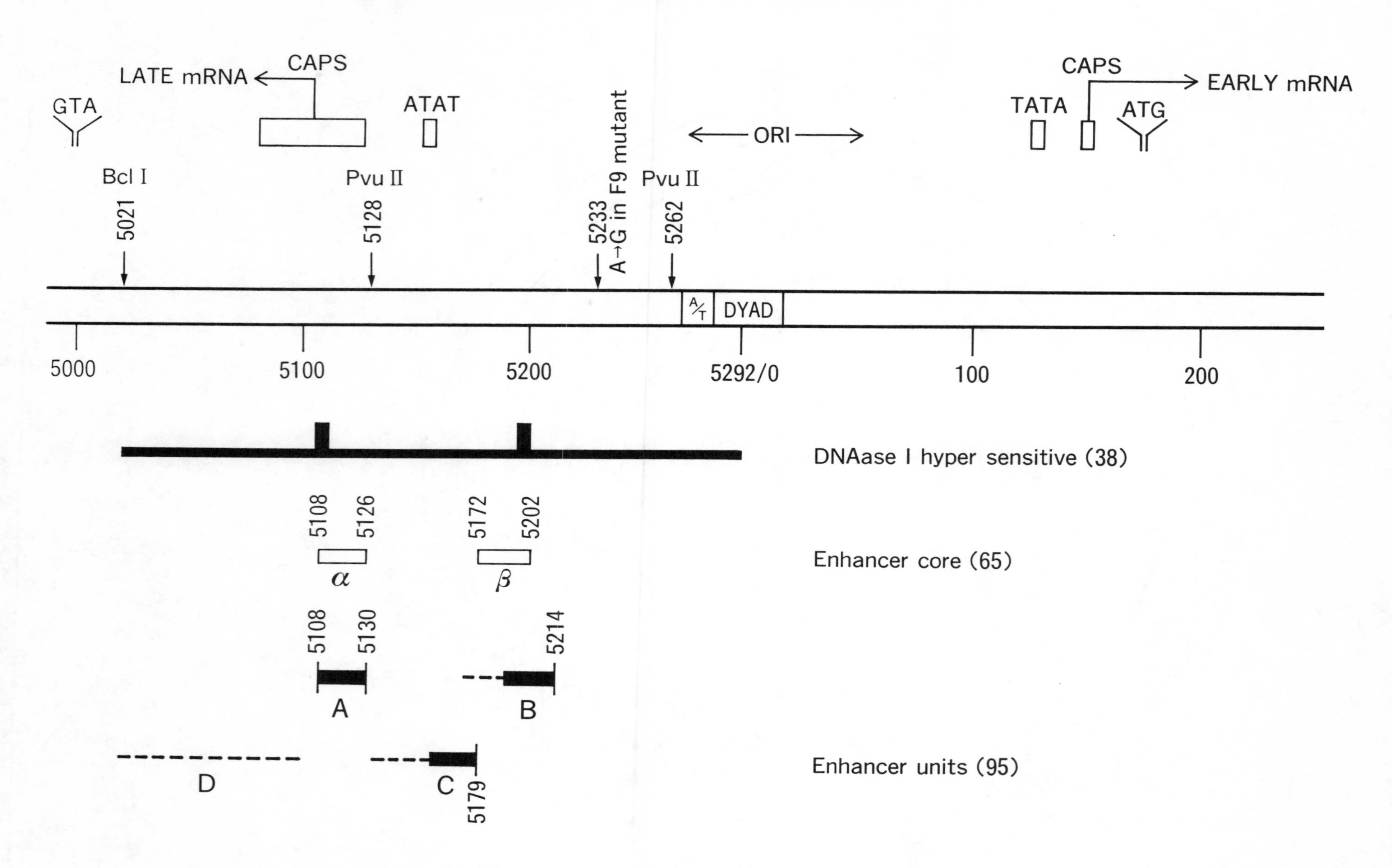
Fig. 2. Regulatory Region of Polyomavirus Genome
GTA
LATE mRNA
CAPS
ATAT
ORI
TATA
CAPS
EARLY mRNA
ATG
Bcl I
5021
Pvu II
5128
5233
A→G in F9 mutant
Pvu II
5262
A/T
DYAD
5000
5100
5200
5292/0
100
200
DNAase I hyper sensitive (38)
5108
5126
α
5172
5202
β
Enhancer core (65)
5108
5130
A
5214
B
D
C
5179
Enhancer units (95)

Table 1

Restriction Enzyme Sites

Enzyme	Position
Acc I	368
	1500
	1809
	4944
	4973
Acy I	84
	4048
Aha I	356
Afl II	2223
Alu I	193
	524
	570
	873
	933
	1145
	1232
	1350
	1362
	1374
	1657
	1668
	2006
	2033
	2086
	2227
	2377
	2894
	3328
	3428
	3762
	3919
	4342
	4430
	4481
	4871
	4919
	4955
	5026
	5129
	5263
Apa I	2611
	3262

Enzyme	Position
Asu I	3
	11
	104
	715
	926
	1103
	1412
	1555
	1574
	2274
	2393
	2611
	2612
	2657
	2985
	3060
	3243
	3262
	3263
	3741
	3944
	3957
	4068
	4177
	4236
	4558
	4675
	5045
Ava I	657
	1016
Ava II	3
	715
	926
	1103
	1412
	1574
	2274
	2393
	2985
	3741
	3957
	4675
Ava III	1910
Avr II	4707

Enzyme	Position
BamH I	4632
Bbv I	262
	277
	1438
	1666
	1725
	2031
	2080
	2304
	2453
	4893
	5127
BbvS I	208
	262
	277
	1363
	1438
	1666
	1725
	2031
	2034
	2080
	2228
	2304
	2453
	3595
	3894
	4224
	4893
	5127
Bcl I	5021
Bgl I	87
Cau II	1100
	1213
	1487
	2614
	2990
	4409
	5291
Cfr I	836
	5132

Enzyme	Position
Dde I	185
	377
	440
	551
	826
	949
	967
	1078
	1267
	1369
	1396
	1618
	1654
	2118
	2487
	3143
	3702
	4323
	4339
	4385
EcoR I	1560
EcoR I	114
	123
	333
	1528
	1561
	1644
	1771
	1868
	1998
	2062
	2283
	2349
	2680
	2699
	2742
	2925
	3348
	3442
	3615
	3831
	4127
	4877
	5009
	5061
	5154

Table 1 (cont'd)

Enzyme	Position
EcoR II	8
	406
	591
	718
	1256
	1571
	1758
	2050
	2457
	2497
	2930
	3033
	3549
	4174
	4441
	4605
	5222
EcoR V	4106
FnuD II	2309
	2972
Fnu4H I	208
	262
	277
	835
	944
	1363
	1438
	1663
	1666
	1725
	2031
	2034
	2080
	2228
	2304
	2307
	2453
	3595
	3894
	4051
	4224
	4893
	4992
	5127

Enzyme	Position
Fok I	427
	721
	2461
	3389
Gdi II	5132
Hae I	1760
	2297
	3108
	3432
	4017
	4434
	4551
	5224
Hae II	84
	95
Hae III	11
	105
	778
	837
	1555
	1761
	2298
	2612
	2657
	3060
	3109
	3243
	3263
	3433
	3944
	4018
	4069
	4178
	4236
	4435
	4552
	4558
	5046
	5133
	5225
	5289

Enzyme	Position
Hga I	2254
	4048
	5124
HgiA I	569
	725
	1373
	2816
	3286
	4276
	4341
HgiC I	84
	594
	2176
	4355
	4693
	4724
HgiJ II	569
	794
	1080
	1307
	1373
	2611
	3262
	3680
	4341
Hha I	85
	96
	2310
	2971
Hinc II	2962
	3466
Hind III	1656
	3918
Hinf I	385
	436
	960
	1269
	1287
	1335
	2039

Enzyme	position
Hinf I cont'd	3447
	3634
	3807
	4154
	4206
	4931
	5073
Hpa II	399
	1101
	1213
	1488
	2615
	2991
	4409
	5291
Hph I	461
	3148
	3736
Kpn I	2176
	4693
Mbo I	1192
	1318
	1323
	1857
	2763
	3384
	4633
Mbo II	34
	255
	1054
	1057
	1794
	2415
	5033
Mnl I	139
	149
	160
	494
	1763
	1940
	1988

Table 1 (cont'd)

Enzyme	Position	Enzyme	Position	Enzyme	Position	Enzyme	Position
Mnl I	2117	NspB II	835	Sac I	569	Sna I	368
cont'd	2140		944		1373		1500
	2202		1663		4341		
	3024		2307			Sph I	696
	3027		4051	Sac III	1482		1594
	3246		4992		3813		1970
	3338						
	3690	NspC I	421	Sca I	601	Stu I	1760
	3711		633		886		2297
	3744		696		4787		4017
	3839		1178				5224
	3863		1594	ScrF I	8		
	3867		1970		406	Taq I	1316
	3966		3078		591		1774
	4078				718		4620
	4198	Pst I	484		1100		4766
	4283		2356		1213		4965
	4359		2451		1256		
	4678		3313		1487	Tth111 I	3454
	4807		4225		1571		3469
	4810				1758		
	4858	Pvu II	1144		2050	Tth111 II	693
	4963		2032		2457		1523
	5048		5128		2497		2444
	5067		5262		2614		2601
	5146				2930		3579
	5174	Rsa I	444		2990		4774
	5213		602		3033		
			887		3549	Xba I	2477
Mst II	2117		1176		4174		2522
			1225		4409		
Nar I	84		2074		4441	Xho II	1191
			2177		4605		3383
Nco I	1273		2575		5222		4632
	3051		2994		5291		
	3111		3358			Xmn I	2586
	4554		4025	SfaN I	587		5231
	4640		4115		1330		
			4334		2145		
Nde I	2735		4570		2219		
	2882		4694		3390		
			4788				

The following sites do not appear:

Asu II	Mst I	Bal I	Sac II	Cla I	Sal I
BssH I	Pvu I	BstE II	Xho I	Hpa I	Xma III
Hgi E II	Sma I	Nae I	Bgl II	Nru I	Mlu I

Table 2

Landmarks on Polyoma Virus (A2 Strain)

Nucleotide no.	Sequence	Significance	Ref.
A. Early region (from about 1-2920 nucleotides)			
5263-45 (5267-18) 5281-20	Hairpin loop	Region around viral origin of replication dyad symmetry	29,77,78
39-43 50-54 60-64	GAGGC	Large T-antigen binding sites	17,32,70
142-138 152-148 163-159	opposite strand GAGGC		
5078-5085 5151-5158 120-127 272-279	TTAAAATA TTTAATTA TATAATTA TATAAGCA	"Hogness boxes"--potential early promoters (>75% homology with canonical TATAAATA)	78
148-153		Major early cap sites	48
173-175	ATG	Initiation codon for known early proteins (T antigens)	78
406-412 790-795	CCAG/GTA TACAG/G	Splice joint for large T-antigen messenger	79,94
743-749 790-795	CCAA/GTA TACAG/G	Splice joint for small T-antigen messenger	79,94
743-749 804-809	CCAA/GTA CCTAG/A	Splice joint for middle T-antigen messenger	79,94
806-809	TAG	Termination codon for small T-antigen	50,79
1476-1481	AATAAA	Possible processing signal for early mRNA (function unknown)	50,79
1498-1500	TAG	Termination codon for middle T-antigen	78
2913-2915	TGA	Termination codon for large T-antigen	79
2915-2920	AATAAA	Processing signal for early mRNA(s)	50,79

Table 2 (Cont'd)

Nucleotide no.	Sequence	Significance	Ref.
B. Late region (from about 5292-2908 nucleotides)			
5283-5268	$TTAATACTA_8$	A, T-rich region near viral origin of replication	78
5262-5021		Enhancer activity,also required for DNA replication	15,16
5103-5117	TAAGCAGGAAGTGAC	Homologous to Ad5Ela enhancer	39
5188-5203	GTGTGGTTTTGCAAGA	Homologous to SV40,MSV and AD-2 Ela enhancers	39
-5100-		DNase I hypersensitive region, HS1	38
-5210-		DNase I hypersensitive region, HS2	38
5267-5021		Sequence changes observed in the genome of mutants capable of growing in embryonal carcinoma and trophoblast cells	31,51 52,75 87
5155-5148	TTAAAAGA	"Hogness box"--potential late promoter	1
5128< >5021		Major late cap sites	25
5023-5018 5081-5076	TCAA/GTA TTAAG/A	"Leader-to-leader" splice which gives rise to tandem repeats in late mRNAs	92
5023-5018 5019-5014	TCAA/GTA GTAGA/T	Possible "leader-to-body" splice for 19S late mRNA (alternative, no splice)	1
5023-5018 4712-4705	TCAA/GTA CCTAG/G	Splice joint for 18S late mRNA	80,92
5023-5018 4127-4122	TCAA/GTA TCTGA/G	Splice joint for 16S late mRNA	79,82 92
5002-5000	ATG	Initiation codon for VP2	1,40,81
4657-4655	ATG	Initiation codon for VP3	1,40,81
4074-4072	ATG	Initiation codon for VP1	1,40,81
4045-4043	TAA	Termination codon for VP2/VP3	1
4074-4046		Overlap region between genes coding for VP2/VP3 and VP1	1,81
2925-2923	TAA	Termination codon for VP1	81
2913-2908	AATAAA	Processing signal for late mRNAs	79,81

Table 3

Map Positions of Mutants of Polyoma Virus

Mutant	Class	Map Position	Ref.
ts a	ts A	2172	13,90
ts 609	ts A	1656< >2310	64
ts 25 E	ts A	2862	13,90
ts 48	ts A	2320	13,22
ts 52	ts A	2862	13
ts 616	ts A	2310< >2971	64
ts 697	ts A	2310< >2971	64
ts 25D	ts A	2310< >2615	24
ts 10	late ts	2971< >3918	64
ts 1260	late ts	2971< >3918	64
ts c	late ts	2971< >3918	64
208	late ts	2971< >3918	64
ts 59	late ts	2971< >3918	34
ts 3	late ts	4409< >5291	24
NG 18	hrt	del. 512-698	37,81
6 B5	hrt	del. 447-625	37
A 8	hrt	del. 456-582	37
B 2	hrt	del. 489-729	37
II 5	hrt	del. 634-734	7
SD 15	hrt	del. 420-560 (plus multiple base changes)	7
NG59	hrt	707 (one base change	7
HA 33	hrt	plus insertion of	7
3A-1	hrt	3 bp)	7
dl 8	mlt	del. 990-1080	75
dl 8/LT	splice del.	410-794	83
dl 8/MT	splice del.	747-808	83
dl 23	mlt	del. 1138-1240	75
dl 23/LT	splice del.	410-794	83
dl 23/MT	splice del.	747-808	83
45	mlt	del. 1075-1141	4
dl 1013	mlt	del. 1191-1212	57
dl 1014	mlt	del. 1198-1206	57
dl 1015	mlt	del. 1244-1274	57
dl 22	mlt	del. 1092-1191	18
dl 27	mlt	del. 1093-1150	18
dl 2208	mlt	del. 1126-1183	66
P155	mlt	del. 1348-1360	33
PyECF9-1 }	PyEC		5,52
Pyf441 }	PyEC	5233 A → G	31
PyECA	PyEC		63
PyEC500	PyEC	5021< >5265	63
PyEC5000	PyEC		63
M206/ev1001	PyEC		58,63

Table 3 (Cont'd.)

Mutant	Class	Map Position	Ref.
PyECF9-5	PyEC		52
PyECPCC4-97	PyEC		51
PyECPCC4-204	PyEC		51
Pyf101	PyEC	5021< >87	31
Pyf111	PyEC		31
Pyhr N-2	PyEC		74
Pyhr N-5	PyEC		74
Py-Mgl-Py-Mg5	PyEC		62
PyTr91	PyTr	del. (5131-5156) and insertion	86
PyTr92	PyTr	del. (5131-5156) and insertion	86
PyNB	growth in NB (neuroblastoma cells)	5050< >5295	59
PyFL78	growth in friend cells	insertion at 5139	14
Py34BXK100	Py-SV hybrid	del. 5021-5262; insertion SV 100-300	6
Py34BXSV	Py-SV hybrid	del. 5021-5262; insertion SV 5172-300	6
B mutants	deletions	243< >5247	12
cs 3	Cold sensitive for	1408	88
cs 5	transformation	1227,1320	88
MOP 1033		1017-1018	87
Py1387-T		1387	8
am 1294	Termination codon	1278	88
am 1292	for middle T	1276	88
am 1045	(non-transforming)	1029	88
bc 1051	Large T splice site	410 (G-C to A-T)	65
LT virus	del.	410-794	84
ST virus	del. ST splice	747-794	89
MT virus	del. MT splice	747-808	89
dl 1061, 1062	Middle, small T splice site	763< >711	65
Py808 A	Middle T splice site	808	55
din21	del.-insertion (mouse)	49-98	19
mutant 44	del.-insertion (foreign, mouse?)	92-95	10
RMI	Py-mouse hybrid	insertion in VP1 region	85
RMII	Py-mouse hybrid	insertion in VP1 region	41
Ad175	del.-insertion (ad-2)	93-170	23
pAS131(Py1178T)	point mutant	1178 A → T	9,68
pAS132-137	point mutants	1188< >1159	68
Rx	MT hydrophobic domain	1490< >1380	60
pLT	del. in LT	1656< >960	2
r9,rB,rP	frame-shift(c deletion)	1247< >1239	95
ins 1-4	AATAAA insertions	2976< >2821	53
PT250	point mutants	983 A → T	61
pTH	point mutants	983, 1178 A → T	61
Py1178T/dl 1014	double mutant	1178 A → T;del.1198-1206	57,73

REFERENCES

1. ARRAND, J., et al. 1980. J. Virol. 33:606-618.
2. ASSELIN, C., and M. BASTIN. 1985. J. Virol. 56.958-968.
3. BENJAMIN, T.L. 1970. Proc. Natl. Acad. Sci. USA 67:394-399.
4. BENDIG, M.M., et al. 1980. J. Virol. 33:1215-1220.
5. BOHNLEIN, E., et al. 1985. Nucl. Acids Res. 13:4789-4809.
6. CAMPBELL, B.A., and L.P.VILLARREAL. 1985. Mol. Cell. Biol. 5:1534-1537.
7. CARMICHAEL, G.G., and T.L. BENJAMIN. 1980. J. Biol. Chem. 255:230-235.
8. CARMICHAEL, G.G., et al. 1982. Proc. Natl. Acad. Sci. USA 79:3579-3583.
9. CARMICHAEL, G.G., et al. 1984. Proc. Natl. Acad. Sci. USA 81:679-683.
10. CLARK, K.L., et al. 1984. J. Virol. 52:1032-1035.
11. COGEN, B. 1978. Virology 85:222-230.
12. DAILY, L., and C. BASILICO. 1985. J. Virol. 54:739-749.
13. DEININGER, P.L., et al. 1981. J. Virol. 37:871-875.
14. de SIMONE, V., et al. 1985. Mol. Cell. Biol. 5:2142-2146.
15. de VILLIERS, J., and W. SCHAFFNER. 1981. Nucl. Acids Res. 9:6251-6264.
16. de VILLIERS, J., et al. 1984. Nature 312:242-246.
17. DILWORTH, et al. 1984. Proc. Natl. Acad. Sci. USA 81:1941-1945.
18. DING, D.M., et al. 1982a. J. Virol. 44:1080-1083.
19. DING, D.M., et al. 1982b. EMBO J. 1:461-466.
20. ECKHART, W. 1969. Virology 38:120-125.
21. ECKHART, W., and R. DULBECCO. 1974. Virology 60:359-369.
22. ECKHART, W., et al. 1981. Virology 109:35-46.
23. FARMERIE, W.G. and W.R. FOLK. 1984. Proc. Natl. Acad. Sci. USA 81:6919-6923.
24. FEUNTEUN, J., et al. 1976. Proc. Natl. Acad. Sci. USA 73:4169-4173.
25. FLAVELL, A.J., et al. 1980. J. Virol. 33:902-908.
26. FRANCKE, B., and W. ECKHART. 1973. Virology 55:127-135.
27. FRIED, M. 1965. Proc. Natl. Acad. Sci. USA 53:486-491.
28. FRIED, M., and B.E. GRIFFIN. 1977. Adv. Cancer Res. 24:67-113.
29. FRIEDMANN, T., et al. 1978. J. Biol. Chem. 253:6561-6567.
30. FRIEDMANN, T.R., et al. 1979. Cell 17:715-724.
31. FUJIMURA, F.K., et al. 1981. Cell 23:809-814.
32. GAUDRAY, P., et al. 1981.Nucl. Acids Res. 9:5697-5710.
33. GELINAS, C., et al. J. Virol. 1982. 43:1072-1081.
34. GIBSON, W., et al. 1977. Virology 80:21-41.
35. GRIFFIN, B.E., and C. MADDOCK. 1979. J. Virol. 31:645-656.
36. GRIFFIN, B.E., et al. 1974. Proc. Natl. Acad. Sci. USA 71:2077-2081.
37. HATTORI, J., et al. 1979. Cell 16:505-513.
38. HERBOMEL, P., et al. 1981. Cell 25:651-657.
39. HERBOMEL, P., et al. 1984. Cell 39:653-662.
40. HEWICK, R.M., et al. 1980. J. Virol. 33:631-636.
41. HUBERDEAU, D., et al. 1985. Mol. Cell. Biol. 5:2608-2612.
42. HUNTER, T., et al. 1978. Proc. Natl. Acad. Sci. USA 75:5917-5921.
43. HUTCHINSON, M.A., et al. 1978. Cell 15:65-77.
44. ITO, Y., et al. 1977. Proc. Natl. Acad. Sci. USA 74:1259-1263.
45. ITO, Y., et al. 1977. Proc. Natl. Acad. Sci. USA 74:4666-4670.
46. ITO, Y. 1979. Virology 98:261-266.
47. ITO, Y. 1980. Viral Oncology (ed. G. Klein). Raven Press,NY,pp.447-480.
48. JAT, P., et al. 1982. Nucl. Acids Res. 10:871-887.
49. KAMEN, R., et al. 1974. Cold Spring Harbor Symp. Quant. Biol.39:187-198.
50. KAMEN, R., et al. 1980. Cold Spring Harbor Symp. Quant. Biol. 44:63-75.
51. KATINKA, M., et al. 1980. Cell 20:393-399.
52. KATINKA, M., et al. 1981. Nature 290:720-722.
53. LANOIX, J., et al. 1986. J. Virol. 58:733-742.
54. LEGON, S., et al. 1979. Cell 16:373-388.
55. LIANG, T.J., et al. 1984. Mol. Cell Biol. 4:2774-2785.
56. MAGNUSSON, G., and P. GERG. 1979. J. Virol. 32:523-529.
57. MAGNUSSON, G., et al. 1981. J. Virol. 39:673-683.
58. MAGNUSSON, G., and M.G. NILSSON. 1982. Virology 119:12-21.
59. MAIONE, R., et al. 1985. EMBO J. 4:3215-3221.
60. MARKLAND, W., et al. 1986. J. Virol. 59:82-89.
61. MARKLAND, W., et al. 1986. J. Virol. 59:384-391.
62. MELIN, F., et al. 1985. J. Virol. 53:862-866.
63. MELIN, F., et al. 1985. EMBO J. 4:1799-1803.
64. MILLER, L.K., and M. FRIED. 1976. J. Virol. 18:824-832.
65. MULLER, W.J., et al. 1988. Mol. Cell. Biol. 8:5000-5015.
66. NILSSON, S.V., and MAGNUSSON, G. 1983. EMBO J. 2:2095-2101.
67. NILSSON, S.V., et al. 1983. J. Virol. 46:284-287.
68. NODA, T., et al. 1986. J. Virol. 60:105-113.
69. OOSTRA, B.A., et al. 1983. Nature 304:456-459.
70. POMERANTZ, B.J., and J.A. HASSELL. 1984. J. Virol. 49:925-937.
71. RASSOULZADEGAN, M., et al. 1982. Nature 300:713-718.
72. RASSOULZADEGAN, M., et al. 1983. Proc.Natl.Acad.Sci.USA 80:4354-4358
73. SCHAFFHAUSEN, B.S., et al. 1978. Proc.Natl.Acad.Sci.USA 75:79-83.
74. SCHAFFHAUSEN, B.S., et al. 1985. Virology 143:671-675.
75. SEKIKAWA, K., and A.J.LEVINE. 1981. Proc.Natl.Acad.Sci.USA 78:1100-1104
76. SMOLAR, N., and B.E. GRIFFIN. 1981. J. Virol. 38:958-967.
77. SOEDA, E., et al. 1978. Proc. Natl. Acad. Sci. USA 75:162-166.
78. SOEDA, E., et al. 1979. Cell 17:357-370.
79. SOEDA, E., et al. 1979. Nucl. Acids Res. 7:839-858.
80. SOEDA, E., et al. 1980. Nature 283:445-453.
81. SOEDA, E., et al. 1980. J. Virol. 33:619-630.
82. SOEDA, E., and B.E. Griffin. 1978. Nature 276:294-298.
83. SRIVATSAN, E.S., et al. 1981. J. Virol. 37:244-247.
84. STRAUSS, M., et al. 1986. Gene 49:331-340.
85. STREULI, C. Personal communication.
86. SYLLA, B.S., et al. 1984. Cell 37:661-667.
87. TANAKA, K., et al. 1982. EMBO J. 1:1521-1527.
88. TEMPLETON, D., and W. ECKHART. 1982. J. Virol. 41:1014-1024.
89. TEMPLETON, D., and W. ECKHART. 1984. J. Virol. 49:799-805.
90. TEMPLETON, D., et al. 1986. J. Virol. 57:367-370.
91. THOMAS, T., et al. 1981. J. Virol. 37:1094-1098.
92. TREISMAN, R. 1980. Nucl. Acids Res. 8:4867-4888.
93. TREISMAN, R., et al. 1981. Nature 292:595-600.
94. TREISMAN, R., et al. 1981. J. Mol. Appl. Genet. 1:83-92.
95. VELDMAN, G.M., et al. 1985. Mol. Cell. Biol. 5:649-658.
96. WILSON, J.B., et al. 1986. Cell 44:477-487.

GENETIC MAP OF ADENOVIRUS

Göran Akusjärvi[1] and Göran Wadell[2]

1) Department of Microbial Genetics, Karolinska Institute, Box 60400, S-104 01 Stockholm, Sweden.
2) Department of Virology, University of Umeå, S-901-85 Umeå, Sweden.

July 1989

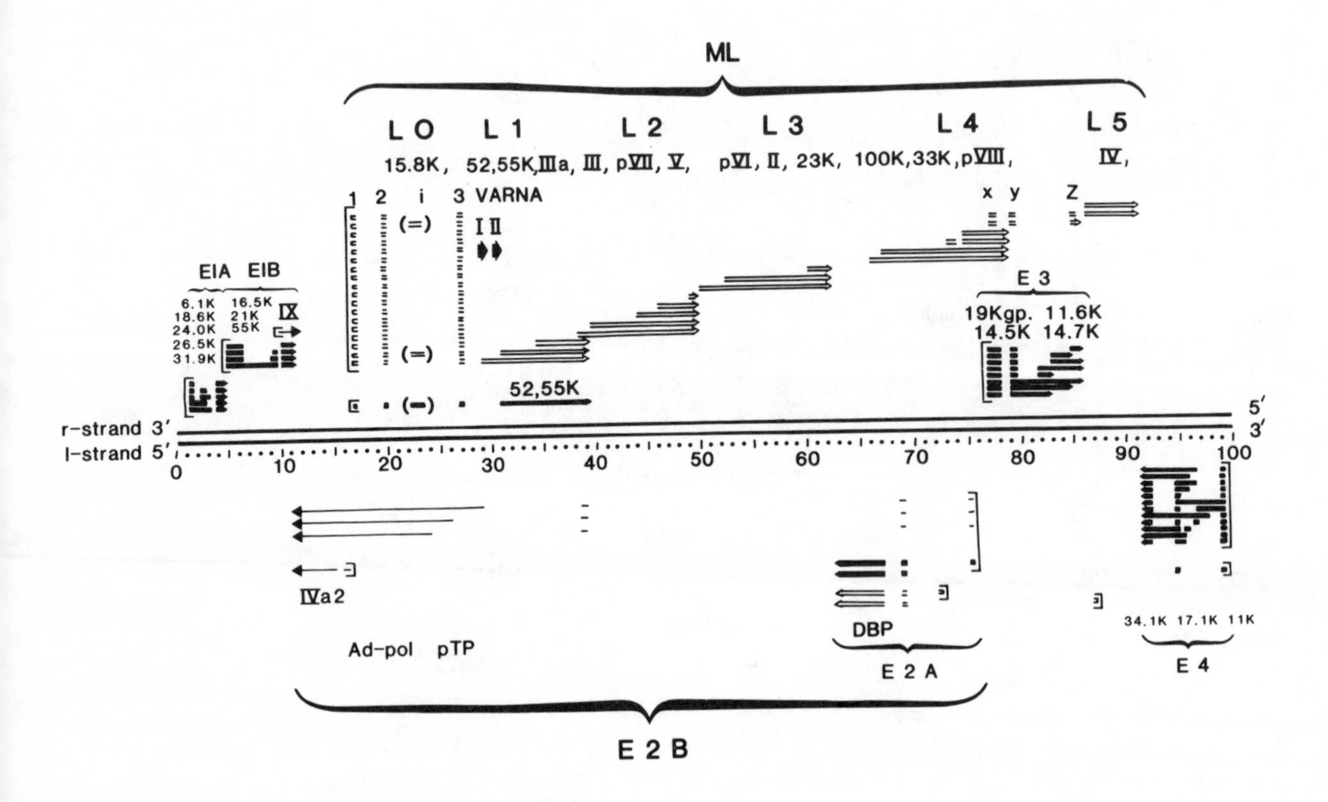

Table 2: NUCLEOTIDE SEQUENCES DETERMINED FROM HUMAN ADENOVIRUSES OF SUBGENERA A THROUGH F. The entire Ad2 sequence (subgenus C) is available (72) and used as a reference in this table.

	E1, pIX	IVa2	Major late promoter	E2B	VA RNA	L3 Hexon	L3 Protease	E2A DBP	L4 100K	E3	L5 Fiber
Approximate Coordinates:	0-11	11-16	16-22	15-29.5	29.5-30.5	52-60.5	60.5-62.5	62.5-67	67-74	76.5-86	86-91
Subgenus A:	Ad12 (55,84)			Ad12 (78)	Ad12 (40)			Ad12 (60)			
B:	Ad3 (33) Ad7 (30)	Ad7 (34)	Ad3 (35) Ad7 (35)	Ad7 (36)				Ad7 (71)		Ad3 (81) Ad35 (38)	Ad3 (80) Ad7 (49)
C:	Ad5 (96)	Ad5 (96)	Ad5 (96)	Ad5 (96)	Ad5 (96)	Ad5 (56)	Ad5 (59)	Ad5 (61)	Ad5 (60)	Ad5 (28)	Ad5 (27)
E:	Ad4 (88)						Ad4 (50)	Ad4 (57)			
F:	Ad40 (12, 95) Ad41 (51, 95)					Ad41 (90)	Ad40 (104) Ad41 (104)	Ad40 (104) Ad41 (104)			

Table 1. Genetic map of the adenovirus type 2 genome. (Modified and updated from ref. 72).

Feature	Base#	Mapunit	Strand	Reference
End of inverted terminal repetition	102	0.28		18, 77
Start of packaging sequence/E1A enhancer region	194	0.54		45, 48
End packaging sequence/E1A enhancer region	358	1.00		45, 48
Cap site for E1A mRNAs	498	1.39	r	20, 47
Initiator AUG for E1A proteins	559	1.56	r	32
5' splice site for E1A 9S 10S and 11S mRNAs	636	7.77	r	100, 93, 82
3' splice site for E1A 10S and 11S mRNAs	853	2.37	r	93, 82
5' splice site for E1A 12S mRNA	973	2.71	r	68
5' splice site for E1A 13S mRNA	1111	3.09	r	68
3' splice site for all E1A mRNAs	1226	3.41	r	68, 100
Terminator UGA for E1A 9S mRNA polypeptide	1313	3.65	r	
Terminator UAA for E1A 10S, 11S, 12S and 13S mRNA polypept.	1540	4.29	r	
AAUAAA for E1A mRNAs	1608	4.47	r	
Poly-A addition site for E1A mRNAs	1630	4.54	r	68
Cap site for E1B mRNAs	1699	4.73	r	20, 47
Initiator AUG for E1B 21K (175R) polypeptide	1711	4.76	r	15
Initiator AUG for E1B 55K (495R), 155R and 82R polypeptides	2016	5.61	r	16, 64
Terminator UGA for E1B 21K (175R) polypeptide	2236	6.22	r	
5' splice site for E1B 13S, 14S and 14.5S mRNAs	2249	6.26	r	69,101
3' splice site for E1B 14.5S mRNA	3212	8.94	r	101
Terminator UAA for E1B 14.5S mRNA	3254	9.05	r	
3' splice site for E1B 14S mRNA	3270	9.10	r	101
Terminator UGA for E1B 55K (495R) and 155R proteins	3501	9.74	r	
5' splice site for E1B 14S, 14.5S and 22S mRNAs	3504	9.75	r	69, 101
Cap site for polypeptide IX mRNA	3576	9.95	r	13, 20
3' splice site for all E1B mRNAs	3589	9.99	r	69
Initiator AUG for polypeptide IX	3600	10.02	r	15
Terminator UGA for 82R polypeptide	3601	10.02	r	64
Terminator UAA for polypeptide IX	4020	11.18	r	
AAUAAA for E1B and polypeptide IX mRNAs	4029	11.21	r	
Poly-A site for the E2B and IVa2 mRNAs	4050	11.27	l	13
Poly-A site for E1B and polypeptide IX mRNA	4061	11.30	r	13
Terminator UAA for IVa2	4083	11.36	l	
AAUAAA for IVa2 mRNA	4085	11.37	l	
Terminator UAG for Ad-DNA polymerase	5189	14.44	l	
3' splice site for IVa2 mRNA	5417	15.07	l	94 (Ad5)
5' splice site for IVa2 mRNA	5696	15.85	l	94 (Ad5)
1st AUG in IVa2 reading frame	5708	15.88	l	
Cap site for IVa2 mRNA	5826	16.21	l	20
Cap site for "major late" promoter	6039	16.80	r	20
5' splice site for 1st late leader	6079	16.92	r	8, 107
Initiator AUG 13.5K polypeptide (11.6K ORF)	6280	17.48	r	63
Terminator UAG 13.5K polypeptide (11.6K ORF)	6598	18.36	r	63
3' splice site for 2nd late leader	7101	19.76	r	8, 107
5' splice site for 2nd late leader	7172	19.96	r	8, 107
3' splice site for "i" leader	7942	22.10	r	37, 98
Initiator AUG for 15.8K "i" leader protein	7968	22.17	r	37, 63, 87, 98
1st AUG in DNA polymerase reading frame	8357	23.25	l	
5' splice site for "i" leader	8381	23.32	r	37, 98
Terminator UAG for Terminal protein	8575	23.86	l	
3' splice site for 3rd late leader	9634	26.81	r	8, 107
Terminator UGA for 15.8K 'i' leader protein	9655	26.87	r	37, 98
5' splice site for 3rd late leader	9723	27.06	r	8, 107
3' splice site for L1$_a$ mRNA	10434	29.03	r	3
1st AUG in Terminal protein reading frame	10534	29.31	l	
5' end of VA RNAI (minor A start)	10607	29.52	r	25, 66, 97
5' end of VA RNAI (major G start)	10610	29.52	r	25, 66, 97
3' end of VA RNAI	10766	29.96	r	24, 66
3' end of V200	10805	30.07	r	106
5' end of VA RNAII	10866	30.24	r	9
3' end of VA RNAII	11023	30.67	r	9
3' splice site for 52,55K mRNA	11040	30.72	r	6, 58
Initiator AUG for the 52,55K polypeptide	11040	30.72	r	65
Terminator UAA for 52,55K polypeptide	12285	34.18	r	
3' splice site for IIIa mRNA	12308	34.25	r	58
1st AUG in polypeptide IIIa reading frame	12308	34.25	r	
Terminator UAA for polypeptide IIIa	14063	39.13	r	
Variable run of A's (19-20)	14064	39.14		73
AAUAAA for L1 mRNAs	14092	39.21	r	
Major poly-A addition site for L1 mRNAs	14113	39.27	r	62
Minor poly-A addition site for L1 mRNAs	14118	39.29	r	4, 44
3' splice site for polypeptide III mRNA	14150	39.37	r	62
1st AUG in polypeptide III reading frame	14151	39.38	r	
Terminator UGA for polypeptide III	15864	44.14	r	
Initiator AUG for pVII	15873	44.17	r	85
Gly/Ala cleavage site for pVII to VII	15945	44.37	r	85
Terminator UAG for pVII	16467	45.82	r	
3' splice site for polypeptide V mRNA	16516	45.96	r	14
1st AUG in polypeptide V reading frame	16539	46.02	r	
Terminator UAA for polypeptide V	17646	49.10	r	
Initiator AUG for pmu	17676	49.19	r	14, 17
Gly/Met cleavage site for pmu to mu	17772	49.45	r	17, 52, 102
Gly/Ile cleavage site for pmu to mu	17829	49.61	r	17, 52, 102
Terminator UGA for pmu	17916	49.85	r	14, 17
AAUAAA for L2 mRNAs	17949	49.95	r	
Poly-A addition site for L2 mRNAs	17969	50.00	r	5,62
3' splice site for pVI mRNA	18000	50.09	r	5
Initiator AUG for pVI	18001	50.09	r	86
Gly/Ala cleavage site for pVI to VI	18100	50.37	r	86
Terminator UAA for pVI	18751	52.18	r	
3' splice site for hexon mRNA	18802	52.32	r	7
Initiator AUG for hexon (polypeptide II)	18838	52.42	r	15, 54
3' splice site for 23K viral endoprotease mRNA	21650	60.24	r	10
Terminator UAA for hexon (polypeptide II)	21742	60.50	r	53
1st AUG in the 23K viral endoprotease reading frame	21778	60.60	r	10, 105
Terminator UAA for the 23K viral endoprotease	22390	62.30	r	10, 105
AAUAAA for L3 mRNAs	22418	62.38	r	
Poly-A addition site for E2A mRNAs	22420	62.39	l	10, 62
AAUAAA for E2A mRNAs	22439	62.44	l	
Poly-A addition site for L3 mRNAs	22443	62.45	r	10, 62
Terminator UAA for DBP (72K protein)	22492	62.59	l	
1st AUG in 100K reading frame	23526	65.46	r	
1st AUG in DBP (72K protein) reading frame	24079	67.00	l	
3' splice site for main body of 72K mRNA	24088	67.03	l	43, 61
3' splice site for 100K mRNA	24095	67.05	r	60

Feature	Base#	Mapunit	Strand	Reference
1st AUG in 100K polypeptide mRNA	24108	67.08	r	60
5' splice site for 2nd leader in 72K mRNA	24715	68.77	l	43, 61
3' splice site for 2nd leader in 72K mRNA	24791	68.98	l	43, 61
5' splice site 1st leader of late 72K mRNA	25886	72.03	l	61
Cap site for 72K mRNA late	25954	72.22	l	20
Alternative cap site for 72K mRNA late	25956	72.23	l	20
1st AUG in 33K polypeptide reading frame	26239	73.01	r	
Terminator UAG for 100K polypeptide	26523	73.80	r	
Likely 5' splice site for 33K mRNA	26551	73.88	r	67
Likely 3' splice site for 33K mRNA	26754	74.45	r	67
5' splice site for 1st leader of 72K mRNA	27025	75.20	l	19, 43
Cap site for 72K mRNA early	27091	75.38	l	20
Cap site for 72K mRNA early	27092	75.39	l	20
Terminator UAG for 33K polypeptide	27125	75.48	r	
1st AUG in pVIII reading frame	27215	75.73	r	
Cap site for E3 mRNAs	27609	76.83	r	20
3' splice site "x" leader	27713	77.12	r	4
Terminator UGA for pVIII	27896	77.62	r	
5' splice site for "x" leader	27980	77.86	r	4, 83
AAUAAA for L4 mRNAs	28205	78.48	r	
Poly-A addition site for L4 mRNAs	28223	78.53	r	62
3' splice site for "y" leader	28376	78.96	r	83, 92, 107
5' splice site for "y" leader	28559	79.47	r	83, 92, 107
Initiator AUG for E3 19K glycoprotein	28812	80.17	r	70
Terminator UGA for E3 19K glycoprotein	29289	81.50	r	
3' splice site for E3 11.6K mRNA	29348	81.67	r	28 (Ad5)
1st AUG in E3 11.6K reading frame	29468	82.00	r	103
3' splice site for 10.4K mRNA	29765	82.83	r	28 (Ad5)
ATTAAA for E3A mRNAs	29769	82.84	r	2, 23, 83
Terminator UAA for E3 11.6K polypeptide	29771	82.84	r	103
Probable start AUG in 10.4K EGF-receptor regul. protein	29781	82.87	r	22
Poly-A addition site for E3A mRNAs	29792	82.90	r	2
Poly-A addition site for E3A mRNAs	29799	82.92	r	2, 83
Poly-A addition site for E3A mRNAs	29801	82.93	r	2
Poly-A addition site for E3A mRNAs	29804	82.93	r	2
Terminator UAA in 10.4K EGF-receptor regulating protein	30053	83.63	r	22
1st AUG in 14.5K ORF	30059	83.64	r	
3' splice site for "z" leader	30438	84.70	r	83, 89
1st AUG in 14.7K protein which has anti-TNF activity	30444	84.71	r	42, 89
Terminator UGA for 14.5K ORF	30449	84.73	r	
5' splice site for "z" leader	30582	85.10	r	83, 92
Terminator UAA for 14.7K protein with anti-TNF activity	30828	85.78	r	42, 89
AAUAAA for E3B mRNAs	30842	85.82	r	
Poly-A addition site for E3B mRNAs	30864	85.88	r	83
3' splice site for fiber mRNA	31030	86.35	r	92, 108
Initiator AUG for fiber (polypeptide IV)	31030	86.35	r	15
AAUAAA for L5 mRNAs	32774	91.20	r	
Terminator UAA for fiber (polypeptide IV)	32776	91.20	r	
Poly-A addition site for L5 mRNAs	32798	91.27	r	62
Poly-A addition site for E4 mRNAs	32802	91.28	l	39, 99
AAUAAA for E4 mRNAs	32821	91.33	l	
Terminator UGA for 19.5K polypeptide (ORF 6/7)	32916	91.59	l	29
3' splice site for large E4 intron (ORF 6/7 mRNA)	33192	92.36	l	39, 91, 99
Terminator UAG for 34.1K polypeptide (ORF 6)	33195	92.37	l	29, 76
Secondary 3' splice site for large E4 intron	33283	92.61	l	39
5' splice site for large intron in E4 (ORF 6/7 mRNA)	33904	94.34	l	39, 91, 99
Terminator UAG in 13.3K reading frame (ORF 4)	34000	94.61	l	
Initiator AUG for E4 ORF 6 and ORF6/7 polypeptides	34077	94.82	l	29, 76
3' splice site in E4	34082	94.84	l	39
3' splice site in E4	34241	95.28	l	39
5' splice site in E4	34288	95.41	l	99
3' splice site in E4	34329	95.53	l	3
1st AUG in 13.3K reading frame (ORF 4)	34342	95.56	l	
Variable run of A's (13-16) in E4	34357	95.60		41
Terminator UAA for E4 11K protein (ORF 3)	34358	95.61	l	75
3' splice site in E4	34379	95.66	l	91
3' splice site in E4	34435	95.82	l	39
5' splice site in E4	34606	96.30	l	39
Terminator UGA for 15.3K reading frame (ORF 2)	34705	96.57	l	
Initiator AUG for E4 11K protein (ORF 3)	34706	96.57	l	31, 75
3' splice site in E4	34735	96.66	l	39
2nd AUG in 15.3K reading frame (ORF 2)	35095	97.66	l	
3' splice site in E4	35107	97.69	l	99
Terminator UAA for 14.3K reading frame (ORF 1)	35148	97.80	l	
1st AUG in 14.3K reading frame (ORF 1)	35532	98.87	l	
5' splice site for 1st leader of E4 mRNAs	35548	98.92	l	39, 99
Cap sites E4 mRNAs	35609-35614	99.10	l	20, 46
End of inverted terminal repetition	35836	99.71	l	18, 77

Principal organization of mRNA and protein encoding regions on the human adenovirus type 2 (Ad2) genome (adapted from ref. 26). The Ad2 chromosome is a linear double-stranded DNA molecule of 35, 937 ± 9 base pairs (72) which has a protein, the terminal protein TP, covalently linked to the 5' end of each DNA strand (74). The ambiguity in length is due to two regions of heterogeneity (at map units 39.14 and 95.60; Table 1) which is found within Ad2 stocks and between strains held in different laboratories. The genome is divided into 100 map units. Thick lines represent mRNAs expressed from transcription units which are activated early after infection (1 to 8 hrs) and thin lines represent mRNAs expressed from transcription units which are activated at intermediate times of infection (7 to 15 hrs). Open arrows show sequences present in mRNAs expressed late after infection (12 to 48 hrs). Vertical brackets indicate position of promoter sites and arrow heads location of 3' ends. Gaps in the arrows represent intervening sequences (introns) that are removed by RNA splicing. Polypeptides that have been assigned to different regions are also shown. The position of two RNA polymerase III transcriptions units, encoding VA RNA_I and VA RNA_{II} are also shown. For a review of the structural and functional organization of the adenovirus genome see ref. 11.

Restriction endonuclease cleavage maps of DNAs from several human and animal adenoviruses have been presented in the 1982, 1984 and 1987 editions of Genetic Maps.

		Serotype	Published year
HUMAN	Subgenera A:	12 (strain Huie)	1982
		31 (strain 1315)	1982
	B:	3	1982, 1987
		7 (strain Gomen)	1982
		16 (strain Chang 79)	1982
		35	1987
	C:	2	1984
	D:	8	1984
	E:	4	1984
	F:	40	1984, 1987
		41	1987

The genomic structure of a number of adenovirus-SV40 hybrid viruses were presented in the 1984 edition of Genetic Maps.

ANIMAL	CELO virus (fowl adenovirus type 1)	1987
	Mouse adenovirus FL	1987
	Simian adenovirus type 7	1987
	Bovine adenovirus type 3, 4 and 7	1987

Furthermore, a useful reference catalogue showing the restriction endonuclease cleavage pattern (for BamHI, BglII, BstEII, HindIII and SmaI) of DNAs from 41 human adenovirus serotypes has also been published (1).

REFERENCES

1. ADRIAN, T. et al. 1986. Arch. Virol. 9: Nr 3+4
2. AHMED, C.M. I. et al. 1982. Gene 20: 339-346
3. AKUSJÄRVI, G. 1985. J. Virol. 56: 879-886
4. AKUSJÄRVI, G. unpublished
5. AKUSJÄRVI, G. and H. PERSSON. 1981. J. Virol. 38: 469-482
6. AKUSJÄRVI, G. and H. PERSSON. 1981. Nature 292: 420-426
7. AKUSJÄRVI, G. and U. PETTERSSON. 1978. Proc. Natl. Acad. Sci. USA 75: 5822-5826
8. AKUSJÄRVI, G. and U. PETTERSSON. 1979. J. Mol. Biol. 134: 143-158
9. AKUSJÄRVI, G., et al. 1980. Proc. Natl. Acad. Sci. USA 77: 2424-2428
10. AKUSJÄRVI, G., et al. 1981. Nucleic Acids Res. 9: 1-17
11. AKUSJÄRVI, G., et al. 1986. In W. Doerfler (Ed.) Developments in molecular virology. Vol. 8 Adenovirus DNA. pp. 51-98
12. ALLARD, A. and WADELL, G. 1988. Virology 164: 220-229
13. ALESTRÖM, P., et al. 1980. Cell 19: 671-681
14. ALESTRÖM, P., et al. 1984. J. Biol. Chem. 259: 13980-13985
15. ANDERSSON, C.W. and J.B. LEWIS. 1980. Virology 104: 27-41
16. ANDERSSON, C.W., et al. 1984. J. Virol. 50: 387-396
17. ANDERSON, C.W. et al. 1989. Virology in press.
18. ARRAND, J.R. and R.J. ROBERTS. 1979. J. Mol. Biol. 128: 577-594
19. BAKER, C.C., et al. 1979. Cell 18: 569-580
20. BAKER, C.C. and E.B. ZIFF. 1981. J. Mol. Biol. 149: 189-221
21. BALL, A.O. et al. 1989. Virology 170: 523-536.
22. CARLIN, C.R. et al. 1989. Cell 57: 135-144
23. BHAT, B.M. and W.S.M. WOLD. 1985. Mol. Cell. Biol. 5: 3183-3193
24. CELMA, M.L., et al. 1977. J. Biol. Chem. 252: 9032-9042
25. CELMA, M.L., et al. 1977. J. Biol. Chem. 252: 9043-9046
26. CHOW, L.T., et al. 1979. J. Mol. Biol. 134: 265-303
27. CHROBOCZEK, J. and JACROT, B. 1987. Virology 161: 549-554
28. CLADARAS, C. and W.S.M. WOLD. 1985. Virology 140: 28-43
29. CUTT, J.R. et al. 1987. J. Virol. 61: 543-552
30. DIJKEMA, R. et al. 1982. Gene 18: 143-156
31. DOWNEY, J.F., et al. 1983. J. Virol. 45: 514-523
32. DOWNEY, J.F., et al. 1984. J. Virol. 50: 30-37
33. ENGLER, J. 1981. Gene 13: 387-394
34. ENGLER, J. and vanBREE, 1982. Gene 19: 71-80
35. ENGLER, J., et al. 1981. Gene 13: 133-143
36. ENGLER, J., et al. 1983. Gene 21: 145-159
37. FALVEY, E. and E.B. ZIFF. 1983. J. Virol. 45: 185-191
38. FLOMENBERG, P.R., et al. 1988. J. Virol. 62: 4431-4437
39. FREYER, G.A., et al. 1984. Nucleic Acids Res. 12: 3503-3519
40. FÖHRING, B., et al. 1979. Virology 95: 295-302
41. GINGERAS, T.R., et al. 1982. J. Biol. Chem. 257: 13475-13491
42. GOODING, L.R. et al. 1988. Cell 53: 341-346
43. GOLDENBERG, C.J. and S.D. HAUSER. 1983. Nucleic Acids Res. 11: 1337-1348
44. HALES, K.H., et al.1988. J. Virol. 62: 1464-1468
45. HAMMARSKJÖLD, M.L. and G. WINBERG. 1980. Cell 20: 787-795
46. HASHIMOTO, S., et al. 1981. Nucleic Acids Res. 9: 1675-1689
47. HASHIMOTO, S., et al. 1981. Biochemistry 20: 6640-6647
48. HEARING, P. and T. SHENK. 1983. Cell 33: 695-703
49. HONG, J.S. et al. 1988. Virology 167: 545-553
50. HOUDE, A and WEBER, J.M. 1987. Gene 54: 51-56
51. ISHINO, M. et al. 1988. Virology 165: 95-102
52. HOSOKAWA, K. and M.T. SUNG. 1976. J. Virol. 17:924-934
53. JÖRNVALL, H. et al. 1974. Eur. J. Biochem. 48: 179-192
54. JÖRNVALL, H. et al. 1979. Biophys. Res. Comm. Biochem. 56: 4439-4457
55. KIMURA, T. et al. 1981. Nucleic Acids Res. 9: 6571-6580
56. KINLOCH, R. et al. 1984. J. Biol. Chem. 259: 6431-6436
57. KITCHINGHAM, G.R. 1985. Virology 146:90-101
58. KREIVI, J.P. and G. AKUSJÄRVI. 1989. Unpublished results.
59. KRUIJER, W. et al. 1980. Nucleic Acids Res. 8: 6033-6042
60. KRUIJER, W., et al. 1983. Virology 128: 140-153
61. KRUIJER, W., et al. 1981. Nucleic Acids Res. 9: 4439-4457
62. LEMOULLEC, J.M., et al. 1983. J. Virol. 48: 127-134
63. LEWIS, J.B. and C.W. ANDERSSON. 1983. Virology 127: 112-123
64. LEWIS, J.B. and C.W. ANDERSON. 1987. J. Virol. 61: 3879-3888
65. LEWIS, J.B., et al. 1985. Virology 143: 452-466
66. OHE, K. and S.M. WEISSMAN. 1971. J. Biol. Chem. 246: 6991-7009
67. OOSTEROM-DRAGON, E.A. and C.W. ANDERSSON. 1983. J. Virol. 45: 251-263
68. PERRICAUDET, M., et al. 1979. Nature 281: 694-696
69. PERRICAUDET, M., et al. 1980. Proc. Natl. Acad. Sci. USA 77: 3778-3782
70. PERSSON, H., et al. 1980. Proc. Natl. Acad. Sci. USA 77, 6349-6353
71. QUINN, C.O. and KITCHINGHAM, G.R. 1984. J. Biol.Chem. 259: 5003-5009
72. ROBERTS, R.J., et al. 1986. In W. Doerfler (Ed.) Developments in molecular virology. Vol. 8. Adenovirus DNA pp. 1-51
73. ROBERTS, R.J., et al. 1984. j. Biol. Chem. 259: 13965-13975
74. ROBINSON, A.J., et al. 1973. Virology 56: 54-69
75. SARNOW, P. et al. 1982. J. Mol. Biol. 162: 565-583
76. SARNOW, P. et al. 1984. J. Virol. 49: 692-700
77. SHINAGAWA, M. and R. PADMANABHAN. 1979. Biochem. Biophys. Res. Com. 87: 671-67
78. SHU, L.M. et al. 1986. Gene 46: 187-195
79. SHU, L.M. et al. 1988. Virology 165: 348-356
80. SIGNÄS, C. et al. 1985. J. Virol. 53: 672-678
81. SIGNÄS, C. et al. 1986. Gene 50: 173-184
82. STEPHENS, C. and HARLOW, E. 1987. EMBO J. 6: 2027-2035
83. STÅLHANDSKE, P., et al. 1983. Gene 22: 157-165
84. SUGISAKI, et al. 1980. Cell 20: 777-786
85. SUNG, M.T., et al. 1983. Proc. Natl. Acad. Sci. USA 80: 2902-2906
86. SUNG, M.T., et al. 1983. J. Biol. Chem. 258: 8266-8272
87. SYMINGTON, J.S., et al. 1986: J. Virol. 57: 848-856
88. TOKUNAGA, O. et al. 1986. Virology 155: 418-433
89. TOLLEFSON, A.E. and W.S.M. WOLD. 1988. J. Virol. 62: 33-39
90. TOOGOOD, C. and HAY, R. 1988. J. Gen. Virol. 69: 2291-2301
91. TIGGES, M.A. and H.J. RASKAS. 1984. J. Virol. 50: 106-117
92. UHLEN, M., et al. 1982. EMBO J. 1:249-254
93. ULFENDAHL, P.J. et al. 1987. EMBO J. 6:2037-2044
94. VAN BEVEREN, C.P., et al. 1981. Gene 16: 179-189
95. Van LOON, A. et al. 1987. Gene 58: 109-126
96. Van ORMONDT, H. and GALIBERT,, F. 1984. Curr. Top. Microbiol. Immunol. 110: 73-142
97. VENNSTRÖM, B., et al. 1978. Nucleic Acids Res. 5: 195-204
98. VIRTANEN, A., et al. 1982. Nucleic Acids Res. 10: 2539
99. VIRTANEN, A., et al. 1984. J. Virol. 51: 822-831
100. VIRTANEN, A. and U. PETTERSSON. 1983. J. Mol. Biol. 165: 496-499
101. VIRTANEN, A. and U. PETTERSSON. 1985. J. Virol. 54: 383-391
102. WEBER, J.M. and C.W. ANDERSON. 1988. J. Virol. 62: 1741-1745
103. WOLD, W.S.M. et al. 1984. J. Virol. 52: 307-313
104. VOS, H.L. et al. 1988. Virology 163: 1-10
105. YEH-KAI, L. et al. 1983. J. Mol. Biol. 167: 217-222
106. WEINMANN, R., et al. 1976. Cell 7: 557-566
107. ZAIN, S., et al. 1979. J. Mol. Biol. 135: 413-433
108. ZAIN, B.S. and R.J. ROBERTS. 1979. J. Mol. Biol. 131: 341-352

EPSTEIN-BARR VIRUS (B95-8 STRAIN)

P. J. Farrell
Ludwig Institute for Cancer Research
St. Mary's Branch
St. Mary's Hospital Medical School
Norfolk Place
London W2 1PG, England

Data revised to 1st July 1989

This map is an update of that given in [44]. The original DNA sequence [5] base 359 was deleted so the revised sequence around that position reads TCAGTCTTT [72]. To avoid renumbering the entire sequence, position 1 was moved 1 base to the left of the Eco RI site separating Eco RI Dhet from Eco RI I (ie the first A of AGAATTC). Reading frames are named according to the Bam HI fragment in which they begin, followed by L or R for left or rightward on the map, then F for frame and a number. For example, the third leftward frame starting in Bam HI G would be BGLF3. Reading frames are shown as pointed boxes and are shaded according to their expression class: black for latent cycle, diagonally cross-hatched for early productive cycle and horizontal dashes for late productive cycle. No RNA has been assigned to the unshaded reading frames. mRNAs are shown as horizontal arrowed lines; rightward RNAs are shown above the reading frames and leftward RNAs are below. RNA ends that have been mapped precisely are marked with a vertical dash, other termini shown are only deduced from Northern blotting experiments and inspection of the DNA sequence. RNAs whose structure remains uncertain are indicated by dashed lines. The diagram and feature table now only show TATA boxes and poly A sites that seem to be relevant to EBV gene expression. The table gives detailed information on the features shown in the diagram. Features on the complementary strand (ie. leftward on the map) are indicated by (C) in the table. Sequence relationships between EBV reading frames and those of the herpesviruses HSV[83], CMV[71], VZV[37] and HVS[52] were mostly identified by the above authors or in the original EBV sequence [5] and are not referenced individually in the feature table.

From	To	Description	
58	272		Exon 2 terminal protein RNAs [72,113]
360	458		Exon 3 terminal protein RNAs
540	788		Exon 4 terminal protein RNAs
871	951		Exon 5 terminal protein RNAs
1026	1196		Exon 6 terminal protein RNAs
1280	1495		Exon 7 terminal protein RNAs
1574	1682		Exon 8 terminal protein RNAs
1691	1691	TATA:	TATTAAA BN-R1 late promoter before BNRF1, gives 4.1kb late RNA.

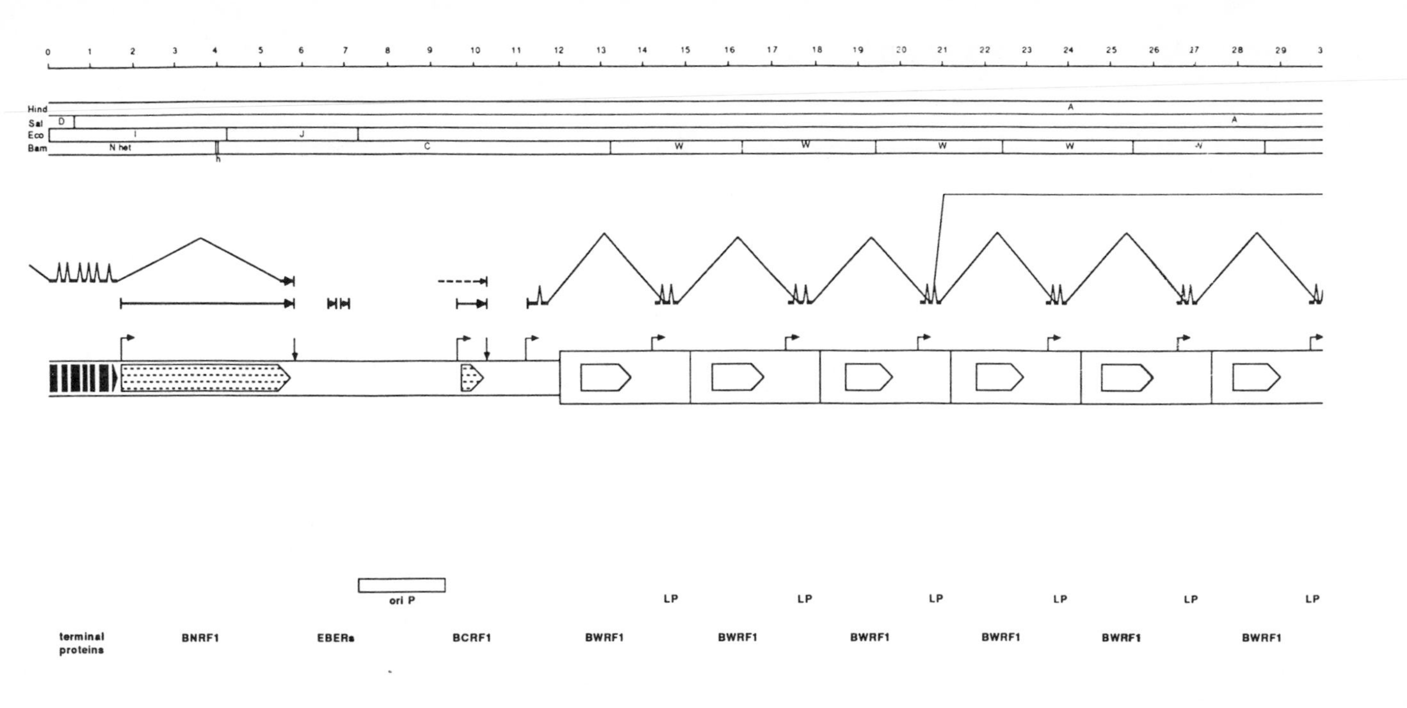
Hind
Sal
Eco
Bam
N het
ori P
LP
terminal proteins
BNRF1
EBERs
BCRF1
BWRF1

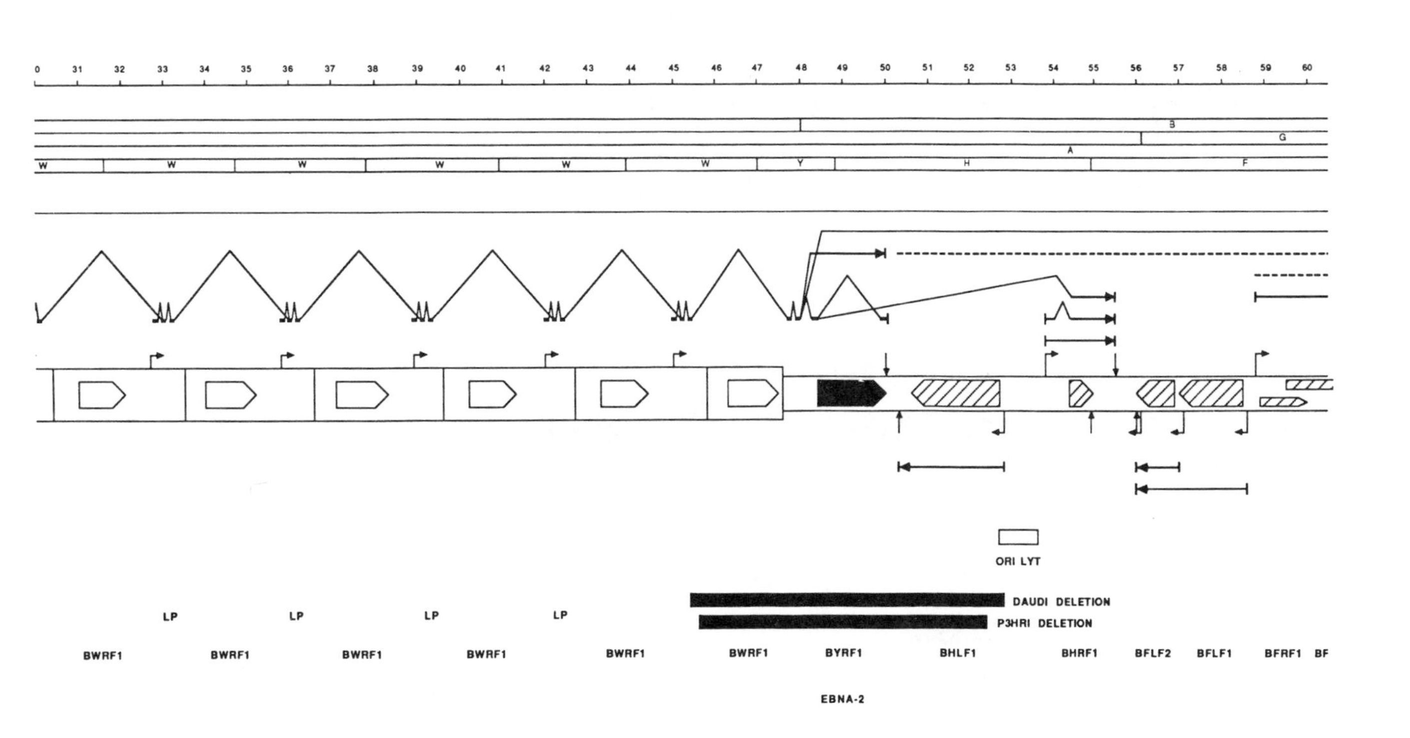
ORI LYT
DAUDI DELETION
P3HRI DELETION
LP
BWRF1
BYRF1
BHLF1
BHRF1
BFLF2
BFLF1
BFRF1
EBNA-2

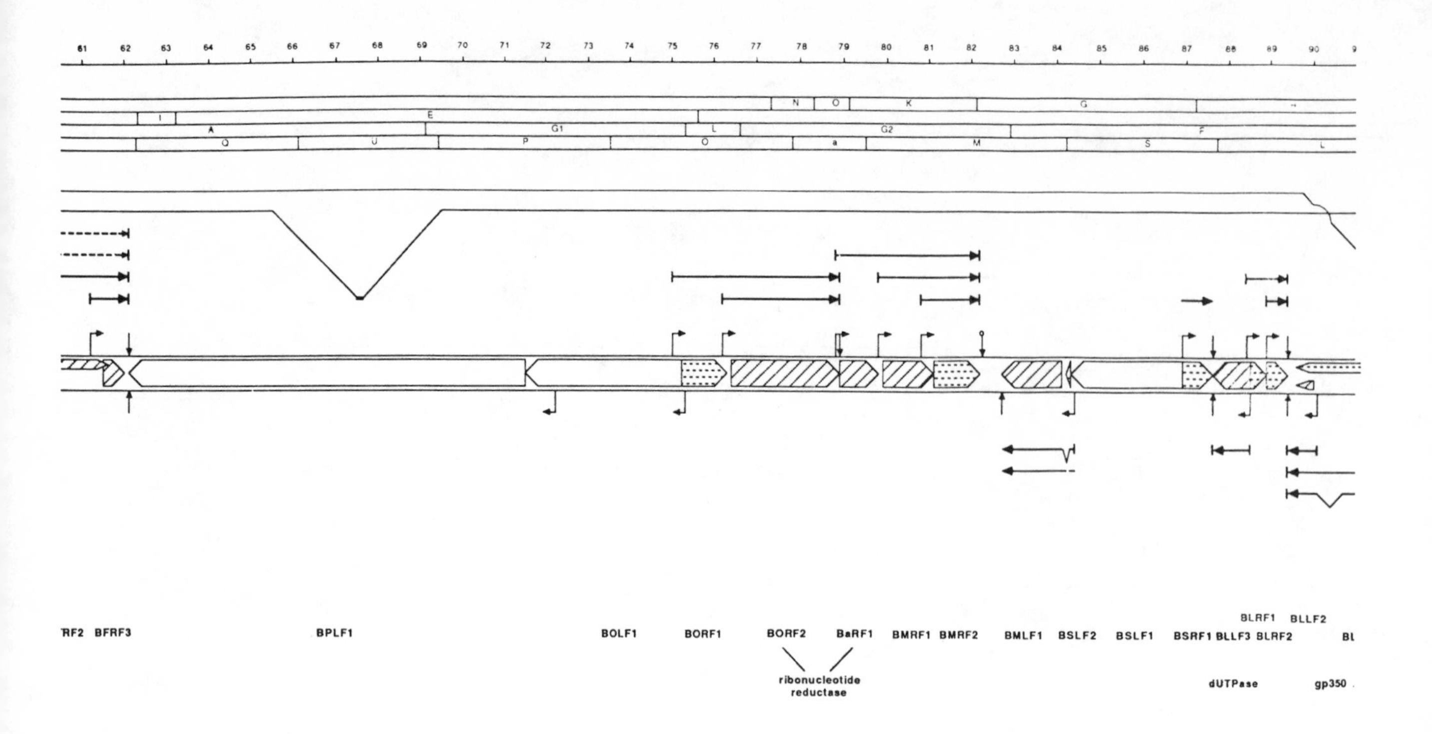

RF2 BFRF3
BPLF1
BOLF1
BORF1
BORF2
BaRF1
BMRF1 BMRF2
BMLF1
BSLF2
BSLF1
BSRF1 BLLF3 BLRF2
BLRF1 BLLF2
ribonucleotide reductase
dUTPase
gp350

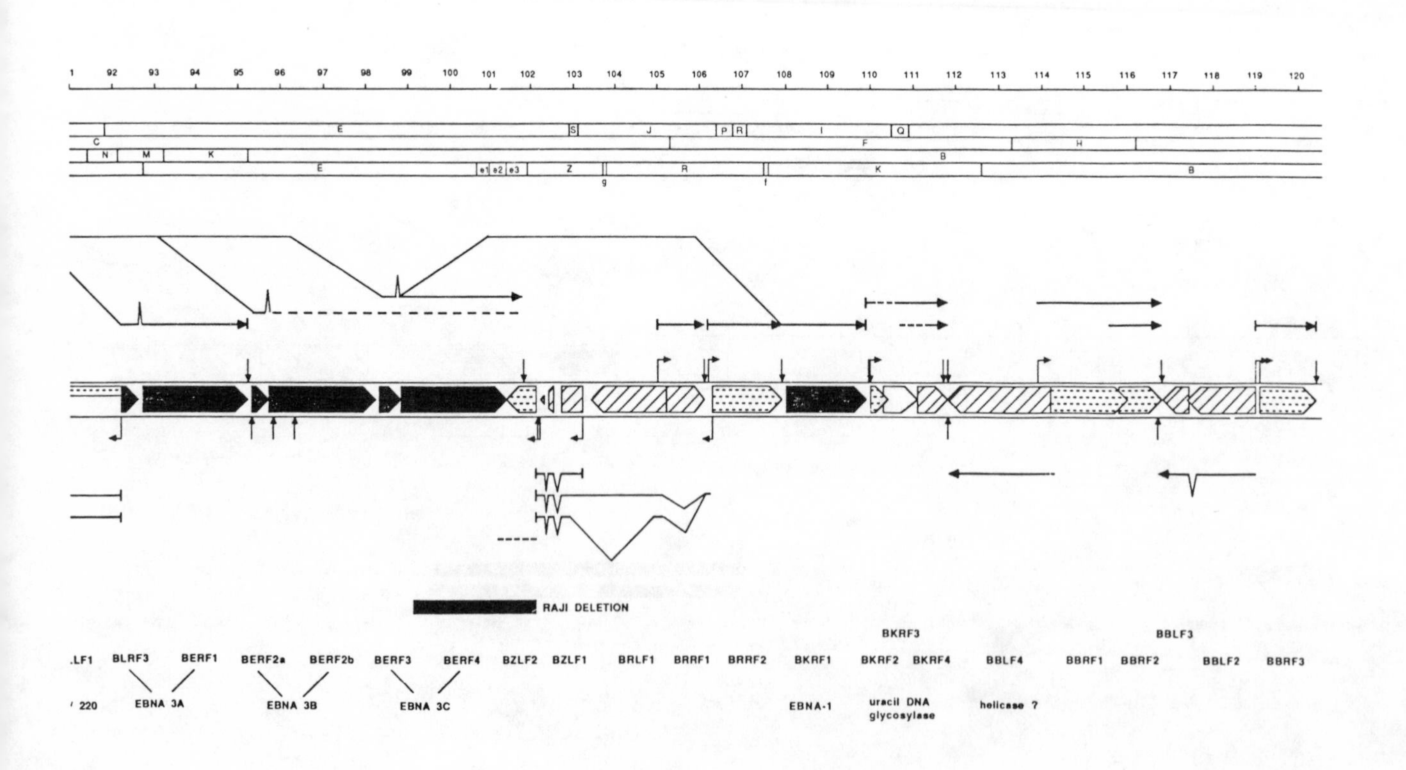

RAJI DELETION
.LF1 BLRF3 BERF1 BERF2a BERF2b BERF3 BERF4 BZLF2 BZLF1 BRLF1 BRRF1 BRRF2 BKRF1 BKRF2 BKRF4 BBLF4 BBRF1 BBRF2 BBLF2 BBRF3
BKRF3
BBLF3
220
EBNA 3A
EBNA 3B
EBNA 3C
EBNA-1
uracil DNA glycosylase
helicase ?

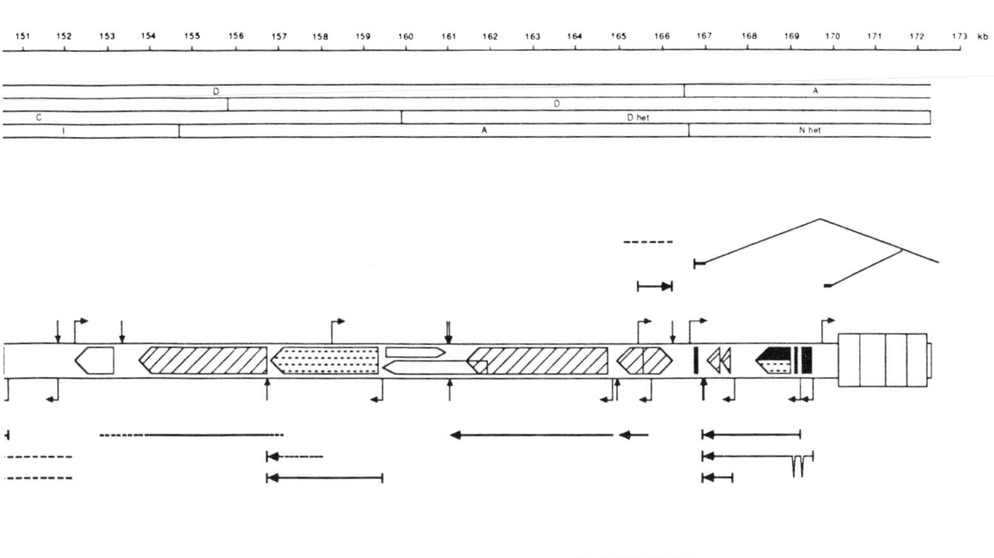

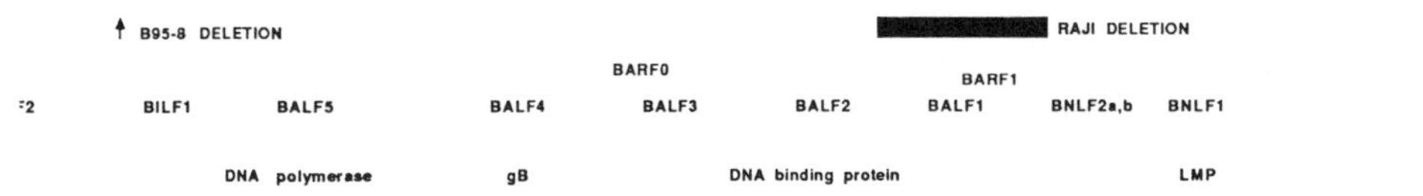

151 152 153 154 155 156 157 158 159 160 161 162 163 164 165 166 167 168 169 170 171 172 173 kb
D
A
D
C
D het
I
A
N het
B95-8 DELETION
RAJI DELETION
BARF0
BARF1
F2
BILF1
BALF5
BALF4
BALF3
BALF2
BALF1
BNLF2a,b
BNLF1
DNA polymerase
gB
DNA binding protein
LMP

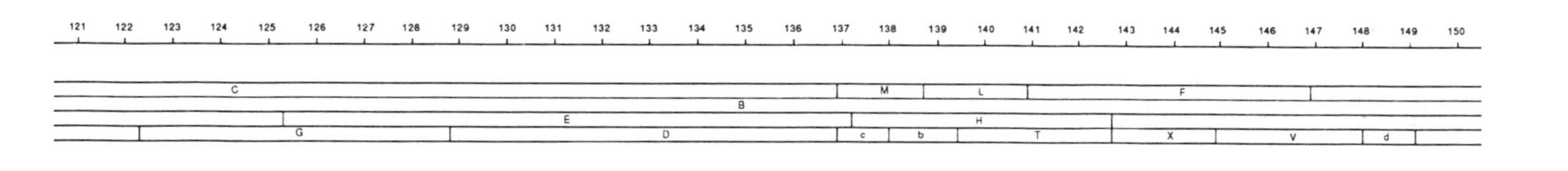
121 122 123 124 125 126 127 128 129 130 131 132 133 134 135 136 137 138 139 140 141 142 143 144 145 146 147 148 149 150
C
M
L
F
B
E
H
G
D
c
b
T
X
V
d

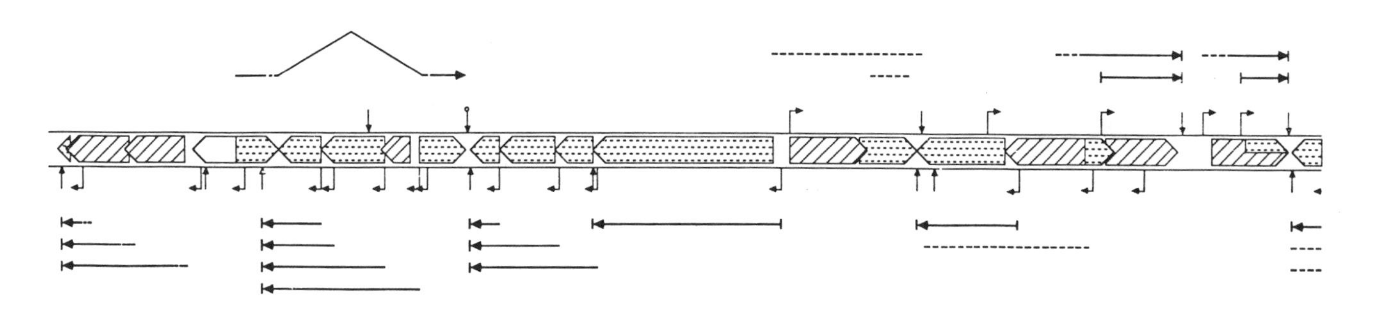

BdRF1
BBLF1 BGLF5 BGLF4 BGLF3 BGRF1 BGLF2 BGLF1 BDLF4 BDRF1 BDLF3 BDLF2 BDLF1
BcLF1
BcRF1
BTRF1
BXLF2
BXLF1 BXRF1 BVRF1
BVRF2
BILI
alkaline exonuclease
protein kinase?
capsid
gH
TK

Start	End		Type	Description
				Probably encodes non glycosylated 140kd protein in membrane antigen [21,63]
1736	5689			BNRF1 reading frame, 5 NXT/S
1795	1795	(C)	POLYA:	AATAAA
3955	3955		BAM:	Bam H1 Nhet/h
3994	3994		BAM:	Bam H1 h/C
5408	5856			Exon 9 terminal protein RNAs
5863	5863			alternative 3' end to TP RNAs
5841	5841		POLYA:	AATAAA, end of 4.1kb late RNA and TP latent RNA.
6629	6795			Pol III RNA EBER 1 [3,13,50]
6956	7128			Pol III RNA EBER 2. EBER 1 and EBER 2 are pol III transcripts but also have upstream control regions that resemble pol II promoter elements [62]
7315	9312			Origin of replication, ori P [25,102,121,139,140]
7421	8042			21x30bp repeats, binding sites for EBNA-1 (site I, [100,138]). Tandem repeat part of oriP. Also functions as a cell type specific enhancer [79,101]
9021	9133		HPN:	Dyad symmetry, site II for EBNA-1 binding [100]. Dyad symmetry part of oriP [102].
9631	9631		TATA:	TATAAAT BC-R1 late promoter before BCRF1
9675	10184			BCRF1 reading frame
10257	10257		POLYA:	AATAAA, end of 0.8kb late RNA from BC-R1 and end of 1.6 kb late RNA, start unknown
11305	11305		TATA:	TACAAAA; BC-R2 promoter for highly spliced EBNA latent RNAs [20]. Regulated by enhancer in oriP [122]
11336	11480			Exon C1 of EBNA RNAs [16,20]
11626	11657			Exon C2 of EBNA RNAs [16,20]
11649	11649		DONOR:	Alternative splice donor at 3' end of C2 exon in an EBNA 3C RNA [115].
12001	15072			3072 repeat 1
12541	13689			BCRF2 (same as BWRF1) repeat reading frame 1
13215	13215		BAM:	BamH1 C/W1
*14352	14352		TATA:	TATAAAG BWR1 one of the promoters for highly spliced EBNA and LP RNAs [104,105,111,120]
*14384	14410			Exon W0 of EBNA/LP RNAs [111,120]
*14554	14619			Exon W1 (also W66) part of leader protein (LP) gene [17,64,111,114,119,120,132,134]. LP is also called EBNA-5 [38] and EBNA-4 [107,119]
*14559	14619			Exon W1' (also W61) of EBNA/LP RNAs forms initiator met [111] when fused to exon W0 or exon C2.
*14701	14832			Exon W2 (also W132) part of LP gene [17,111]

* it is not known to what extent these features may be active in particular copies of the major internal repeat region so they are indicated at these positions as an example. Sequencing of cDNAs shows that many alternatively spliced forms of RNA containing different combinations of W1,W2 etc leaders are produced [18,19]. Only two independent clones of the major internal repeat (Bam HI W) have been sequenced so the possibility of minor sequence variation between repeat copies has not been excluded. 11 full copies of the major internal repeat were put into the B95-8 sequence [5] as a realistic but arbitrary number of repeats. Subsequent more accurate measurement indicates that 8 complete copies is more typical in B95-8 cells [1].

Start	End	Type	Description
15073	18144		3072 repeat 2
15613	16761		BWRF1 reading frame 2 [132]
16287	16287	BAM:	BamH1 W1/W2
17424	17424	TATA:	TATAAAG
17626	17691		Exon W1, LP gene [17,111,119]
17773	17904		Exon W2, LP gene [17,111,119]
18145	21216		3072 repeat 3
18685	19833		BWRF1 reading frame 3
19359	19359	BAM:	BamH1 W2/W3
20496	20496	TATA:	TATAAAG
20698	20763		Exon W1, LP gene [17,111,119]
20845	20976		Exon W2, LP gene [17,111,119]
21217	24288		3072 repeat 4
21757	22905		BWRF1 reading frame 4
22431	22431	BAM:	BamH1 W3/W4
23568	23568	TATA:	TATAAAG
23771	23835		Exon W1, LP gene [17,111,119]
23917	24048		Exon W2, LP gene [17,111,119]
24289	27360		3072 repeat 5
24829	25977		BWRF1 reading frame 5
25503	25503	BAM:	BamH1 W4/W5
26640	26640	TATA:	TATAAAG
26842	26907		Exon W1, LP gene [17,111,119]
26989	27120		Exon W2, LP gene [17,111,119]
27361	30432		3072 repeat 6
27901	29049		BWRF1 reading frame 6
28575	28575	BAM:	BamH1 W5/W6
29712	29712	TATA:	TATAAAG
29914	29979		Exon W1, LP gene [17,111,119]
30061	30192		Exon W2, LP gene [17,111,119]
30433	33504		3072 repeat 7
30973	32121		BWRF1 reading frame 7
31647	31647	BAM:	BamH1 W6/W7
32784	32784	TATA:	TATAAAG
32986	33051		Exon W1, LP gene [17,111,119]
33133	33264		Exon W2, LP gene [17,111,119]
33505	36576		3072 repeat 8
34045	35193		BWRF1 reading frame 8

Start	End	Strand	Type	Description
34719	34719		BAM:	BamH1 W7/W8
35856	35856		TATA:	TATAAAG
36058	36123			Exon W1, LP gene [17,111,119]
36205	36336			Exon W2, LP gene [17,111,119]
36577	39648			3072 repeat 9
37117	38265			BWRF1 reading frame 9
37791	37791		BAM:	BamH1 W8/W9
38928	38928		TATA:	TATAAAG
39130	39195			Exon W1, LP gene [17,111,119]
39277	39408			Exon W2, LP gene [17,111,119]
39649	42720			3072 repeat 10
40189	41337			BWRF1 reading frame 10
40863	40863		BAM:	BamH1 W9/W10
42000	42000		TATA:	TATAAAG
42202	42267			Exon W1, LP gene [17,111,119]
42349	42480			Exon W2, LP gene [17,111,119]
42721	45792			3072 repeat 11
43261	44409			BWRF1 reading frame 11
43935	43935		BAM:	BamH1 W10/W11
45072	45072		TATA:	TATAAAG
45274	45339			Exon W1, LP gene [17,111,119]
45415	52824		DEL:	DAUDI deletion [68]
45421	45552			Exon W2, LP gene [17,111,119]
45644	52450		DEL:	P3HR1 deletion [66]
45793	47643			partial 3072 repeat 12
46333	47481			BWRF1 reading frame 12
47007	47007		BAM:	BamH1 W11/Y
47761	47793			Exon Y1 [17]
47878	47999			Exon Y2 [17] and EBNA-1 RNA [119]
48386	48444			Exon Y3 [17]
48386	50032			Coding exon for EBNA-2 [111]
48429	49964			BYRF1, encodes EBNA-2 [34,35,39,40,41,59,87,108,109,134,143] EBNA-2 required for immortalization by EBV
48678	48800			14 x "CCCCCACCA" repeats
48848	48848		BAM:	BamH1 Y/H
49525	49578			9 x "GGGGCA" repeats
49852	50032			Exon H1 [17]
50003	50003		POLYA:	AATAAA, end of T1 cDNA [17] and EBNA-2 RNA
50317	50317	(C)	POLYA:	AATAAA, end of 2.5kb early RNA containing BHLF1
50578	52115			12 x "125bp" repeats
52557	50578	(C)		BHLF1 early reading frame
52589	53581		ORI:	Ori Lyt, lytic origin of replication [54]
52589	52944		ORI:	Part of Ori Lyt, upstream element [54]
52654	53697			Region homologous to Eco RI C of Raji. Duplicated left and right regions contain DL and DR promoters [22,23] and ori lyt
52817	52817	(C)	TATA:	GATAAAA promoter for 2.5kb early RNA containing BHLF1 [47,66].
53207	53581		ORI:	Part of Ori Lyt, enhancer element, palindrome
53759	53759		TATA:	TATTAAC likely promoter for class III and IV early RNAs encoding BHRF1 [92]
53895	**53895**		**DONOR:**	**CGGGTAACT donor for splice to 54335 in class IV early RNAs encoding BHRF1 [92]**
54335	54335		ACCEPT:	TTTTCTAG acceptor from 48444 in class I, 47999 in class II, and 53895 in class IV early RNAs encoding BHRF1 [4,92,98]
54376	54948			BHRF1 reading frame, limited homology to bcl-2 gene [28]. Early gene in B95-8 cells and part of restricted EA complex.
54853	54853		BAM:	BamH1 H/F
55518	55518		POLYA:	AATAAA, 3' end of 2.5kb, 1.9kb, 1.7kb and 0.6kb early RNAs
55990	55990	(C)	POLYA:	AATAAA, 3' end of 2.3kb and 1.1kb early RNAs from 58568 and 57081
56935	55982	(C)		BFLF2 reading frame, 4 NXT/S, homologous to RF 27 in VZV and HFRF2 in CMV
57081	57081	(C)	TATA:	TATTTAA before BFLF2; BFL2 promoter gives 1.1kb early RNA
58525	56951	(C)		BFLF1 reading frame, 2 NXT/S homologous to RF 26 in VZV and HFRF1 in CMV
58568	58568	(C)	TATA:	TATTAAA before BFLF1, BFL1 promoter gives 2.3kb early RNA
58832	58832		TATA:	TATAAAA before BFRF1
58891	59898			BFRF1 early reading frame, 1 NXT/S, homologous to HFLF4 in CMV
59610	61580			BFRF2 early reading frame, homologous to HFLF5 in CMV
61344	61344		TATA:	TATTTAA before BFRF3
61456	62034			BFRF3 early reading frame
62068	62068	(C)	POLYA:	AATAAA
62069	62069		POLYA:	AATAAA, 3' end of 10, 6.5, 3.7, 3.4, 3.1, 2.5 and 0.8kb early RNAs
62249	62249		BAM:	BamH1 F/Q
62430	62477			Site III for EBNA-1 binding [100]
66121	66121		BAM:	BamH1 Q/U
67477	67649			Exon in EBNA-1 RNA [119] and cDNA clone T4 [18]
69410	69410		BAM:	BamH1 U/P
69684	69930			5 x 51bp repeats
70387	70521			9 x 15bp repeat
71527	62081	(C)		BPLF1 reading frame, 1 NXT/S, analogous to VZV RF22
72192	72192	(C)	TATA:	TATTAAA before BPLF1
73468	73468		BAM:	BamH1 P/O
75017	75017		TATA:	TATTTAA BO-R1 late promoter before BORF1, gives 3.9kb late RNA
75238	76329			BORF1 late reading frame, 2 NXT/S homologous to

Start	End	Strand	Feature	Description
				VZV RF20
75239	71523	(C)		BOLF1 reading frame, 1 NXT/S analogous to VZV RF 21
75322	75322	(C)	TATA:	TATTTAG before BOLF1
76169	76169		TATA:	TACATAT BO-R2 early promoter before BORF2, gives 2.8kb RNA
76407	78884			BORF2 early reading frame, 2 NXT/S. Homology HSV 140K ribonucleotide reductase [48,49,51] and RF 19 VZV
77835	77835		BAM:	Bam H1 O/a
78804	78804		TATA:	TATAAGT Ba-R1 early promoter before BaRF1, gives 3.5kb RNA
78883	78883		POLYA:	AATAAA, end of 3.9kb late RNA from 75017 and 2.8kb early RNA from 76169
78900	79805			BaRF1 early reading frame, homologous to HSV 38K ribonucleotide reductase [48,49,51] and RF 18 VZV
79537	79537		BAM:	Bam H1 a/M
79840	79840		TATA:	CATAAAT BM-R1 early promoter before BMRF1, gives 2.5kb RNA
79899	81110			BMRF1 early reading frame. Early antigen protein recognised by R3 monoclonal [26,42,90,91].
80779	80779		TATA:	TATTTAA BM-R2 late promoter before BMRF2
80832	80832		TATA:	GATAAAA, possible promoter for 1.4kb late RNA encoding BMRF2
81118	82188			BMRF2 early reading frame
82180	82180		POLYA:	ATTAAA, end of 3.5kb early RNA from 78804, 2.5kb early RNA from 79840 and 1.4kb late RNA [97].
82319	82461			2x71bp repeats
82747	82747	(C)	POLYA:	AATAAA
83640	83729			10x9bp repeats
84122	84122	(C)	ACCEPT:	CTCCCCTCTGCAG acceptor in spliced form of BMLF1 RNA
84122	82746	(C)		BMLF1 early reading frame. Diffuse early antigen [27,136,137]. Also homologous to RF 4 VZV and IE63 of HSV.(BSLF2 + BMLF1) is also called EB2 [24]. Post-transcriptional activator of gene expression [24,74,90]
84227	84227	(C)	DONOR:	CAGGTAAGA donor in spliced form of BMLF1 RNA [112]
84233	84233		BAM:	Bam H1 M/S
84288	84229	(C)		BSLF2 early reading frame in 5' exon of spliced RNA encoding BMLF1 [112]
84356	84356	(C)	TATA:	CATAAAT before BSLF2 and BMLF1. Two RNAs start here; one is spliced and the other is unspliced, both traverse BMLF1.
86881	84260	(C)		BSLF1 reading frame, homologous to RF 6 VZV
86882	86882		TATA:	TATTTAA BS-R1 late promoter before BSRF1
86924	87577			BSRF1 reading frame
87599	87599		POLYA:	AATAAA
87613	87613	(C)	POLYA:	AATAAA, end 1.0kb early RNA from BL-L3
87650	87650		BAM:	Bam H1 S/L
88474	87641	(C)		BLLF3 early reading frame (BLLF2 in [5]). Homologous to RF 8 VZV and dUTPase HSV.
88507	88507		TATA:	TATATAT BL-R1 late promoter before BLRF1, gives 1.0kb late RNA
88514	88514	(C)	TATA:	TATATAT BL-L3 early promoter before BLLF3, gives 1.0kb early RNA
88547	88852			BLRF1 late reading frame
88863	88863		TATA:	TATTTAA BL-R2 late promoter before BLRF2, gives 0.6kb late RNA
88925	89410			BLRF2 late reading frame, 2 NXS/T
89412	89412		POLYA:	AATAAA, end of 1.0kb and 0.6kb late RNAs
89425	89425	(C)	POLYA:	AATAAA, end of 0.7kb early, 2.2kb late and 2.8kb late RNAs
90013	89569	(C)		BLLF2 early reading frame (BLLF3 in [5])
90051	90051	(C)	TATA:	TATAACA BL-L2 early promoter before BLLF2, gives 0.7kb early RNA
90177	90639			21 copies of 21bp approximate repeat
90652	90062	(C)		intervening sequence in gp220 gene
92153	89433	(C)		BLLF1b, late reading frame gp220 membrane antigen, spliced form of BLLF1a [12,15,65]
92153	89433	(C)		BLLF1a, late reading frame, gp350 membrane antigen, 36 NXT/S [8,12,15,36,65,127,130,135]
92192	92192	(C)	TATA:	TATTAAA BL-L1 late promoter before BLLF1a,b. Gives 2.8 and 2.2kb late RNAs
92238	92581			Exon in EBNA 3A RNA [18]
92243	92581			BLRF3 reading frame
92670	95248			Exon in EBNA 3A RNA [18]
92670	95162			BERF1 frame, homology with BERF2b and BERF4. A fusion of BLRF3 with BERF1 encodes EBNA-3A, also known as EBNA 3, latent cycle gene [16,60,67,69,103,110,115,117].
92703	927C3		BAM:	Bam H1 L/E
94208	94277			repeat type A
94281	94306			repeat type B
94307	94381			repeat type C
94386	94411			repeat type B
94412	94489			repeat type C
94490	94560			repeat type A
94571	94648			repeat type C
94649	94719			repeat type A
94896	94982			repeat type D
94983	95069			repeat type D
95221	95221		POLYA:	AATAAA
95272	95272	(C)	POLYA:	AATAAA
95353	95721			BERF2a reading frame
95725	98244			BERF2b frame, homology with BERF1 and BERF4.

Start	End	Strand	Type	Description
				BERF2a and BERF2b are spliced together to make EBNA3B latent protein [95], also known as EBNA 4 [2,103]
95819	95819	(C)	POLYA:	AATAAA
96276	96276	(C)	POLYA:	AATAAA
97522	97698			3x60bp repeat
98323	98766			BERF3 reading frame
98364	98730			Exon in EBNA-1 RNA [119]
98731	98731		DONOR:	AAGGTGAGT donor
98805	99050			Exon in T4 cDNA [18] 99050 is not the end of the RNA.
98805	101420			BERF4 frame, homology with BERF1 and BERF2b. BERF3 and BERF4 are spliced together to make the EBNA3C [2,96,115] latent protein, also known as EBNA 6 [2,103]
99126	102118		DEL:	Deletion in Raji [56]
100122	100304			10 x "15bp" repeat
100613	100613		BAM:	Bam H1 E/e1
100665	100781			3x39bp repeat
100919	100919		BAM:	Bam H1 e1/e2
101426	101426		BAM:	Bam H1 e2/e3
101947	101947		BAM:	Bam H1 e3/Z
102116	101448	(C)		BZLF2 reading frame 3x NXT/S. 2.5kb late RNA traverses BZLF2, ends unknown.
102153	102153		TATA:	TATTAAT
102156	102156	(C)	POLYA:	AATAAA 3' end of 0.9kb and 2.8kb RNAs encoding BZLF1 and BRLF1
102160	102160	(C)	TATA:	TATTAAT
102341	102126	(C)		3' terminal exon of 0.9kb and 2.8kb early RNAs
102380	102380	(C)	TATA:	CATAAAT
102420	102420	(C)	TATA:	TATATAC
102504	102504	(C)	POLYA:	AATAAA, apparently not functional
102530	102425	(C)		Exon of 0.9kb and 2.8kb early RNAs
102581	102652			semi-repetitive sequence, homologous to human c-fos 3' sequence
102918	102918	(c)		splice acceptor used in RZ fusion gene [80]
103155	102213	(C)		BZLF1 reading frame, modified from [5]. Has two splices within frame. 2xNXT/S. Immediate early gene which disrupts latency [14,30,31,53,70,80,106,116,124,126,128,131], also called EB1 [24] and ZEBRA [31]. Sequence specific DNA binding protein [45] and transcription factor.
103194	102655	(C)		First exon of 0.9kb early RNA encoding BZLF1
103231	103231	(C)	TATA:	TTTAAA of BZL1 immediate early promoter gives 0.9kb RNA
103311	103256	(C)		Upstream of BZL1, homology to 106243 to 106188
103462	103453	(C)		TAATGAAATC sequence
103741	103741		BAM:	Bam H1 Z/g
103816	103816		BAM:	Bam H1 g/R
104989	104927	(C)		BRLF2 poss. small 5' exon
105016	105016		TATA:	TATAAAT before BRRF1, possible promoter for 1.1kb early RNA encoding BRRF1
105182	106111			BRRF1 early reading frame
105183	103369	(C)		BRLF1 reading frame, (immediate?) early gene, acts as transcription activator.
105185	104926	(C)		exon in RZ fusion RNA [80]
105185	105185	(C)	ACCEPT:	splice acceptor in 2.8kb early RNA encoding BRLF1 and RZ fusion RNA [14,80]
106110	106110		POLYA:	AATAAA, 3' end of early 1.1kb RNA encoding BRRF1
106181	106126	(C)		5' leader exon for 2.8kb early RNA encoding BRLF1 and RZ fusion RNA [14,80]
106213	106213	(C)	TATA:	CATAAAA
106243	106243		TATA:	TATAAAA before BRRF2, possible promoter for 1.8kb RNA encoding BRRF2
106243	106188	(C)		Homology to upstream region of BZL1
106302	107912			BRRF2 reading frame
106385	106385	(C)	TATA:	GATAAAA
107457	107457		BAM:	Bam H1 R/f
107565	107565		BAM:	Bam H1 f/K
107914	107914		POLYA:	AATAAA, 3' end of 1.8kb RNA encoding BRRF2
107942	107942		ACCEPT:	splice acceptor for EBNA-1 RNA [119]
107950	109872			BKRF1 encodes EBNA-1 protein, latent cycle gene. [46,59,123,125]
108217	108924			EBNA triplet repeat GGA,GCA,GGG.
109809	109866			6.5 x 9bp repeats, encodes gly every 4th residue
109905	109905		TATA:	TATTAAA before BKRF2, possible start for 2.3kb late RNA
109937	109937		POLYA:	AATAAA 3' end of EBNA-1 RNA
109958	110368			BKRF2 reading frame
110271	110271		DONOR:	TCCGTGAGT possible donor at end of BKRF2
110275	111117			BKRF3 reading frame, homologous to RF 59 VZV homologous to uracil DNA glycosylase [88]
111098	111098		DONOR:	TCGGTGAGA possible donor at end BKRF3
111107	111784			BKRF4 reading frame, contains complex repetitive sequence
111272	111272		DONOR:	GACGTGAGT poss.donor before rpt.seq. in BKRF4
111719	111719		POLYA:	AATAAA : currently unknown which
111787	111787		POLYA:	AATAAA : is 3' end of the 2.3kb late and 1.1kb early RNAs
111830	111830	(C)	POLYA:	AATAAA
112620	112620		BAM:	Bam H1 K/B
113876	113876		TATA:	TATTTAT before BBRF1
114204	116042			BBRF1 late reading frame, homologous to RF 54 VZV
114259	111833	(C)		BBLF4 early reading frame, very good homology to RF55 VZV, likely helicase [33]
115843	116781			BBRF2 late reading frame, homologous to RF 53 VZV
116696	116696	(C)	POLYA:	AATAAA

116785	116785		POLYA:	AATAAA
117386	116784	(C)		BBLF3 early reading frame, spliced to BBLF2. BBLF3 contains a consensus nucleotide binding site
117515	117386	(C)		intron spliced out in RNA linking BBLF2 and BBLF3
118981	118981		TATA:	TATAAAA BBR1 late promoter before BBRF3
119080	117515	(C)		BBLF2 early reading frame, spliced to BBLF3
119098	119098		TATA:	TATTTAA BBR3 late promoter before BBRF3
119137	120351			BBRF3 late reading frame
120358	120358		POLYA:	AATAAA
120764	120764	(C)	POLYA:	AATAAA, 3' end of 0.6kb late, 1.6kb early, 3.0kb early RNAs
120974	120750	(C)		BBLF1 late reading frame, possibly homologous to RF 49 VZV
121331	121331	(C)	TATA:	TATTAAA BBL1 late promoter before BBLF1
122313	122313		BAM:	Bam H1 B/G
122341	120932	(C)		BGLF5 early reading frame, homologous to RF 48 VZV and alkaline exonuclease of HSV [85,142]
123506	123506	(C)	DONOR:	AAGGTGACT possible donor
123692	122328	(C)		BGLF4 early reading frame, homologous to RF 47 VZV and HSV UL13. Possible protein kinase [118]
124117	124117	(C)	TATA:	TATAAAA
124219	124219	(C)	POLYA:	AATAAA
124938	125912			BGRF1 reading frame, homologous to RF 45 VZV and spliced HSV gene [29]. Spliced to BDRF1. Northern blots in BGRF1 detect 2.7, 2.6, 2.1kb late and 1.9kb early RNAs. 2.6, 2.1kb RNAs very weak.
124939	123944	(C)		BGLF3 reading frame
125113	125113	(C)	TATA:	TATAAAT before BGLF3
125484	125484	(C)	POLYA:	AATAAA, 3' end of 1.6kb late, 1.8kb late, 3.0kb late and 3.7kb early RNAs
125873	125873			Predicted donor in RNA coding BGRF1 and BDRF1
126873	125866	(C)		BGLF2 late reading frame, poor homology to RF44 VZV
126929	126929	(C)	TATA:	TATTAAA late promoter before BGLF2, gives 1.6kb late RNA
127237	127237	(C)	TATA:	TATAAAA, potential promoter for 1.8kb late RNA
128029	128029		POLYA:	AATAAA
128374	126854	(C)		BGLF1 late reading frame
128432	128432	(C)	TATA:	TATTTAA before BGLF1, potential promoter for 3.0kb late RNA
128848	128848		BAM:	Bam H1 G/D
129021	128347	(C)		BDLF4 early reading frame
129054	129054	(C)	TATA:	TATTTGC before BDLF4, potential promoter for 3.7kb early RNA
129188	130348			BDRF1 reading frame, homologous to RF 42 VZV and spliced gene in HSV [29]. Spliced from BGRF1. Northern blots in BDRF1 detect 2.7,2.6 kb late and 1.9kb early RNAs. Possibly 1.8kb early RNA.
129214	129214			Predicted acceptor in RNA encoding BGRF1 and BDRF1
129377	129377	(C)	TATA:	TATAAAG
130347	130347		POLYA:	ATTAAA
130359	130359	(C)	POLYA:	AATAAA, 3' end of 0.9kb late RNA, 2.3kb late RNA and 3.2kb late RNA
131066	130365	(C)		BDLF3 late reading frame 9xNXT/S, possible relation to HSV1 gC [8]
131104	131104	(C)	TATA:	TATAAAA late promoter before BDLF3, gives 0.9kb late RNA
132389	131130	(C)		BDLF2 late reading frame
132476	132476	(C)	TATA:	TATTTAA before BDLF2, likely promoter for 2.3kb late RNA
133305	132403	(C)		BDLF1 late reading frame, poor homology to RF41 VZV
133312	133312	(C)	POLYA:	AATAAA, 3' end of 4.5kb late RNA
133332	133332	(C)	DONOR:	AAGGTGGTT possible donor
133352	133352	(C)	TATA:	TATTAAA before BDLF1
133386	133386	(C)	TATA:	TATATAA
136868	136868		BAM:	Bam H1 D/c
137466	133324	(C)		BcLF1 late reading frame, homologous to RF 40 VZV and major capsid protein of HSV [37,83,86]
137710	137710	(C)	TATA:	TATTAAA EHL1 promoter before BcLF1, gives 4.5kb late RNA
137857	137857		TATA:	CATAAAC
137862	139715			BcRF1 reading frame
138019	138019		BAM:	Bam H1 c/b
139352	139352		BAM:	Bam H1 b/T
139642	140916			BTRF1 reading frame. Northern blots detect 0.95kb late and 3.8kb early RNA
140902	140902	(C)	POLYA:	AATAAA, 3' end of 2.5kb late RNA
140970	140970		POLYA:	AATAAA
141286	141286	(C)	POLYA:	AATAAA
142589	142589		TATA:	GATAAAA
142740	142740		BAM:	Bam H1 T/X
143036	140919	(C)		BXLF2 late reading frame, encodes gp85. Homologous to RF 37 VZV and glycoprotein H of HSV (gpIII of VZV) [32,57,84,89]
143310	143310	(C)	TATA:	TATAAGA late promoter before BXLF2, gives 2.5kb late RNA
144791	144791		ACCEPT:	TCTTTCGTTTTCAGG poss. acceptor before BXRF1
144860	145603			BXRF1 late reading frame, homologous to RF 35 VZV. Basic (core?) protein.
144861	143041	(C)		BXLF1 early reading frame, thymidine kinase [77,78]. Weak homology to RF 36 VZV and HSV thymidine kinase. 4.0kb early RNA presumably encodes the TK. Also a 2.2kb late RNA here.
144862	144862		BAM:	Bam H1 X/V

145135	145135	(C)	TATA:	TATAACA before BXLF1
145302	145302		TATA:	TATTTAA before BVRF1, potential promoter for 1.9kb early RNA
145416	147125			BVRF1 early reading frame, homologous to RF 34 VZV
146273	146273	(C)	TATA:	CATAAAA
147170	147170		POLYA:	AATAAA, 3' end of 2.4kb late and 1.9kb early RNAs
147721	147721		TATA:	TATTTAT before BVRF2, potential promoter for 2.1kb early RNA
147927	149741			BVRF2 early reading frame, N-terminus homologous to RF 33 VZV
148007	148007		BAM:	Bam H1 V/d
148620	148620		TATA:	TATTTAA late promoter before BdRF1, gives 1.2kb late RNA
148707	149741			BdRF1 reading frame; the C terminus of BVRF2
149115	149115		BAM:	Bam H1 d/I
149727	149727		POLYA:	AATAAA, 3' end of 2.1kb early and 1.2kb late RNAs
149758	149758	(C)	POLYA:	AATAAA, 3' end of 1.0kb late, 1.5kb late and 1.8kb late RNAs
150525	149782	(C)		BILF2 late reading frame 11xNXT/S
150571	150571	(C)	TATA:	TATTTAG before BILF2. Potential promoter for 1.0kb late RNA.
151236	151618			repetitive sequence 3X
151767	151767		POLYA:	AATAAA
151780	151780	(C)	TATA:	CATAAAA
152012	152013		DEL:	B95-8 deletion with respect to other EBV strains, 12 kb deleted relative to Raji. Deletion contains DR region, second ori lyt.
152230	152230		TATA:	CATAAAA
153099	152164	(C)		BILF1 reading frame, membrane protein?, 3xNXS/T [11]
153259	153259		POLYA:	AATAAA
153637	153637		HPN:	22bp 2-fold symmetry
154747	154747		BAM:	Bam H1 I/A
156707	156707	(C)	POLYA:	AATAAA; 3' end of 2.5kb late (gB) RNA and 1.8kb late RNA
156746	153702	(C)		BALF5 DNA polymerase [5], homologous to many DNA polymerases, CMV HFLF2 and RF 28 VZV. 4.5kb early RNA encodes BALF5, RNA ends unknown [73].
158204	158204		TATA:	TATAAAA
159322	156752	(C)		BALF4 late reading frame 9xNXT/S homologous to HSV1 glycoprotein B, CMV HFLF1 and RF 31 VZV (gpII) [8,9,43,93,94]
159370	159370	(C)	TATA:	TATTTAA late promoter before BALF4, gives 2.5kb late RNA
159579	160991			BARF0 reading frame
160966	160966		POLYA:	AATAAA
160990	160990		POLYA:	ATTAAA
161013	161013	(C)	POLYA:	AATAAA, presumed end of 3.9kb early RNA
161678	159312	(C)		BALF3 reading frame
163978	166635		DEL:	deletion in Raji [56]
164770	161387	(C)		BALF2 early reading frame, homologous to RF 29 VZV and major DNA binding protein HSV [141]. 3.9kb RNA
164814	164814	(C)	TATA:	CATTTAA before BALF2, presumed promoter for 3.9kb early RNA
164851	164851	(C)	POLYA:	AATAAA
165442	165442		TATA:	GATAAAA
165466	165466		TATA:	TATAAGA early promoter before BARF1, gives 0.8kb early RNA
165504	166166			BARF1 reading frame
165517	164858	(C)		BALF1 early reading frame, 0.7kb early RNA
165713	165713	(C)	TATA:	TATAAAG before BALF1
166165	166165		POLYA:	AATAAA 3' end of 0.8kb early RNA. Also 1kb late RNA in this region.
166469	166475		TATA:	TTATTTT promoter for terminal protein 1 (TP1)
166498	166916			Exon 1 of TP1 RNA [72,113]. TP1 also called IMP-2A [113]
166561	166563			Likely initiator met of terminal protein 1
166614	166614		BAM:	Bam H1 A/Nhet
166946	166946	(C)	POLYA:	AATAAA) 3' end of 0.8kb early, 2.5kb
166950	166950	(C)	POLYA:	AATAAA) late and 2.5kb latent RNAs
167303	167001	(C)		BNLF2b reading frame
167320	167320		TATA:	CATAAAA
167486	167307	(C)		BNLF2a reading frame
167525	167525	(C)	TATA:	TATAAAA early promoter before BNLF2a,b. Gives 0.8kb RNA
168399	168574			5 x "33bp" repeats
168965	168163	(C)		BNLF1 coding part of exon c of latent membrane protein
169128	169042	(C)		BNLF1 exon b of latent membrane protein mRNA
169201	169201	(C)	TATA:	TATTACA ED-L1A late promoter, gives 2.5kb late RNA N terminal half of LMP is sufficient for transforming function in 3T3 cells and is toxic at high levels [5,6,55,56,58,75,76,81,108,109,129,133]
169474	169207	(C)		BNLF1 exon a of latent membrane protein mRNA
169546	169546	(C)	TATA:	TACATAAGC EDL1 promoter before BNLF1 gives 2.5kb latent RNA (LMP)
169740	169906			5' exon of TP2 mRNA (exon 1'). Mapping of 5' end to 169780 is uncertain [113]. TP2 also called LMP-2B [113]
170094	170631			terminal repeat 1 538bp [10,56,99]
170632	171154			terminal repeat 2 523bp
171155	171692			terminal repeat 3 538bp
171693	172231			terminal repeat 4 538bp

REFERENCES

1. Allan GJ and Rowe DT (1989) Virology : in press.
2. Allday MJ, Crawford DH and Griffin BE (1988) Nucleic Acids Res. 16: 4353-4369.
3. Arrand JR, Young LS and Tugwood JD (1989) J. Virol. 63: 983-986.
4. Austin PJ, Flemington E, Yandava CN, Strominger JL and Speck SH (1988) Proc. Natl. Acad. Sci USA 85: 3678-3683.
5. Baer R, Bankier AT, Biggin MD, Deininger PL, Farrell PJ, Gibson TJ, Hatfull G, Hudson GS, Satchwell SC, Seguin C, Tuffnell PS and Barrell BG (1984) Nature 310: 207-211.
6. Baichwal VR and Sugden B (1987) J. Virol. 61: 866-75.
7. Baichwal VR and Sugden B (1988) Oncogene. 2: 461-469.
8. Balachandran N, Oba DE and Hutt-Fletcher LM (1987) J. Virol. 61: 1125-35.
9. Balachandran N, Pittari J and Hutt-Fletcher LM (1986) J. Virol. 60: 369-75.
10. Bankier AT, Deininger PL, Satchwell SC, Baer R, Farrell PJ and Barrell BG (1983) Mol. Biol. Med. 1: 425-445.
11. Becker Y, Tabor E and Asher Y (1988) Leukemia. 2: 178-192.
12. Beisel C, Tanner J, Matsuo T, Thorley-Lawson D, Kedsay F and Keiff E (1985) J. Virol. 54: 665-674.
13. Bhat RA and Thimmappaya B (1985) J. Virol 56: 750-756.
14. Biggin M, Bodescot M, Perricaudet M and Farrell P (1987) J. Virol. 61: 3120-3132.
15. Biggin M, Farrell PJ and Barrell BG (1984) EMBO J. 3: 1083-1090.
16. Bodescot M, Brison O and Perricaudet M (1986) Nucleic Acids Res. 14: 2611-2620.
17. Bodescot M, Chambraud B, Farrell P and Perricaudet M (1984) EMBO J. 3: 1913-1917.
18. Bodescot M and Perricaudet M (1986) Nucleic Acids Res. 17: 7103-14.
19. Bodescot M and Perricaudet M (1987) Nucleic Acids Res. 15: 5887-8.
20. Bodescot M, Perricaudet M and Farrell PJ (1987) J. Virol. 61: 3424-3430.
21. Cameron KR, Stamminger T, Craxton M, Bodemer W, Honess RW and Fleckenstein B (1987) J. Virol. 61: 2063-70.
22. Chavrier P, Gruffat H, Chevallier-Graco A, Buisson M and Sergeant A (1989) J. Virol. 63: 607-614.
23. Chevallier-Greco A, Gruffat H, Manet E, Calender A and Sergeant A (1989) J. Virol. 63: 615-623.
24. Chevallier-Greco A, Manet E, Chavrier P, Mosnier C, Daillie J and Sergeant A (1986) EMBO J. 5: 3243-3250.
25. Chittenden T, Lupton S and Levine AJ (1989) J. Virol. 63: 3016-3025.
26. Cho M-S, Jeang KT and Hayward SD (1985a) J. Virol. 56: 852-859.
27. Cho M-S, Milman G and Hayward SD (1985b) J. Virol. 56: 860-866.
28. Cleary ML, Smith SD and Sklar J (1986) Cell 47: 19-28.
29. Costa RH, Draper KG, Kelly TJ and Wagner EK (1985) J. Virol. 54: 317-328.
30. Countryman J, Jenson H, Seibl R, Wolf H and Miller G (1987) J. Virol. 61: 3672-3679.
31. Countryman J and Miller G (1985) Proc. Natl. Acad. Sci. USA 82: 4085-4089.
32. Cranage MP, Smith GL, Bell SE, Hart H, Brown C, Bankier AT, Tomlinson P, Barrell BG and Minson TC (1988) J. Virol. 62: 1416-1422.
33. Crute JJ, Mocarski ES and Lehman IR (1988) Nucleic Acids Res 16: 6585-6596.
34. Dambaugh T, Hennessy K, Chamnaukit L and Keiff E (1984) Proc. Natl. Acad. Sci. USA 81: 7632-7636.
35. Dambaugh T, Wang F, Hennessy K, Rickinson A and Kieff E (1986) J. Virol. 59: 453-62.
36. David EM and Morgan AJ (1988) J. Immunol. Meth. 108: 231-237.
37. Davison AJ and Taylor P (1987) J. Gen. Virol. 68: 1067-81.
38. Dillner J, Kallin B, Alexander H, Ernberg I, Uno M, Ono Y, Klein G and Lerner RA (1986) Proc. Natl. Acad. Sci. USA 83: 6641-6645.
39. Dillner J, Kallin B, Klein G, Jornvall H, Alexander H and Lerner R (1985) EMBO J. 4: 1813-1818.
40. Dillner J, Sternas L, Kallin B, Alexander H, Ehlin-Henriksson B, Jornvall H, Klein G and Lerner R (1984) Proc. Natl. Acad. Sci. USA. 81: 4652-4656.
41. Dillner J, Wendel-Hansen V, Kjellstrom G, Kallin B and Rosen A (1988) Int. J. Cancer. 42: 721-728.
42. Dolken G, Hecht T, Rockel D and Hirsch FW (1987) Virology. 157: 460-472.
43. Emini EA, Luka J, Armstrong ME, Keller PM, Ellis RW and Pearson GR (1987) Virology. 157: 552-6.
44. Farrell PJ (1989) The Epstein-Barr Virus Genome. In G. Klein (ed): Advances in Viral Oncology. New York: Raven Press, pp. 103-132.
45. Farrell PJ, Rowe DT, Rooney CM and Kouzarides T (1989) EMBO J. 8: 127-133.
46. Fischer D, Robert M, Shedd D, Summers W, Robinson J, Wolak J, Stefano J and Miller G (1984) Proc. Natl. Acad. Sci USA 81: 43-47.
47. Freese UK, Laux G, Hudenwentz J, Schwarz E and Bornkamm G (1983) J. Virol. 48: 731-743.
48. Gibson T, Stockwell P, Ginsburg M and Barrell B (1984) Nucleic Acids Res. 12: 5087-5099.
49. Gibson TJ, Barrell BG and Farrell PJ (1986) Virology. 152: 136-48.
50. Glickman JN, Howe JG and Steitz JA (1988) J. Virol. 62: 902-911.
51. Goldschmidts W, Luka J and Pearson GR (1987) Virology. 157: 220-7.
52. Gompels UA, Craxton MA and Honess RW (1988) J. Virol. 62: 757-767.
53. Grogan E, Jenson H, Countryman J, Heston L, Gradoville L and Miller G (1987) Proc. Natl. Acad. Sci. USA 84: 1332-1337.
54. Hammerschmidt W and Sugden B (1988) Cell. 55: 427-433.
55. Hammerschmidt W, Sugden B and Baichwal VR (1989) J. Virol. 63: 2469-2475.
56. Hatfull G, Bankier T, Barrell BG and Farrell PJ (1988) Virology. 164: 334-341.

57. Heineman T, Gong M, Sample J and Kieff E (1988) J. Virol. 62: 1101-7.
58. Hennessy K, Fennewald S, Hummel M, Cole T and Keiff E (1984) Proc. Natl. Acad. Sci. USA 81: 7207-7211.
59. Hennessy K, Heller M, van Santen V and Kieff E (1983) Science 220: 1396-1398.
60. Hennessy K and Keiff E (1985) Science. 227: 1238-1240.
61. Hennessy K, Wang F, Woodland Bushman E and Kieff E (1986) Proc. Natl. Acad. Sci. USA. 83: 5693-5697.
62. Howe JG and Shu M-D (1989) Cell 57: 825-834.
63. Hudson GS, Bankier AT, Satchwell SC and Barrell BG (1985) Virology. 147: 81-98.
64. Huen DS, Grand RJ and Young LS (1988) Oncogene. 3: 729.
65. Hummel M, Thorley-Lawson D and Kieff E (1984) J. Virol. 49: 413-417.
66. Jeang K-T and Hayward SD (1983) J. Virol. 48: 135-148.
67. Joab I, Rowe DT, Bodescot M, Nicolas J-C, Farrell PJ and Perricaudet M (1987) J. Virol. 61: 3340-4.
68. Jones M, Foster L, Sheedy T and Griffin BE (1984) EMBO J. 3: 813-821.
69. Kallin B, Dillner J, Ernberg I, Ehlin-Henriksson B, Rosen A, Henle W, Henle G and Klein G (1986) Proc. Natl. Acad. Sci. 83: 1499-1503.
70. Kenney S, Kamine J, Holley-Gutherie E, Lin J-G, Mar E-C and Pagano J (1989) J. Virol. 63: 1729-1736.
71. Kouzarides T, Bankier AT, Satchwell SC, Weston K, Tomlinson P and Barrell BG (1987) Virology 157: 397-413.
72. Laux G, Perricaudet M and Farrell PJ (1988) EMBO J. 7: 769-775.
73. Li J-S, Zhou BS, Dutschman GE, Grill SP, Tan R-S and Cheng Y-C (1987) J. Virol. 61: 2947-9.
74. Lieberman PM, O'Hare P, Hayward GS and Hayward SD (1986) J. Virol. 60: 140-148.
75. Liebowitz D, Kopan R, Fuchs E, Sample J and Kieff E (1987) Mol. Cell.Biol. 7: 2299-2308.
76. Liebowitz D, Wang D and Kieff E (1986) J. Virol. 58: 233-7.
77. Littler E and Arrand JR (1988) J. Virol. 62: 3892-3895.
78. Littler E, Zeuthen J, McBride A, Sorensen E, Powell KL, Walsh-Arrand JE and Arrand JR (1986) EMBO J. 5: 1959-66.
79. Lupton S and Levine AJ (1985) Mol. Cell. Biol. 5: 2533-2542.
80. Manet E, Gruffat H, Trescol-Biemont MC, Moreno N, Chambard P, Giot JF and Sergeant A (1989) EMBO J 8: 1819-1826.
81. Mann KP and Thorley-Lawson D (1987) J. Virol. 61: 2100-8.
82. Marschall M, Leser U, Seibl R and Wolf H (1989) J. Virol. 63: 938-942.
83. McGeoch DJ, Dalrymple MA, Davison AJ, Dolan A Frame MC, McNab D, Perry LJ, Scott JE and Taylor P (1988) J.Gen.Virol. 69: 1531-1574.
84. McGeoch DJ and Davison AJ (1986) Nucleic Acids Res. 14: 4281-4292.
85. McGeoch DJ, Dolan A and Frame MC (1986) Nucleic. Acids Res. 14: 3435-3448.
86. Middeldorp JM and Meloen RH (1988) J. Virol. Meth. 21: 147-153.
87. Mueller-Lantsch N, Lenoir GM, Sauter M, Takaki K, Bechet JM, Kuklik-Roos C, Wunderlich D and Bornkamm GW (1985) EMBO J. 4: 1805-1811.
88. Mullaney J, Moss HWM and McGeoch DJ (1989) J. Gen. Virol. 70: 449-454.
89. Oba DE and Hutt-Fletcher LM (1988) J. Virol. 62: 1108-1114.
90. Oguro MO, Shimizu N, Ono Y and Takada K (1987) J. Virol. 61: 3310-3313.
91. Pearson G, Vroman B, Chase B, Sculley T, Hummel M and Keiff E (1983) J. Virol. 47: 193-201.
92. Pearson GR, Luka J, Petti L, Sample J, Birkenbach M, Braun D and Kieff E (1987) Virology. 160: 151-162.
93. Pellett PE, Biggin MD, Barrell BG and Roizman B (1985) J. Virol. 56: 807-813.
94. Pellett PE, Jenkins JF, Ackermann M, Sarmiento M and Roizman B (1986) J. Virol. 60: 1134-40.
95. Petti L and Kieff E (1988) J. Virol. 62: 2173-8.
96. Petti L, Sample J, Wang F and Kieff E (1988) J. Virol. 62: 1330-8.
97. Pfitzner AJ, Strominger JL and Speck SH (1987) J. Virol. 61: 2943-6.
98. Pfitzner AJ, Tsai EC, Strominger JL and Speck SH (1987) J. Virol 61: 2902-9.
99. Raab-Traub N and Flynn K (1986) Cell 47: 883-889.
100. Rawlins DR, Milman G, Hayward SD and Hayward GS (1985) Cell. 42: 859-868.
101. Reisman D and Sugden B (1986) Mol. Cell. Biol. 6: 3838-3846.
102. Reisman D, Yates J and Sugden B (1985) Mol. Cell Biol. 5: 1822-1832.
103. Ricksten A, Kallin B, Alexander H, Dillner J, Fahraeus R, Klein G, Lerner R and Rymo L (1988) Proc. Nat. Acad. Sci. USA. 85: 995-1000.
104. Ricksten A, Olsson A, Andersson T and Rymo L (1988) Nucleic Acids Res. 16: 8391-8411.
105. Ricksten A, Svensonn C, Welinder C and Rymo L (1987) J. Gen. Virol. 68: 2407-2419.
106. Rooney CM, Rowe DT, Ragot T and Farrell PJ (1989) J. Virol. 63: 3109-3116.
107. Rowe D, Farrell PJ and Miller G (1987) Virology. 156: 153-162.
108. Rowe DT, Rowe M, Evans GI, Wallace LE, Farrell PJ and Rickinson AB (1986) EMBO J. 5: 2599-2607.
109. Rowe M, Rowe DT, Gregory CD, Young LS, Farrell PJ, Rupani H and Rickinson AB (1987) EMBO J. 6: 2743-2753.
110. Rowe M, Young LS, Cadwallader K, Petti L, Kieff E and Rickinson AB (1989) J. Virol. 63: 1031-1039.
111. Sample J, Hummel M, Braun D, Birkenbach M and Kieff E (1986) Proc. Natl. Acad. Sci. USA. 83: 5096-5100.
112. Sample J, Lancz G and Nonoyama M (1986) J. Virol. 57: 145-54.
113. Sample J, Liebowitz D and Kieff E (1989) J. Virol. 63: 933-937.
114. Sauter M, Boos H, Hirsch F and Mueller-Lantzsch N (1988) Virology. 166: 586-591.
115. Sawada K, Yamamoto M, Tabata T, Smith M, Tanaka A and Nonoyama M (1989) Virology. 168: 22-31.
116. Seibl R, Motz M and Wolf H (1986) J. Virol. 60: 902-909.
117. Shimizu N, Yamaki M, Sakuma S, Ono Y and Takada K (1988) Int.J. Cancer. 41: 744-752.

118. Smith RF and Smith TF (1989) J. Virol. 63: 450-455.
119. Speck S and Strominger J (1985) Proc. Natl. Acad. Sci. USA 82: 8305-8309.
120. Speck SH, Pfitzner A and Strominger JL (1986) Proc. Natl. Acad. Sci. USA. 83: 9298-9302.
121. Sugden B and Warren N (1988) Mol. Biol. Med. 5: 85-95.
122. Sugden B and Warren N (1989) J. Virol. 63: 2644-2649.
123. Summers W, Grogan E, Shedd D, Robert M, Liu C-R and Miller G (1982) Proc. Natl. Acad. Sci. USA. 79: 5688-5692.
124. Takada K and Ono Y (1989) J. Virol. 63: 445-449.
125. Takada K, Shimizu N, Oguro M and Ono Y (1986) J. Virol. 60: 324-30.
126. Takada K, Shimizu N, Sakuma S and Ono Y (1986) J. Virol. 57: 1016-1022.
127. Tanner J, Whang Y, Sample J, Sears A and Kieff E (1988) J. Virology. 62: 4452-4464.
128. Taylor N, Countryman J, Rooney C, Katz D and Miller G (1989) J. Virol. 63: 1721-1728.
129. Thorley-Lawson DA and Israelsohn ES (1987) Proc. Nat. Acad. Sci. USA. 84: 5384-9.
130. Tosoni-Pittoni E, Joab I, Nicolas JC and Perricaudet M (1989) Biochem. Biophys. Res. Commun. 158: 676-685.
131. Urier G, Buisson M, Chambard P and Sergeant A (1989) EMBO J. 8: 1447-1453.
132. Walls D and Gannon F (1988) EMBO J. 7: 1191-7.
133. Wang D, Liebowitz D and Keiff E (1985) Cell. 43: 831-840.
134. Wang F, Petti L, Braun D, Seung S and Kieff E (1987) J. Virol. 61: 945-954.
135. Whang Y, Silberlang M, Morgan A, Munshi S, Lenny AB, Ellis RW and Kieff E (1987) Virology. 51: 1796-1807.
136. Wong K-M and Levine AJ (1989) Virology. 168: 101-112.
137. Wong KM and Levine AJ (1986) J. Virol. 60: 149-56.
138. Wysokenski DA and Yates JL (1989) J. Virol. 63: 2657-2666.
139. Yates J, Warren N, Reisman D and Sugden B (1984) Proc. Natl. Acad. Sci. USA 81: 3806-3810.
140. Yates JL, Warren N and Sugden B (1985) Nature. 313: 812-815.
141. Zhang CX, Decaussin G, Daillie J and Ooka T (1988) J. Virol. 62: 1862-9.
142. Zhang CX, Decaussin G, de Turenne Tessier M, Daillie J and Ooka T (1987) Nucleic Acids Res. 15: 2707-19.
143. Zimber U, Aldinger HK, Lenoir GM, Vuillaume M, Knebel-Doeberitz LG, Desgranges C, Wittmann P, Freese UK, Schneider U and Bornkamm GW (1986) Virology. 154: 56-66.

Duncan J. McGeoch
Valerie G. Preston

MRC Virology Unit
Institute of Virology
University of Glasgow
Church Street
Glasgow G11 5JR
Scotland

Sandra K. Weller

Department of Microbiology
University of Connecticut
Health Center
Farmington, CT 06032

Priscilla A. Schaffer

Dana-Farber Cancer Institute and
Department of Microbiology and
Molecular Genetics
Harvard Medical School
Boston, MA 02115

The 152 kbp genome of herpes simplex virus type 1 (HSV-1) has been completely sequenced (67, 68, 69, 79), and this description is based on the sequence.

Figure 1 shows a standard representation of the HSV-1 genome, with two unique sequences (U_L and U_S) shown as solid lines and with each flanked by a pair of repeat elements in inverted orientation (TR_L and IR_L; IR_S and TR_S). Scales are indicated in fractional map units (upper scale, 0.0 - 1.0) and in kbp (lower scale, 0 -152).

The genome organization is shown in greater detail in the six panels of Figure 2. At the top of each panel, scales for fractional map units and DNA sequence (kbp) are shown. Below these, restriction sites and fragment names are listed for BamHI (top), EcoRI, HindIII, KpnI and XbaI (bottom). Below these, rightward transcribed RNA species are indicated: reasonably well characterized transcripts are shown as solid lines with arrowheads at their 3' limits, and poorly characterized or hypothetical transcripts are shown as dashed lines. RNA mapping and temporal assignment data are from (4, 25, 26, 41, 45, 47, 88, 89, 105, 106, 107). In the center of each panel of Figure 2 the DNA sequence is represented. Locations of proposed polyadenylation sites are marked by small vertical arrows. Positions of major families of short tandem reiterations are marked as "R" below the DNA. Locations and directions of proposed protein coding open reading frames (ORFs) are indicated by ⊃ shapes, corresponding to gene designations in the lower part of each panel and in Table 1. For genes expressed at immediate-early times (α genes), ORFs are solid black. Early (β) gene ORFS are grey, early-late (γ_1) gene ORFs are hatched, and true-late (γ_2) genes are crosshatched. ORFs for genes of unknown temporal class are white. Many temporal assignments are based on limited data; doubtful cases were included in the early-late class rather than either the early or true-late class. Leftward transcripts are indicated below the DNA, with gene designations and a repeat of the sequence scale.

Table 1 indicates for each gene what types of mutant are known, whether the gene is essential for growth of virus in tissue culture, and what is known of the encoded protein and its function (69, 70). Homologous genes from the two other completely sequenced herpesviruses, varicella-zoster virus (VZV; 30) and Epstein-Barr virus (EBV; 6) are also indicated.

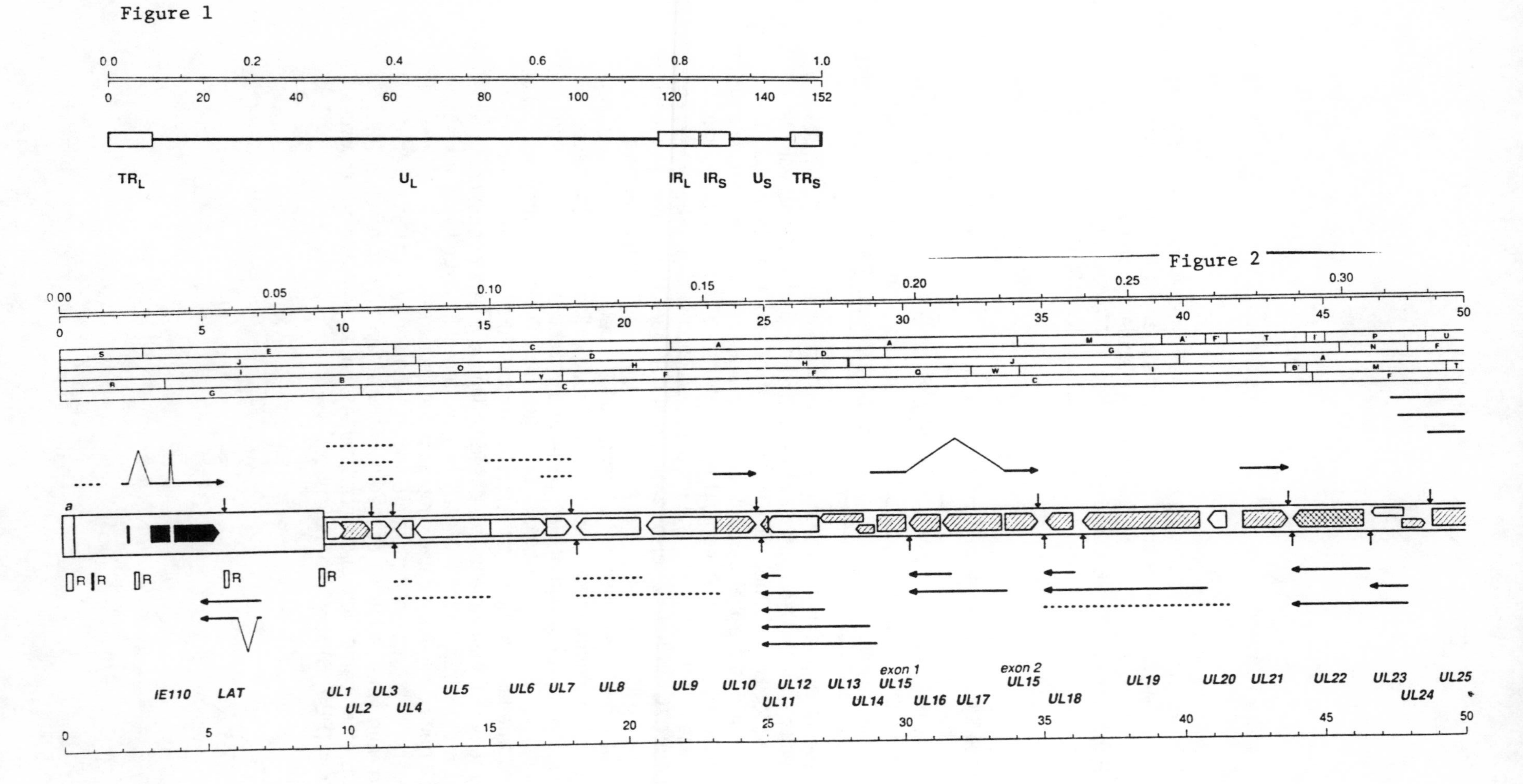
Figure 1
TR_L
U_L
IR_L
IR_S
U_S
TR_S
Figure 2
a
R
IE110
LAT
UL1
UL2
UL3
UL4
UL5
UL6
UL7
UL8
UL9
UL10
UL11
UL12
UL13
UL14
exon 1
UL15
UL16
UL17
exon 2
UL15
UL18
UL19
UL20
UL21
UL22
UL23
UL24
UL25

Figure 2

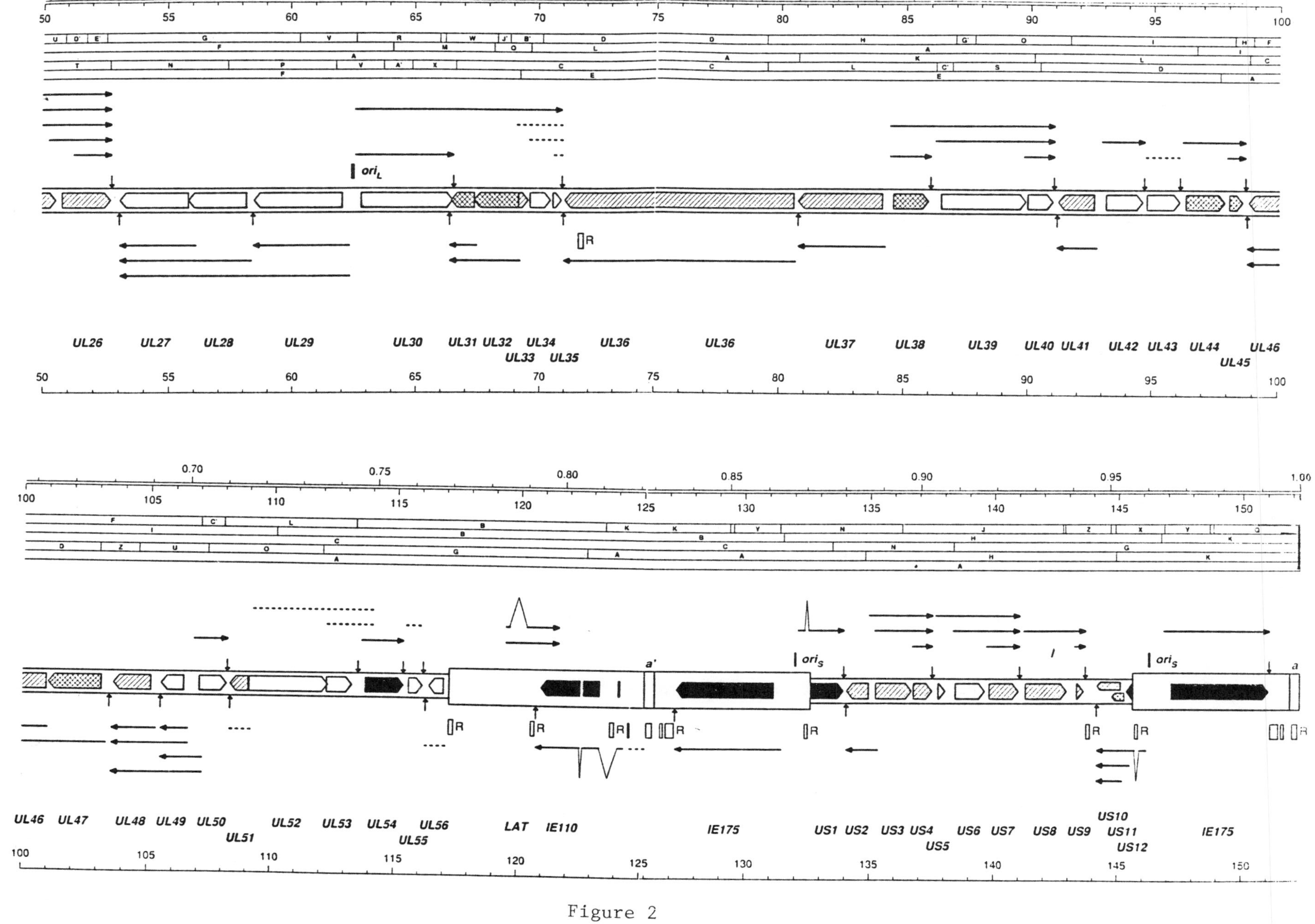
ori_L
UL26 UL27 UL28 UL29 UL30 UL31 UL32 UL33 UL34 UL35 UL36 UL36 UL37 UL38 UL39 UL40 UL41 UL42 UL43 UL44 UL45 UL46
a'
a
ori_S
UL46 UL47 UL48 UL49 UL50 UL51 UL52 UL53 UL54 UL55 UL56 LAT IE110 IE175 US1 US2 US3 US4 US5 US6 US7 US8 US9 US10 US11 US12 IE175

Figure 2

Table 1

HSV-1 Gene	Mutants[a]	Status[b]	Protein/ Function	VZV Gene	EBV Gene	References
IE110	del	ne	IE transcriptional regulatory protein (ICP0, Vmw110)	61		12, 39, 93, 101
UL1	syn?		Unknown	60	BKRF2?	57
UL2	in	ne	Uracil-DNA glycosylase	59	BKRF3	73,115
UL3	-	-	Unknown	58	BKRF4?	
UL4	-	-	Unknown	56		
UL5	ts	e	Required for DNA replication; component of DNA helicase-primase complex	55	BBLF4	27,112,116,117
UL6	ts	e	Presumed virion protein; possible role in DNA packaging	54	BBRF1	97,98,110
UL7	-	-	Unknown	53	BBRF2	
UL8	ts, hr	e	Required for DNA replication; component of DNA helicase-primase complex	52	BBLF3?	15,20,27,112,117
UL9	ts, hr	e	Required for DNA replication; ori-binding protein	51	BBLF2?	14,28,75,113,116
UL10	in	-	Function unknown; possible membrane-inserted protein	50	BBRF3	61
UL11	-	-	Myristylated virion protein	49	BBLF1?	64
UL12	hr	e	Deoxyribonuclease	48	BGLF5	113
UL13	-	-	Predicted protein kinase	47	BGLF4	19,99
UL14	-	-	Unknown	46	BGLF3?	
UL15						
exon 1	-	-	Unknown	45	BGRF1	
exon 2	-	-	Unknown	42	BDRF1	
UL16	-	-	Unknown	44	BGLF2	

Table 1 Cont'd

HSV-1 Gene	Mutants[a]	Status	Protein/ Function	VZV Gene	EBV Gene	References
UL17	-	-	Unknown	43	BGLF1	
UL18	-	-	Capsid protein	41	BDLF1	31
UL19	ts	e	Major capsid protein (VP5, MCP)	40	BCLF1	112
UL20	-	e	Unknown	39		
UL21	-	-	Unknown	38	BcRF1? or BTRF1?	
UL22	ts	e	Glycoprotein H	37	BXLF2	36,110
UL23	various	ne	Thymidine kinase	36	BXLF1	21,29,54,94,103
UL24	del, in, syn	ne	Unknown	35	BXRF1	50,94
UL25	ts	e	Virion protein	34	BVRF1	2
UL26	ts	e	Required for DNA packaging (ICP35, Vmw40 p40, VP22a)	33	BVRF2	9,85,90
UL27	ts, hr, syn, mar	e	Glycoprotein B	31	BALF4	11,46,52,56,95
UL28	ts	e	Probably structural	30	BALF3	77,78
UL29	ts, del	e	ssDNA binding protein required for DNA replication (ICP8, DBP)	29	BALF2	24,76,109,110
Ori_L	del	ne	Origin of DNA replication			81,111
UL30	various	e	Replicative DNA polymerase	28	BALF5	5,17,18,22,23,49
UL31	-	-	Unknown	27	BFLF2	
UL32	ts	e	Probably structural	26	BFLF1	77
UL33	ts	e	Structural; involved in packaging DNA	25		3
UL34	-	-	Possible virion protein	24	BFRF1?	
UL35	-	-	Unknown	23	BFRF2? or BFRF3?	
UL36	ts	e	Very large tegument protein (VP1-3, Vmw273)	22	BPLF1?	7

Table 1 Cont'd

HSV-1 Gene	Mutants[a]	Status	Protein/ Function	VZV Gene	EBV Gene	References
UL37	-	-	Unknown (ICP10)	21	BOLF1?	
UL38	ts	e	Capsid protein	20	BORF1?	31,80
UL39	ts, in	e/ne	Ribonucleotide reductase large subunit (ICP6, Vmw136, RR1)	19	BORF2	42,43
UL40	ts	e/ne	Ribonucleotide reductase small subunit (Vmw38, RR2)	18	BaRF1	86
UL41	vhs	-	Presumed virion protein	17		53
UL42	ts, in, del	e	DNA binding protein, required for DNA replication ($65K_{DBP}$)	16	BMRF1?	16,65,87,116
UL43	in, del	ne	Unknown	15	BMRF2?	
UL44	various	ne	Glycoprotein C	14		38,46,48
UL45	-	-	Unknown			
UL46	-	-	Unknown. May modulate activity of UL48 protein	12		71
UL47	-	-	Unknown. May modulate activity of UL48 protein	11		71
UL48	in	-	Major tegument protein, transactivates IE genes (VP16, Vmw65, αTIF)	10		1,13
UL49	-	-	Unknown	9		
UL50	in	ne	Deoxyuridine triphosphatase	8	BLLF2	40,118
UL51	-	-	Unknown	7	BSRF1?	
UL52	ts, hr	e	Required for DNA replication, function unknown	6	BSLF1	27,44
UL53	syn	-	Possible membrane protein	5		32,82
UL54	ts	e	IE transcriptional regulatory protein (ICP27, Vmw63)	4	BMLF1	66,77,92,95

Table 1 Cont'd

HSV-1 Gene	Mutants[a]	Status	Protein/ Function	VZV Gene	EBV Gene	References
UL55	del	ne	Unknown	3		63
UL56	del	ne	Unknown			63
IE110			see TR_L version, above			
IE175	ts, hr and others	e	IE transcriptional regulatory protein (ICP4, Vmw175)	62		34,35,37,84,91,95,96
Ori_S	del	e	Origin of DNA replication			100,102,114
US1	in	e/ne	IE protein. Function unknown (ICP22, Vmw68)	63		8,83
US2	in, del	ne	Unknown function			60,108
US3	in, del	ne	Protein kinase	66		60
US4	in, del	ne	Glycoprotein G			60,108
US5	in	ne	Unknown; possible glycoprotein			108
US6	hr, mar	e	Glycoprotein D			55,72
US7	in	ne	Glycoprotein I	67		51,58
US8	in, del	ne	Glycoprotein E	68		10,51,59,74
US9	del	ne	Virion protein	65		104
US10	del	ne	Virion protein	64		60,104
US11	del	ne	Function unknown; localizes in nucleolus			62, 104
US12	del	ne	IE protein; function unknown (ICP47, Vmw12)			60,104
Ori_S			See IR_S version, above			
IE175			See IR_S version, above			

a Loci of mutations are indicated: del, deletion; in, insertion; ts, temperature-sensitive; syn, syncytial plaque; mar, monoclonal antibody resistant; vhs, virion host shut-off; hr: host range (virus carrying inactivated gene grows only in a cell line carrying and expressing the gene).

b e, genes known to be essential for growth of virus in tissue culture; e/ne, depends on culture conditions, temperature etc,. ne, non-essential genes;

1. Ace, C.I., et al., 1989. J. Virol. 63:2260-2269.
2. Addison, C., et al., 1984. Virology 138:246-259.
3. M. Al-Kobaisi and V.G. Preston (unpublished)
4. Anderson, K.P., et al., 1981. J. Virol. 37:1011-1027.
5. Aron, G.M., et al., 1975. J. Virol. 16:498-507.
6. Baer, R., et al., 1984. Nature 310:207-211.
7. Batterson, W., et al., 1983. J. Virol. 45:397-407.
8. C. Bogard and P.A. Schaffer (unpublished)
9. Braun, D.K., et al., 1984. J. Virol. 49:142-153.
10. R. Byrn and P.A. Schaffer (unpublished)
11. Cai, W., et al., 1987. J. Virol. 61:714-721.
12. Cai, W., and Schaffer, P.A. 1989. J. Virol. 63: (in press).
13. Campbell, M.E.M., et al., 1984. J. Mol. Biol. 180:1-19.
14. Carmichael, E.P., et al., 1988. J. Virol. 62:91-99.
15. Carmichael, E.P., and Weller, S.K. 1989. J. Virol. 63:591-599.
16. P. Carmillo et al. (unpublished)
17. Chartrand, P., et al., 1979. J. Virol. 31:265-276.
18. Chartrand, P., et al., 1980. Virology 103:311-326.
19. Chee, M.S., et al., 1989. J. Gen. Virol. 70:1151-1160.
20. Chu, C.-T. et al., 1979. Virology 98:168-181.
21. Coen, D.M. and Schaffer, P.A. 1980. Proc. Natl. Acad. Sci. USA 77:2265-2269.
22. Coen, D.M., et al., 1982. J. Virol. 41:909-918.
23. Coen, D.M., et al., 1984. J. Virol. 49:236-247.
24. Conley, A.J., et al., 1981. J. Virol. 37:191-206.
25. Costa, R.H., et al. 1983. J. Virol. 48:591-603.
26. Costa, R.H., et al. 1985. J. Virol. 54:317-328.
27. Crute, J.J., et al., 1989. Proc. Natl. Acad. Sci. USA. 86:2186-2189.
28. C.A. Dabrowski and P.A. Schaffer (unpublished).
29. Darby, G., et al., 1981. Nature 289:81-83.
30. Davison, A.J. and Scott, J.E. 1986. J. Gen. Virol. 67:1759-1816.
31. M. Davison et al., (submitted)
32. Debroy C., et al., 1985. Virology 145:36-48.
33. DeLuca, N.A., et al. 1984. J. Virol. 52:767-776.
34. DeLuca, N.A., et al.1985. J. Virol. 56:558-570.
35. DeLuca, N.A., et al. 1988. J. Virol. 62:732-743.
36. Desai, D.J., et al., 1985. J. Gen. Virol. 69:1147-1156.
37. Dixon, R.A.F., et al., 1980. J. Virol. 36:189-203.
38. Draper, K.G., et al., 1984. J. Virol. 51:578-585.
39. Everett, R.D. 1989. J. Gen. Virol. 70:1185-1202.
40. Fisher, F.B. and Preston, V.G. 1986. Virology 148:190-197.
41. Frink, R.J., et al., 1981. J. Virol. 39:559-572.
42. Goldstein, D.J. and Weller, S.K. 1988a. J. Virol. 62:196-205.
43. Goldstein, D.J. and Weller, S.K. 1988b. Virology 166:41-51.
44. Goldstein, D.J. and Weller, S.K. 1988c. J. Virol. 62:2970-2977.
45. Hall, L.M., et al., 1982. J. Virol. 43:594-607.
46. Holland, T.C., et al., 1983. J. Virol. 46:649-652.
47. Holland, L.E., et al., 1984. J. Virol. 49:947-959.
48. Homa, F.L., et al., 1986. J. Virol. 58:281-289.
49. Honess, R.W. et al., 1984. J. Gen. Virol. 65:1-17.
50. Jacobson, J.G., et al., 1989. J. Virol. 63:1839-1843.
51. Johnson, D.C. et al., 1988. J. Virol. 62:1347-1354.
52. Kousoulas, K.G., et al., 1984. Virology 135:379-394.
53. Kwong, A.D., et al., 1988. J. Virol. 62:912-921.
54. Larder, B.A., et al. 1983. J. Gen. Virol. 64:523-532.
55. Ligas, M.W. and Johnson, D.C. 1988. J. Virol. 62:1486-1494.
56. Little, S.P., et al., 1981. Virology 115:149-160.
57. Little, S.P and Schaffer, P.A. 1981. Virology 112:686-702.
58. Longnecker, R. et al., 1987. Proc. Natl. Acad. Sci. USA 84:4303-4307.
59. Longnecker, R. and Roizman, B. 1986. J. Virol. 583-591.
60. Longnecker, R. and Roizman, B. 1987. Science 236:573-576.
61. C.A. MacLean et al., (unpublished).
62. MacLean, C.A., et al., 1987. J. Gen. Virol. 68:1921-1937
63. MacLean, A.R. and Brown, S.M. 1987. J. Gen. Virol. 68: 1339-1350.
64. MacLean, C.A., et al., 1989. J. Gen. Virol. 70:3147-3157.
65. Marchetti, M., et al. 1988. J. Virol. 62:715-721.
66. McCarthy, A.M., et al. 1989. J. Virol. 63:18-27.
67. McGeoch, D.J., et al., 1985. J. Mol. Biol. 181:1-13.
68. McGeoch, D.J., et al., 1986. Nucl. Acids. Res. 14:1727-1745.
69. McGeoch, D.J., et al., 1988. J. Gen. Virol. 69:1531-1574.
70. McGeoch, D.J. 1989. Ann. Rev. Microbiol. 43:235-265.
71. McKnight, J.L.C., et al., 1987. J. Virol. 61:992-1001.
72. Minson, A.C., et al., 1986. J. Gen. Virol. 67:1001-1013.
73. Mullaney, J., et al., 1989. J. Gen. Virol. 70:449-454.
74. Neidhardt, H., et al., 1987. J. Virol. 61:600-603.
75. Olivo, P.D., et al., 1988. Proc. Natl. Acad. Sci. USA 85:5414-5418.
76. Orberg, P.K. and Schaffer, P.A. 1987. J. Virol. 61:1136-1146.
77. Pancake, B.A., et al. 1983. J. Virol. 47:568-585.
78. Pellett, P.E., et al., 1986. J. Virol. 60:1134-1140.
79. Perry, L.J. and McGeoch, D.J. 1988. J. Gen. Virol. 69:2831-28
80. Pertuiset, B., et al., 1989. J. Virol. 63:2169-2179.
81. Polvino-Bodnar, M., et al., 1987. J. Virol. 61:3528-3535.
82. Pogue-Geile, K.L. and Spear, P.G. 1987. Virology 157:67-74.
83. Post, L.E. and Roizman, B. 1981. Cell 25:227-232.
84. Preston, V.G 1981. J. Virol. 39:150-161.
85. Preston, V.G., et al., 1983. J. Virol. 45:1056-1064.
86. Preston, V.G., et al., 1988. Virology 167:458-467.
87. V.G. Preston, (unpublished).
88. Rixon, F.J. and McGeoch, D.J. 1984. Nucl. Acids. Res. 12:2473-2487.
89. Rixon, F.J. and McGeoch, D.J. 1985. Nucl. Acids. Res. 13:953-973.
90. Rixon, F.J., et al., 1988. J. Gen. Virol. 69:2879-2891.
91. Russell, J., et al., 1987. J. Gen. Virol. 68:2397-2406.
92. Sacks, W.R., et al., 1985. J. Virol. 55:796-805.
93. Sacks, W.R. and Schaffer, P.A. 1987. J. Virol. 61:829-839.
94. Sanders, P.G., et al., 1982. J. Gen. Virol. 63:277-295.
95. Schaffer, P.A., et al., 1973. Virology 52:57-71.
96. Schroder, C.H., et al., 1985. J. Gen. Virol. 1589-1593.
97. Sherman, G. and Bachenheimer, S.L. 1987. Virology 158:427-430.
98. Sherman, G. and Bachenheimer, S.L. 1988. Virology 163:471-480.
99. Smith, R.F. and Smith, T.F. 1988. J. Virol. 63:450-455.
100. Stow, N.D. 1985. J. Gen. Virol. 66:31-42.
101. Stow, N.D. and Stow, E.C. 1986. J. Gen. Virol. 67:2571-2585.
102. Stow, N.D. and McMonagle, E.C. 1983. Virology 130:427-438.
103. Summers, W.P., et al., 1975. Proc. Natl. Acad. Sci. USA 72:4081-4084.
104. Umene, K. 1987. J Gen. Virol. 67:1035-1048.
105. Wagner, E.K. 1985. The Herpesviruses, Plenum Press, New York and London. 3:45-104.
106. Wagner, E.K., et al., 1988a J. Virol. 62:1194-1202.
107. Wagner, E.K., et al., 1988b. J. Virol. 62:4577-4585.
108. Weber, P.C., et al., 1987. Science 236:576-579.
109. Weller, S.K., et al. 1983a. J. Virol. 45:354-366.
110. Weller, S.K., et al. 1983b. Virology 130:290-305.
111. Weller, S.K., et al., 1985. Mol. Cell. Biol. 5:930-942.
112. Weller, S.K., et al. 1987. Virology 161:198-210.
113. S.K. Weller et al., (unpublished).
114. S. Wong and P.A. Schaffer (unpublished).
115. Worrad, D.M., et al., 1988. J. Virol. 62:4774-4777.
116. Wu, C.A., et al., 1988. J. Virol. 62:435-443.
117. Zhu, L. and Weller, S.K. 1988. Virology 166:366-378.
118. L. Zhu and S.K. Weller (unpublished).

HEPATITIS B VIRUS

Hend Farza and Pierre Tiollais, Unité de Recombinaison et Expression génétique, INSERM U.163/CNRS U.A.271, Institut Pasteur, 28 rue du Docteur Roux, 75724 PARIS CEDEX 15 (FRANCE)

The hepatitis B viruses, also called Hepadnaviruses (1), represent a small group of primarily hepatrotopic envelopped DNA viruses that replicate via an RNA intermediate (2,3).
Beside the hepatitis B virus (HBV) of man(4), this family include woodchuck hepatitis virus (WHV) of *Marmota monax* (5), ground squirrel hepatitis B virus (GSHV) of *Spermophilus beecheyi* (6), the tree squirrel hepatitis B virus (7), duck hepatitis B virus (DHBV) of *Anas domesticus* (8) and other ducks, heron hepatitis B virus in gray herons (9) and probably others.
The HBV is the causative agent of polymophic liver diseases in man, ranging from an inapparent form to acute or fulminant hepatitis, chronic hepatitis and cirrhosis. Moreover, an association between HBV chronic infection and the development of hepatocellular carcinoma is now clearly demonstrated. HBV can be detected in the hepatocyte either as a free DNA molecules or in an integrated form (for review see ref.10).
Infected plasma contains viral particles of different sizes and forms. The 22nm diameter particles consist of empty viral envelopes that bears the hepatitis B surface antigen (AgHBs). The 42nm HBV particles (Dane particles) contains three different surface proteins which are referred to as major protein, middle and large protein. Inside the viral envelope, there is the 27nm core particle formed by subunits of core proteins referred to as HBcAg. It contains the viral polymerase and the partially double stranded DNA molecule to which a protein is covalently linked (for review see ref.11).
Different system are now available to study HBV gene expression. Virus multiplication and production of surface antigens have been obtained in hepatoma cell lines transfected by cloned HBV DNA (12), in adult (13) and foetal (14) primary human hepatocytes cultures by direct infection. This has been done also <u>in vivo</u> with the transgenic mouse system (15,16,17,18). So the knowledge of the virus cycle and of its molecular biology is in progress.

The viral surface antigen has a common group specific determinant *a* and carries one member of each of the two pairs of mutually exclusive subtype determinants *d* and *y* (19) and *w* and *r* (20). Thus there are four major subtypes of HBsAg: *adw*, *adr*, *ayw* and *ayr* (21).

Table 1. List of HBV genomes analysed after cloning.

groups	subtype	length(bp)	origin	reference
A	*adw*	3221	U.S.A.	22
	adw	3200	U.S.A.	23
B	*adw*	3215	Japan	24
	adw	3215	Japan	24
	adw	3215	Indonesia	24
	adw	3215	Indonesia	25
C	*adw*	3215	Japan	26
	adr	3215	Japan	27
	adr	3188	Japan	23
	adr	3214	Japan	28
	ayr	3215	Japan	29
D	*ayw*	3182	France	30
	ayw	3182	USSR	31

The classification into group A, B, C and D has been recently proposed by Okamoto et al.(24).

The groups A, B and D are homogeneous while group C is not (24). In this later group the *adw* genome is closely related to the *ayr* and *adr* genomes. Two by two analysis of the nucleotide sequences show some degree of divergence. The divergence is about 10% for viruses of different subtypes and about 2% for viruses of the same subtype except for the *ayr* subtype which diverges only 2% from the *adr* subtype.
The variations of length are due to small deletions or insertions as shown in table 2.

Table 2: The genome length heterogeneity with respect to subtype *ayw* (30). The position of an insertion/deletion relative to *ayw* is that of the first base in the *ayw* sequence before the insertion/deletion.

subtype	length	insertion /deletion	position
ayw	3182		
adw	3221	+6	2353-5
		+33	2850
adw	3200	+6	2353-5
		+12	2850
adr	3188	-27	1790-2
		+33	2850
adr	3214	-1	1227
		+33	2850
adr	3215	+33	2850

Note: +6 (2353-5) correlates with the *adw* subtype.+33 or +12 (2850) correlates with the *adw* and *adr* subtypes.
Occasionally mixed subtypes have been reported ie *adwr*, *adyw*, *adywr* and *aywr*. Both apparently excluded determinants map to the same molecule. It is not known if this represents a double infection or an uncommon HBV strain.

Virion DNA: The HBV genome is a small circular partially double-stranded DNA molecule with a single stranded region of variable length. The minus (-) strand is linear and of fixed length of about 3200 nucleotides. It is the coding strand from which the viral mRNA and the viral pregenomic RNA are transcribed. At its 5' end, there is a protein that serves as primer for reverse transcription.
The plus strand (+) is of variable length ranging from 50 to 100% that of the (-) strand.
The maintainance of the circular structure is assured by a short cohesive overlap region of about 200 nucleotides at the 5' end of the two strands. A 12 bp direct repeats (DR1 and DR2) located near the 5' end of both strands seems to serve as a primary site for replication.
The first T of the sequence 5'GAATTC in the (+) strand corresponding to the unique EcoRI site which exists in most genomes is used as reference origin of physical map. Nucleotides numbering is from 5' to 3' in the (+) strand. When the EcoRI site does not exist the base occupying the same position is taken as position 1. Since not two viruses sequenced to date have the same length, no standardized numbering system is proposed.

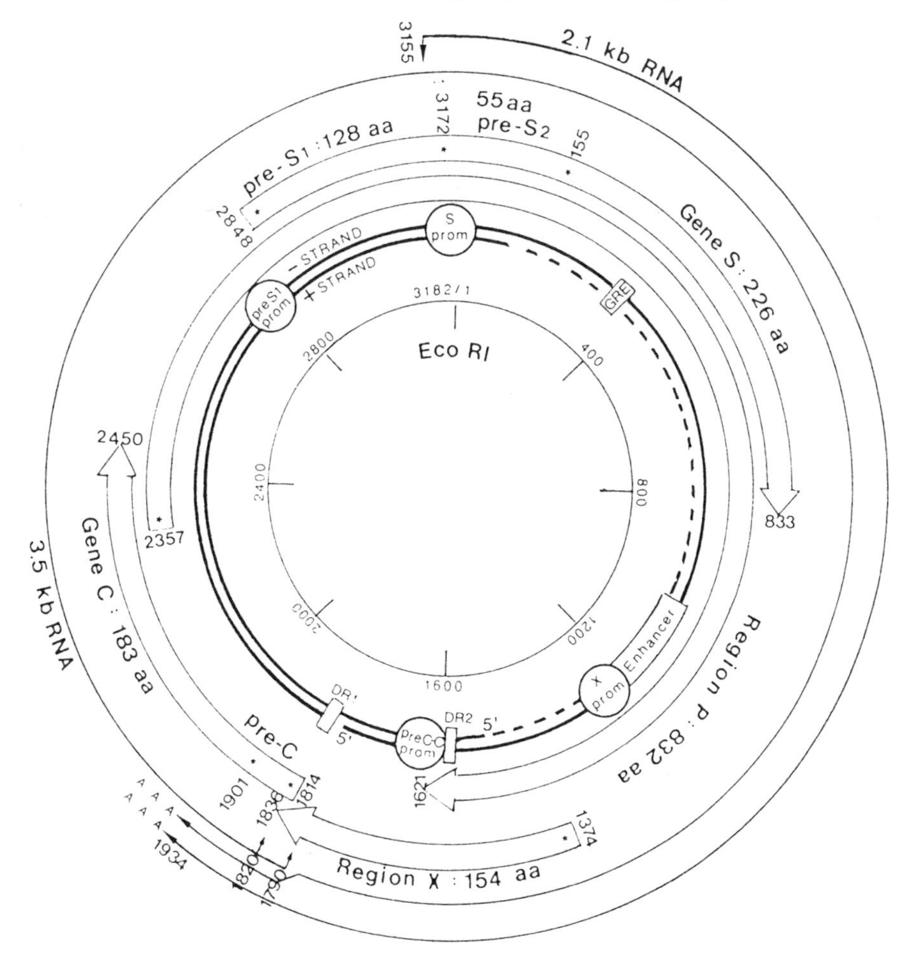

Figure 1

The stars correspond to the AUG codons and the positions given at the end of the arrows correspond to those of the stop codons. Abbreviations: GRE= glucocorticoid responsive element; prom.= promoter.

Physical structure and genetic organization of the HBV genome.
The HBV genome contains four overlapping open reading frames (figure 1).
The envelope open reading frame, also called S region, contains three in phase translation start codons (ATG) defining the N-termini of three envelope polypeptides(11,34).
Referring to the ayw (30) subtype, the shortest polypeptide (226aa)that contains the group a and the subtypes (*d/y, w/r*) determinants is also called major protein because of it relative abundance.
The middle enveloppe polypeptide contains the entire aminoacid sequence of the major polypeptide plus 55aa at the N terminus containing the PreS2 antigen. The large protein is composed by the entire aminoacid sequence of the middle envelope polypeptide plus 128aa at the N-terminus and bears the PreS1 antigen(34).
The two PreS antigens are highly immunogenic neutralizing epitopes, the larger one beeing involved in the virus binding to cell receptors and in the entry in the hepatocytes(34).
The capside open reading frame or region C contains two in frame ATG and encodes a nucleic acid binding protein, HBcAg, that encapsidate the viral nucleic acids and a 29aa longer polypeptide (PreC) that is secreted as HBeAg(35).
The X region encodes a polypeptide expressed during HBV infection and in hepatocellular carcinomas(36). This polypeptide has transactivating properties on HBV and other viral and cellular promoters(37,38,39).
The P region overlaps all the others. It encodes the polymerase and the reverse transcriptase activity (40,41) as well as the protein that binds to the viral DNA 5'terminus of the minus strand(42).
For precise starts and stops for other subtypes consult DNA sequence bank.

Table 3: Some restriction endonuclease sites in different subtypes.

enzyme	ayw(30)	adw(22)	adw(23)	adr(23)	adr(28)
AccI	1	2	2	3	3
AhaII	2	1	1	1	1
ApaI	0	1	1	1	1
AvaI	3	2	2	4	3
BamHI	3	2	2	1	1
BclI	0	0	0	0	0
BglII	3	3	4	2	2
BstEII	2	1	1	1	1
EcoRI*	1	1	1	1	0
EcoRV	0	1	1	0	0
HindIII	0	0	0	0	0
HpaI	0	1	1	0	0
MstI	1	1	1	0	0
MstII	2	2	2	2	1
NciI	1	0	0	1	1
NcoI	1	1	1	1	1
PstI	0	1	1	0	0
XbaI	3	1	1	1	1
XhoI	1	0	0	1	1
XmnI	1	0	0	0	1

Note: There is some heterogeneity in the restriction maps of clones derived from the same sera(33).
*The clone pHBr330 carries an EcoRI site. However two clones derived from the same cloning process did not possess an EcoRI site.

Table 4 gives the sizes of the polypeptides encoded by the regions of the minus strand transcripts.

genome	regions			
	S	C	P	X
ayw	226	212	832	154
adw	226	214	845	154
adw	226	214	475*	154
adr	226	183	834	145
adr	226	212	719	154
ayr	226	183	843	154

*Clone pHBV933 (23) had a codon stop in the P gene that prevented the translation of the putative DNA polymerase.

Table 5: HBV associated proteins. The abbreviation GP means that the protein is glycosylated.

major protein	P25s and GP29s
middle protein	GP33s and GP36s
large protein	P39s and GP42s
Core protein	P22c
PreC/C protein	P25e that is the precursor for the P16e (HBeAg)
P	P90 P70(reverse transcriptase?) P35 (RNaseH?)
X	P17(?)

GENE TRANSCRIPTS:
All the trancripts already described have the same 3' end consisting on a polyadenylation site 1916-TATAAA-1921.
There are four promoters that regulates the expression of the different HBV genes. The given positions correspond to that of the ayw (30) HBV genome.

The PreS1 promoter is a canonical TATA sequence located upstream the ATG of the PreS1 regionand produces 2.4kb transcripts. However this promoter is less functionnal that the other viral promoters(34).
The S promoter is located round position 3155 just upstream to the translation site of the middle protein and produces 2.1kb transcripts(43). The 5' ends of these transcripts are heterogenous and encodes both the major and the middle protein.
The X promoter, which has not been precisely located (44), is weakly active *in vivo* and represent less than 1% of the viral trancripts. In *in vitro* systems, 0.7kb and 0.9kb transcripts have been reported(45). This promoter seems to be more efficient out

of the whole HBV genome context.
The Core promoter, produces transcripts of 3,5kb. Three 5' ends have been located upstream of the C gene: two of them initiating downstream of the ATG of the PreC region, coding for the major capsid antigen and the genomic viral DNA, and another one starting upstream region PreC, coding for the HBeAg(46).
A C gene specific 2,1kb spliced transcript that represent 1/5 to 1/10 of the 2,1kb S transcripts has been recently described(47).
A Polymerase gene promoter has not been identified. The 3,5kb transcripts that have heterogeneous 5' ends(48) could also encode the polymerase gene products.

REGULATORY ELEMENTS:
A transcriptionnal enhancer region is located upstream the X region (positions 1080-1234) within the region P. This enhancer has a limited liver specificity(49).
Many transcriptionnal factors are able to bind to specific sites in the PreS region of HBV DNA and could be somehow responsible for the hepatotropism i.e. HNF1 and AP1(50,51).
A glucocorticoid responsive element (GRE) has been mapped within the S gene between positions 351 and 366(52). This element has no enhancer activity but act synergetically with the viral enhancer(53).
Steroid hormones have been shown to positively regulate the S gene expression in transgenic mice (54).

REFERENCES

1. Gust,I.D. et al. 1986. Intervirology 25: 14-29.
2. Summers J. et W.S. Mason.1982. Cell 29:403-415.
3. Seeger, C et al. 1986. Science 232: 477-484.
4. Bayer, M. et al. 1968. Nature 218: 1057-1059.
5. Summers, J. et al. 1978. Proc. Natl. Acad. Sci. USA 75: 4533-4537.
6. Marion, P.L. et al. 1980. Proc. Natl. Acad. Sci. USA 77: 2941-2944.
7. Feitelson, M.A. et al. 1986. Proc. Natl. Acad. Sci. USA 83: 2994-2997.
8. Mason W.S. et al. 1980. J. Virol. 36: 829-836.
9. Sprengel, R. et al. 1988. J. Virol. 62: 3832-3839.
10. Bréchot C. 1987.Bull.Inst. Pasteur 85: 125-149.
11. Ganem, D and H.E. Varmus. 1987. Ann. Rev. Biochem. 56: 651-693.
12. Sureau, C. et al. 1986. Cell 47: 37-47.
13. Gripon P. et al. 1988. J. Virol. 62: 4136-4143.
14. Ochiya T. et al. 1989. Proc. Natl. Acad. Sci. USA 86: 1875-1879.
15. Chisari F.V. et al. 1985. Science 230: 1157-1160
16. Babinet C. et al. 1985. Science 230: 1160-1163.
17. Farza H. et al. 1988. J.Virol.62: 4144-4152.
18. Araki K. et al. 1988. Proc. Natl. Acad. Sci. USA 86: 207-211.
19. Le Bouvier G.L.1971. J. Inf. Dis.123: 671-675.
20. Bancroft W.H. et al.1972. J. Immunol. 109: 842-848.
21. Le Bouvier G.L. et al. 1972. J. Amer. Med. Ass. 222: 928-930.
22. Valenzuela P. et al. 1980 In: Animal Virus Genetics, B. Fields, R.Jaenisch, C.F. Fox, eds. p.57. Acad. Press, New York.
23. Ono, Y. et al. 1983. Nucl. Acids Res. 11: 1747-1757.
24. Okamoto, H. et al. 1988. J. Gen. Virol. 69: 2575-2583.

25. Sastrosoewignjo, R.I. et al. 1985. ICMR Annals 5: 39-50.
26. Okamoto, H. et al. 1987. J. Virol. 61: 3030-3034.
27. Kobayashi, M. and K. Koike 1984. Gene 30: 227-232.
28. Fujiyama, A. et al. 1983. Nucl. Acids Res. 11: 4601-4610.
29. Okamoto, H. et al. 1986 J. Gen. Virol. 67: 2305-2314.
30. Galibert, F. et al. 1979 Nature 281:646-650.
31. Bichko, V. et al. 1985 FEBS Letters 185: 208-212.
32. Seeger, C and Maragos, J. 1989. J. Virol. 63:1907-1910.
33. Siddiqui, A. et al. 1979. Proc. Natl. Acad. Sci. USA 76: 4664.
34. Neurath A.R. and S.B. Kent. 1988. Adv. Virus Res. 34:65-142.
35. Uy, A. et al. 1986. Virology 155: 89-96.
36. Feitelson M.A. 1986. Hepatology 6: 191-198.
37. Twu J.S. and R.H. Schloemer. 1987. J. Virol.61: 3448-3453.
38. Spandau D.F. and C.H. Lee.1988. J. Virol. 62: 427-434.
39. Twu J.S. and W.S Robinson.1989. Proc. Natl. Acad. Sci. USA 86:2046-2050.
40. Bartenschlager R. and H. Schaller. 1988. EMBO J.7: 4185-4192.
41. Schlicht H.J. et al. 1989. Cell 56: 85-92.
42. Bosch V. et al. 1988. Virology 166: 475-485.
43. Cattaneo et al. 1983. Nature 305: 336-338.
44. Treinin M. and O. Laub. 1987. Mol. Cell. Biol. 7: 545-548.
45. Kaneko S. and R.H. Miller. 1988. J. Virol. 62; 3979-3984.
46. Yaginuma K. et al. 1987. Proc. Natl. Acad. Sci. USA 84: 2678-2682.
47. Su T.S. et al. 1989. Hepatology 9: 180-185.
48. Enders G.H. et al. 1985. Cell 42: 297-308.
49. Shaul Y. et al. 1985. EMBO J. 4: 427-430.
50. Courtois G. et al. 1988. Proc. Natl. Acad. Sci. USA 85: 7937-7941.
51. De Medina T. et al. 1988. Mol. Cell. Biol. 8: 2449-2455.
52. Tur-Kaspa R. et al. 1986. Proc. Natl. Acad. Sci. USA 83: 1627-1631.
53. Tur-Kaspa R. et al. 1988. Virology 167: 630-633.
54. Farza H. et al. 1987. Proc. Natl. Acad. Sci.USA 84:1187-1191.

THE GENOMES OF THE PAPILLOMAVIRUSES

Carl C. Baker and Lex M. Cowsert*
Laboratory of Tumor Virus Biology
National Cancer Institute
National Institutes of Health
Bethesda, Maryland 20892

*Current address:
ISIS Pharmaceuticals
Carlsbad, CA 92008

August, 1989

The papillomaviruses are small DNA viruses originally grouped with the polyomaviruses in the papovavirus family. The genomes of all papillomaviruses are double stranded circular DNA of approximately 8 k base pairs. The viruses are named by host specificity and, in those cases where more than one papillomavirus exists for a given host, by type. Two papillomaviruses are considered to be distinct types if they are less than 50% homologous by stringent hybridization analysis. All major open reading frames (ORFs) are found on one strand (Figure 1) and only one strand is expressed as mRNA (Figure 2).

Figure 1. Genomic Organization of the Papillomaviruses.

The genomes of fifteen papillomaviruses have been sequenced to date. Linearized maps of the genomic organization of fourteen papillomaviruses are depicted. The major ORFs are shown by open boxes and are labeled with numbers and the letters E or L. The E ORFs are located within the 69% transforming segment of the BPV-1 genome or within the analogous regions of the other papillomavirus genomes. The L ORFs map within a region only transcribed for BPV-1 in the differentiated keratinocytes of a bovine fibropapilloma or within the analogous regions of the other papillomavirus genomes. The numbers at the 5' and 3' ends of the ORFs refer to the first nucleotide of the ORF and to the nucleotide preceding the stop codon, respectively. The positions of the first AUG codon are indicated by the dashed vertical line. Probable polyadenylation sites are indicated by triangles labeled with the letter A. The DNA sequences of these papillomavirus genomes are all available from GenBank and the EMBL database. However, investigators should be aware that for some papillomaviruses, the database version of the sequence has not been updated to take into account published corrections to the original sequence. References for the original papillomavirus sequences as well as corrections are as follows: bovine papillomavirus type 1 (BPV-1) (2,12,20,70,73), BPV-2 (D.E. Groff, R. Mitra, and W.D. Lancaster, unpublished results; GenBank locus PPB2CG, accession numbers M20219 and M19551), BPV-4 (57), deer papillomavirus (DPV) (37), European elk papillomavirus (EEPV) (1,29), cotton tail rabbit papillomavirus (CRPV) (32), human papillomavirus type 1a (HPV-1a) (15,20,21,67), HPV-5 (84), HPV-6b (67), HPV-8 (30), HPV-11 (22), HPV-16 (7,10,39,52,68; deletion of "A" at nt 7861 of the sequence of (68), H. Romanczuk, unpublished results), HPV-18 (16), HPV-31 (36), and HPV-33 (17).

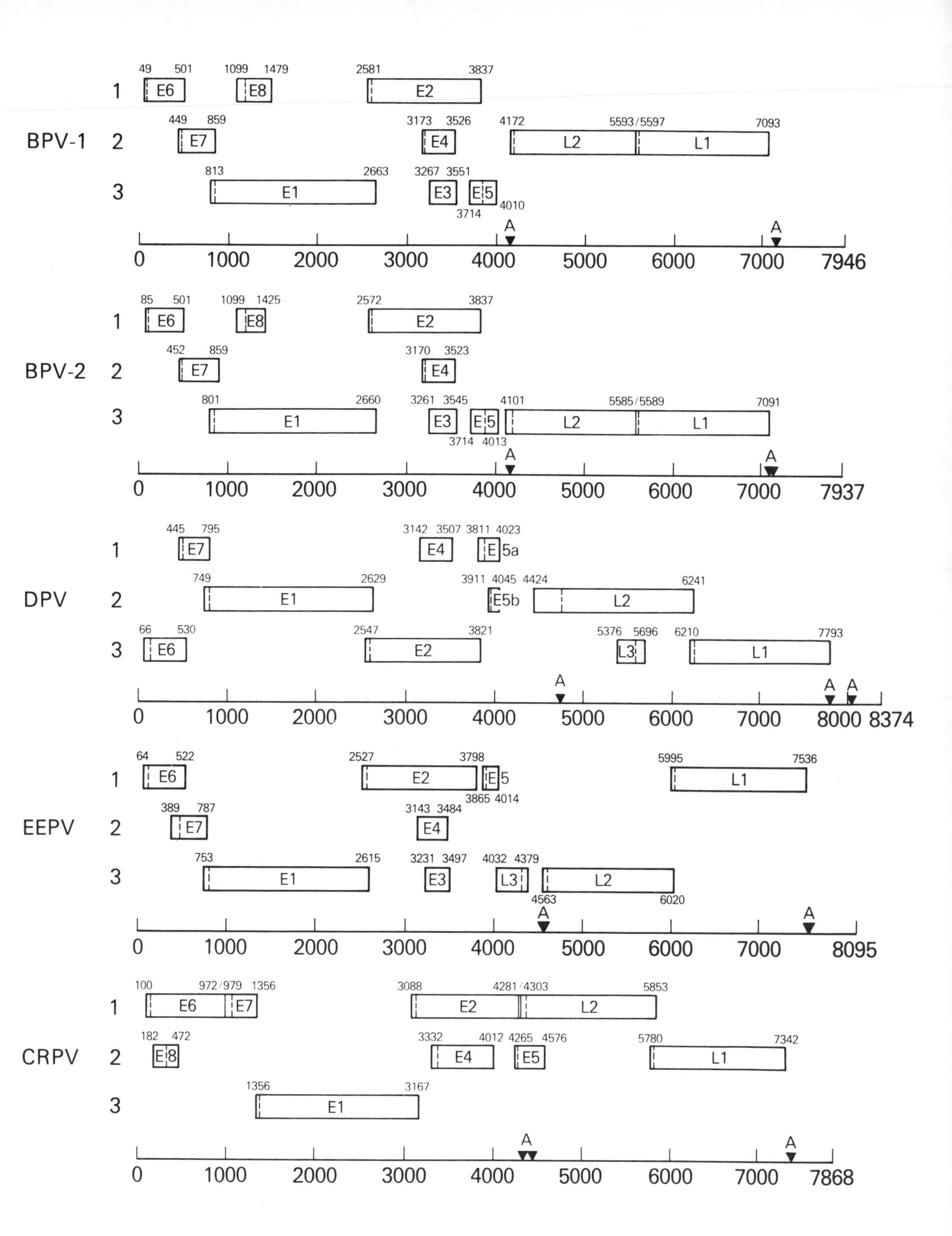

BPV-1
1
49 501 E6
1099 1479 E8
2581 3837 E2
2
449 859 E7
3173 3526 E4
4172 5593/5597 7093 L2 L1
3
813 2663 E1
3267 3551 E3
3714 4010 E5
A A
0 1000 2000 3000 4000 5000 6000 7000 7946
BPV-2
1
85 501 E6
1099 1425 E8
2572 3837 E2
2
452 859 E7
3170 3523 E4
3
801 2660 E1
3261 3545 E3
3714 4013 E5
4101 5585/5589 7091 L2 L1
A A
0 1000 2000 3000 4000 5000 6000 7000 7937
DPV
1
445 795 E7
3142 3507 E4
3811 4023 E5a
2
749 2629 E1
3911 4045 E5b
4424 6241 L2
3
66 530 E6
2547 3821 E2
5376 5696 L3
6210 7793 L1
A A A
0 1000 2000 3000 4000 5000 6000 7000 8000 8374
EEPV
1
64 522 E6
2527 3798 E2
3865 4014 E5
5995 7536 L1
2
389 787 E7
3143 3484 E4
3
753 2615 E1
3231 3497 E3
4032 4379 L3
4563 6020 L2
A A
0 1000 2000 3000 4000 5000 6000 7000 8095
CRPV
1
100 972/979 1356 E6 E7
3088 4281/4303 5853 E2 L2
2
182 472 E8
3332 4012 E4
4265 4576 E5
5780 7342 L1
3
1356 3167 E1
A A
0 1000 2000 3000 4000 5000 6000 7000 7868

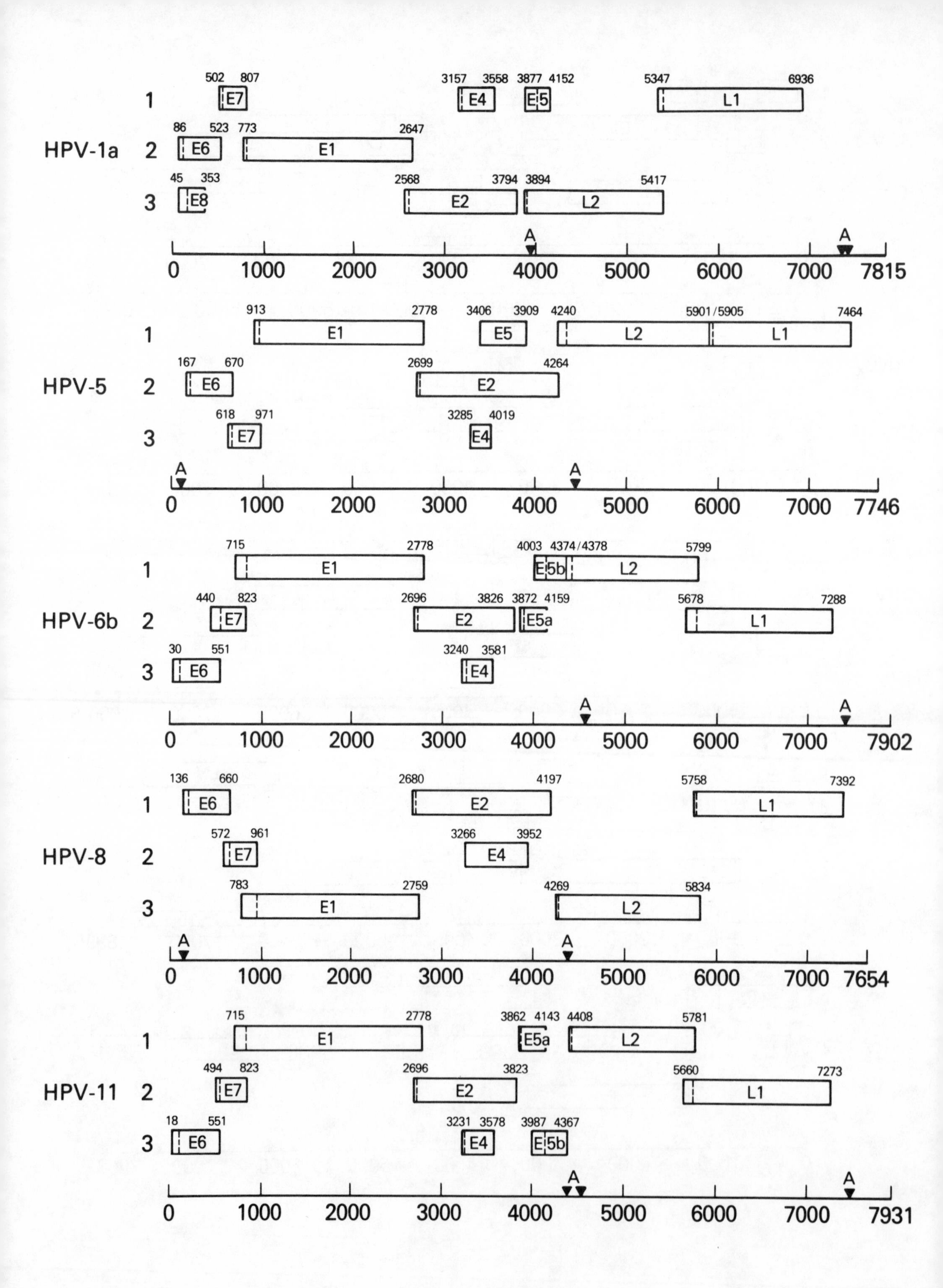
HPV-1a
1
502 807 E7
3157 3558 E4
3877 4152 E5
5347 6936 L1
2
86 523 E6
773 2647 E1
3
45 353 E8
2568 3794 E2
3894 5417 L2
A A
0 1000 2000 3000 4000 5000 6000 7000 7815
HPV-5
1
913 2778 E1
3406 3909 E5
4240 5901/5905 7464 L2 L1
2
167 670 E6
2699 4264 E2
3
618 971 E7
3285 4019 E4
A A
0 1000 2000 3000 4000 5000 6000 7000 7746
HPV-6b
1
715 2778 E1
4003 4374/4378 5799 E5b L2
2
440 823 E7
2696 3826 E2
3872 4159 E5a
5678 7288 L1
3
30 551 E6
3240 3581 E4
A A
0 1000 2000 3000 4000 5000 6000 7000 7902
HPV-8
1
136 660 E6
2680 4197 E2
5758 7392 L1
2
572 961 E7
3266 3952 E4
3
783 2759 E1
4269 5834 L2
A A
0 1000 2000 3000 4000 5000 6000 7000 7654
HPV-11
1
715 2778 E1
3862 4143 E5a
4408 5781 L2
2
494 823 E7
2696 3823 E2
5660 7273 L1
3
18 551 E6
3231 3578 E4
3987 4367 E5b
A A
0 1000 2000 3000 4000 5000 6000 7000 7931

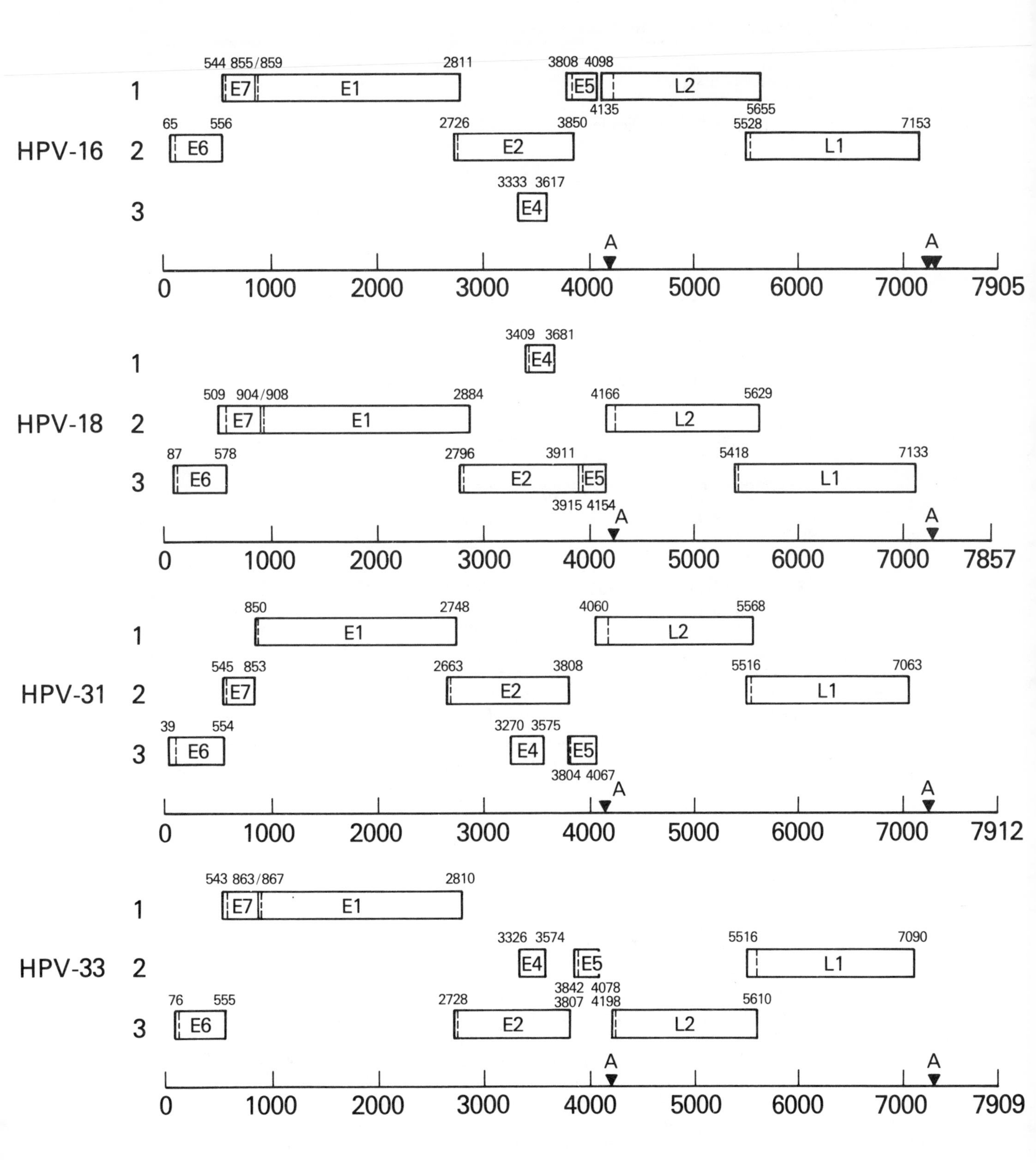
HPV-16
1
544 855/859
E7
E1
2811
3808 4098
E5
L2
4135
5655
2
65 556
E6
2726
E2
3850
5528
L1
7153
3
3333 3617
E4
A
A
0 1000 2000 3000 4000 5000 6000 7000 7905
HPV-18
1
3409 3681
E4
2
509 904/908
E7
E1
2884
4166
L2
5629
3
87 578
E6
2796
E2
3911
E5
3915 4154
5418
L1
7133
A
A
0 1000 2000 3000 4000 5000 6000 7000 7857
HPV-31
1
850
E1
2748
4060
L2
5568
2
545 853
E7
2663
E2
3808
5516
L1
7063
3
39 554
E6
3270 3575
E4
E5
3804 4067
A
A
0 1000 2000 3000 4000 5000 6000 7000 7912
HPV-33
1
543 863/867
E7
E1
2810
2
3326 3574
E4
E5
3842 4078
5516
L1
7090
3
76 555
E6
2728
E2
3807 4198
L2
5610
A
A
0 1000 2000 3000 4000 5000 6000 7000 7909

TABLE 1. PAPILLOMAVIRUS GENE FUNCTIONS

Open Reading Frames	Function(s) Assigned	Virus	Reference
E1 (5' portion)	Modulator of DNA replication	BPV-1	(8,61)
E1 (3' portion)	Replication	BPV-1	(50,63)
E2	Transcriptional regulation,	BPV-1	(23,38,41,60,63,70,71)
	probable indirect effects on	HPV-16	(42,58)
	replication and transformation	HPV-11	(13,42)
(full-length)	Transactivation, DNA binding	BPV-1	(4,54,71)
(3' portion)	Transrepression, DNA binding	BPV-1	(46,53)
E3	None yet assigned	---	---
E4	Abundant cytoplasmic protein in warts	HPV-1	(25)
E5	Transformation	BPV-1	(24,38,60,65,83)
E6	Transformation, plasmid copy number control	BPV-1	(9,64,82)
	Transformation	HPV-16	(55,79)
	Transcriptional regulation	HPV-18	(33)
E7	Plasmid copy number control, transformation	BPV-1	(9,50,56)
	Transcriptional regulation, transformation, and binding to p105-RB	HPV-16	(27,59,74)
E8	Replication, transcriptional regulation (E8/E2 fusion)	BPV-1	(14,a)
L1	Major capsid protein	BPV-1	(18,28)
		HPV-1	(26)
L2	Minor capsid protein	BPV-1	(44)
		HPV-1	(26,45)

[a]M. Lusky, pers. comm.

Figure 2. Transcription map of BPV-1.

The genomic organization of BPV-1 is shown at the top. The solid black line represents the 69% transforming region and the long control region is labelled LCR. The locations of known promoters (6,14,72,73,82) are indicated by arrows and labeled P_n where n is the approximate nucleotide position of the RNA start site for that promoter. P_{7185}, P_{89}, P_{890}, P_{2443}, and P_{3080} are also referred to as P1, P2, P3, P4, and P5 (14,72), respectively. P_L is the major promoter active in the fibropapilloma (6). Polyadenylation sites for the E and the L regions are indicated by A_E and A_L, respectively.

The structures of BPV-1 mRNAs from BPV-1 transformed mouse C127 cells (species A-P) were determined by cDNA cloning as well as electron microscopy, nuclease protection, and primer extension. Numbers indicate the 5' and 3' termini and the splice junctions (first and last nucleotides of exons). Splice junctions in parentheses were deduced from the genomic sequence and have not been confirmed by cDNA sequencing. The 5' most ORF containing a translation initiation codon and a significant coding region is indicated at the right of each mRNA. Although an E6/E7 fusion ORF is the 5' most ORF for species I, the cDNA from which this structure was deduced has been shown to encode the E1 M protein (77). Additional very rare mRNA species from cycloheximide treated BPV-1 transformed C127 cells have been characterized (14), but are not shown here. The structures of mRNAs unique to the BPV-1 fibropapilloma (species Q-V) were determined by cDNA cloning and sequencing (6). A more detailed discussion of BPV-1 transcription can be found in (5).

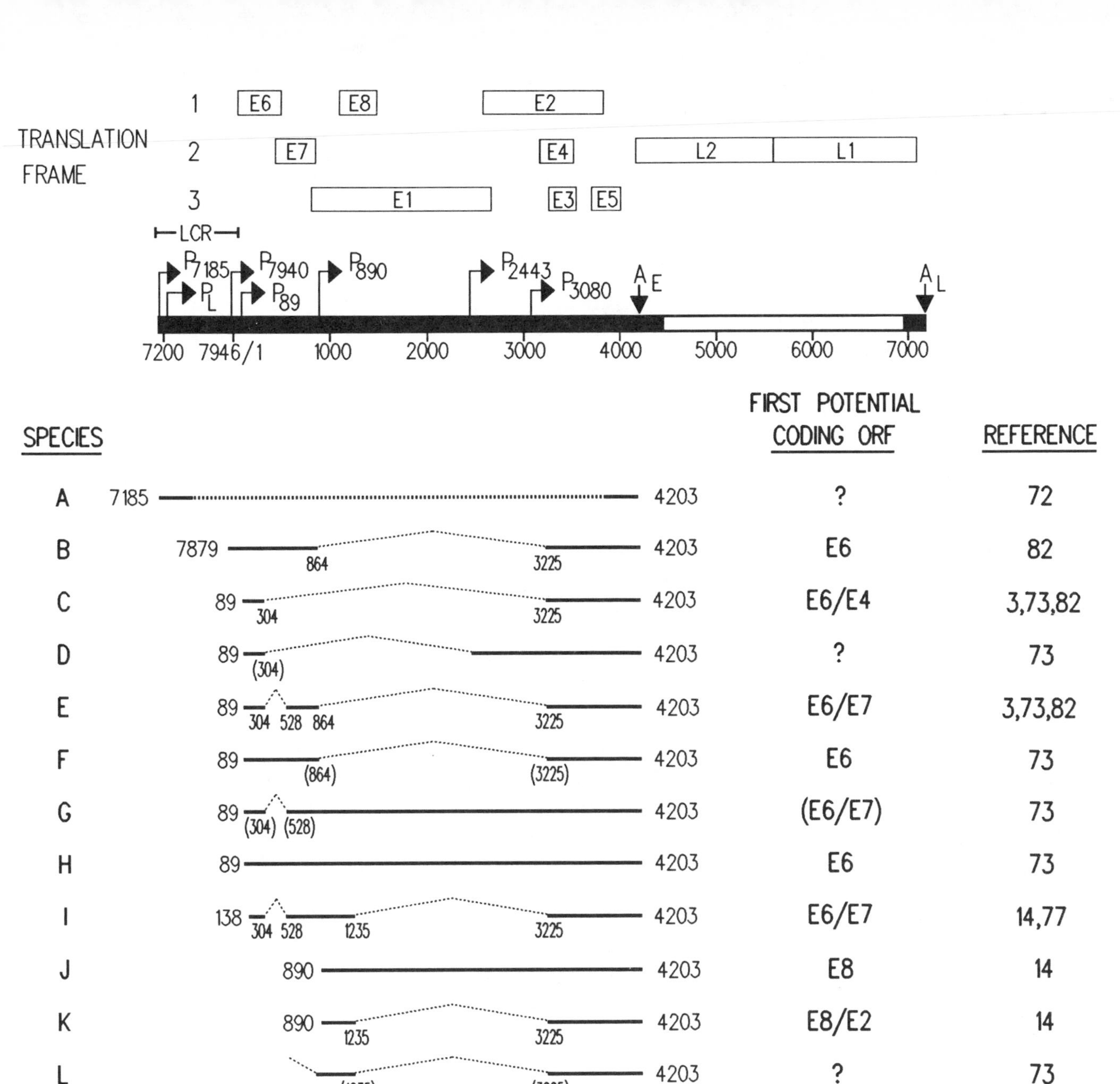

SPECIES	5′ end	Splice sites	3′ end	FIRST POTENTIAL CODING ORF	REFERENCE
A	7185		4203	?	72
B	7879	864, 3225	4203	E6	82
C	89	304, 3225	4203	E6/E4	3,73,82
D	89	(304)	4203	?	73
E	89	304, 528, 864, 3225	4203	E6/E7	3,73,82
F	89	(864), (3225)	4203	E6	73
G	89	(304), (528)	4203	(E6/E7)	73
H	89		4203	E6	73
I	138	304, 528, 1235, 3225	4203	E6/E7	14,77
J	890		4203	E8	14
K	890	1235, 3225	4203	E8/E2	14
L		(1235), (3225)	4203	?	73
M		(2505), (3225)	4203	?	73
N	2443	2505, 3225	4203	E5	3,73,82
O	2443		4203	E2	73,82
P	3080		4203	E2	3,6,46,73
Q	7255		4203	E6	6
R	7232	864, 3225	4203	E6	6
S	7246	7385, 3225	4197	E4	6
T	7217	7385, 3605	4203	E2	6
U	7269	7385, 3605, 3764, 5609	(7175)	L1	6
V	4096		7175	L2	6

TABLE 2. FUNCTIONAL ELEMENTS OF THE BPV-1 GENOME

Element and/or Function	Location[a]	Reference
1. E2 responsive enhancers:		(40,41,70)
$E2RE_1$	nt 7611-7806	
$E2RE_2$	nt 7200-7386	
2. Constitutive enhancer (active in bovine fibroblasts and keratinocytes)	nt 7143-7275	(b)
3. P_{7185} internal promoter element	nt 7194-7217	(72)
4. Distal enhancer	nt 4394-4433	(43,48,80)
5. E2 binding sites ($ACCN_6GGT$ and $ACCN_6GTT$)	nt 16, 855, 1125, 2396, 2921, 3088, 7203, 7365, 7408, 7459, 7510, 7591, 7620, 7634, 7760, 7781, 7896	(4,47,54)
6. RNA start sites:		
P_{7185}	nt 7185	(6,72)
P_L	nt 7214-7256	(6)
P_{7940}	nt 7934-7940	(6)
P_{89}	nt 89	(6,73,82)
P_{890}	nt 886, 896	(14)
P_{2443}	nt 2410, 2436-2447	(3,6,73,82)
P_{3080}	nt 3010, 3070-3080	(3,6,73)
7. Splice donors	nt 304, 864, 1235, 1613, 1940, 2505, 3764, 7385, 7905 [c]	(3,6,14,73,82)
8. Splice acceptors	nt 528, 547, 558, 1024, 1032, 1867, 2558, 3225, 3605, 5609 [c]	(3,6,14,73,82)
9. Polyadenylation sites		
Early region	nt 4197, 4203	(6,14,73,82)
Late region	nt 7175	(6,28)
10. Origin of replication	nt 6940 ± 5%	(78)
11. Plasmid maintenance sequences (PMS)		(49,51)
PMS I	nt 7068-7330 (7103-7171)[d]	
PMS II	nt 1515-1655 (1522-1582)[d]	
12. Negative control of replication (NCOR)		(61)
NCOR I	nt 6950-7355 (7112-7135)[e]	
NCOR II	nt 1475-1515 (1494-1515)[e]	
13. Replication enhancer	nt 6673-6848	(51)

TABLE 2. (continued)

Element and/or Function	Location[a]	Reference
14. DNAse I hypersensitive regions (DHSR)	0.30 m.u. 0.39 m.u. 0.49 m.u. 0.56 m.u. 0.62 m.u. 0.88-0.92 m.u. 0.94-0.98 m.u.	(62)
15. Z-DNA	0.56 m.u.	(62)
16. Sites of static DNA bending in LCR	nt 7477, 7790	(81)

[a]Nucleotide positions refer to the BPV-1 genomic sequence of Chen et al. (12) corrected as documented in the legend to Figure 1.
[b]S. Vande Pol, personal communication.
[c]Nucleotide positions refer to the first or last nucleotide of exons.
[d]Nucleotide region in parentheses indicates region of homology between PMS I and PMS II.
[e]Nucleotide region in parentheses indicates region of homology between NCOR I and NCOR II.

TABLE 3. FUNCTIONAL ELEMENTS OF THE HPV-16 AND HPV-18 GENOMES

Element and/or Function	Location[a]	Reference
1. RNA start sites:		
HPV-16 (P_{97})	nt 97	(69)
HPV-18 (P_{105})	nt 105, 107	(66,76)
2. E2 responsive enhancer:		
HPV-16	nt 7465-57	(19,58)
HPV-18	nt 37-81	(33)
3. Potential E2 binding sites:		
HPV-16	nt 35, 50, 127, 4929, 6196, 6565, 6564, 7452, 7859	(4,47)
HPV-18	nt 42, 58, 4759, 5162, 6538, 6539, 7458, 7822	(4,31,47)
4. Cell-type-specific enhancer:		
HPV-16	nt 7529-7752	(19,34,35)
HPV-18	nt 7510-7739	(31,33,75)
5. Glucocorticoid and progesterone responsive element:		
HPV-16	nt 7642-7656	(11,34)
HPV-18	nt 7839-7853	(11)
6. E6-inducible enhancer:		
HPV-18	nt 7121-7509	(33)
7. AP1 binding sites:		
HPV-18	nt 7600-7615, 7781-7803	(31)

[a]Nucleotide positions for HPV-16 refer to the genomic sequence of Seedorf et al. (68) corrected as documented in the legend to Figure 1.

REFERENCES

1. **Ahola, H., P. Bergman, A.C. Ström, J. Moreno-Lopez, and U. Pettersson.** 1986. Gene **50**:195-205.
2. **Ahola, H., A. Stenlund, J. Moreno-Lopez, and U. Pettersson.** 1983. Nucleic Acids Res. **11**:2639-2650.
3. **Ahola, H., A. Stenlund, J. Moreno-Lopez, and U. Pettersson.** 1987. J. Virol. **61**:2240-2244.
4. **Androphy, E.J., D.R. Lowy, and J.T. Schiller.** 1987. Nature **325**:70-73.
5. **Baker, C.C.** 1990. In Papillomavirus and human cancer. H. Pfister (ed.), CRC Press, Inc., Boca Raton, FL, in press.
6. **Baker, C.C., and P.M. Howley.** 1987. EMBO J. **6**:1027-1035.
7. **Baker, C.C., W.C. Phelps, V. Lindgren, M.J. Braun, M.A. Gonda, and P.M. Howley.** 1987. J. Virol. **61**:962-971.
8. **Berg, L., M. Lusky, A. Stenlund, and M.R. Botchan.** 1986. Cell **46**:753-762.
9. **Berg, L.J., K. Singh, and M. Botchan.** 1986. Mol. Cell. Biol. **6**:859-869.
10. **Bubb, V., D.J. McCance, and R. Schlegel.** 1988. Virology **163**:243-246.
11. **Chan, W.-K., G. Klock, and H.-U. Bernard.** 1989. J. Virol. **63**:3261-3269.
12. **Chen, E.Y., P.M. Howley, A.D. Levinson, and P.H. Seeburg.** 1982. Nature **299**:529-534.
13. **Chin, M.T., R. Hirochika, H. Hirochika, T.R. Broker, and L.T. Chow.** 1988. J. Virol. **62**:2994-3002.
14. **Choe, J., P. Vaillancourt, A. Stenlund, and M. Botchan.** 1989. J. Virol. **63**:1743-1755.
15. **Clad, A., L. Gissmann, B. Meier, U.K. Freese, and E. Schwarz.** 1982. Virology **118**:254-259.
16. **Cole, S.T., and O. Danos.** 1987. J. Mol. Biol. **193**:599-608.
17. **Cole, S.T., and R.E. Streeck.** 1986. J. Virol. **58**:991-995.
18. **Cowsert, L.M., W.P. Pilacinski, and A.B. Jenson.** 1988. Virology **165**:613-615.
19. **Cripe, T.P., T.H. Haugen, J.P. Turk, F. Tabatabai, P.G. Schmid, M. Dürst, L. Gissmann, A. Roman, and L.P. Turek.** 1987. EMBO J. **6**:3745-3753.
20. **Danos, O., L.W. Engel, E.Y. Chen, M. Yaniv, and P.M. Howley.** 1983. J. Virol. **46**:557-566.
21. **Danos, O., M. Katinka, and M. Yaniv.** 1982. EMBO J. **1**:231-236.
22. **Dartmann, K., E. Schwarz, L. Gissmann, and H. zur Hausen.** 1986. Virology **151**:124-130.
23. **DiMaio, D.** 1986. J. Virol. **57**:475-480.
24. **DiMaio, D., D. Guralski, and J.T. Schiller.** 1986. Proc. Natl. Acad. Sci. USA **83**:1797-1801.
25. **Doorbar, J., D. Campbell, R.J. Grand, and P.H. Gallimore.** 1986. EMBO J. **5**:355-362.
26. **Doorbar, J., and P.H. Gallimore.** 1987. J. Virol. **61**:2793-2799.
27. **Dyson, N., P.M. Howley, K. Münger, and E. Harlow.** 1989. Science **243**:934-937.
28. **Engel, L.W., C.A. Heilman, and P.M. Howley.** 1983. J. Virol. **47**:516-528.
29. **Eriksson, A., H. Ahola, U. Pettersson, and J. Moreno-Lopez.** 1988. Virus Genes. **1**:123-133.
30. **Fuchs, P.G., T. Iftner, J. Weninger, and H. Pfister.** 1986. J. Virol. **58**:626-634.
31. **Garcia-Carranca, A., F. Thierry, and M. Yaniv.** 1988. J. Virol. **62**:4321-4330.
32. **Giri, I., O. Danos, and M. Yaniv.** 1985. Proc. Natl. Acad. Sci. USA **82**:1580-1584.
33. **Gius, D., S. Grossman, M.A. Bedell, and L.A. Laimins.** 1988. J. Virol. **62**:665-672.
34. **Gloss, B., H.U. Bernard, K. Seedorf, and G. Klock.** 1987. EMBO J. **6**:3735-3743.
35. **Gloss, B., T. Chong, and H.-U. Bernard.** 1989. J. Virol. **63**:1142-1152.
36. **Goldsborough, M.D., D. DiSilvestre, G.F. Temple, and A.T. Lorincz.** 1989. Virology **171**:306-311.
37. **Groff, D.E., and W.D. Lancaster.** 1985. J. Virol. **56**:85-91.
38. **Groff, D.E., and W.D. Lancaster.** 1986. Virology **150**:221-230.
39. **Halbert, C.L., and D.A. Galloway.** 1988. J. Virol. **62**:1071-1075.
40. **Harrison, S.M., K.L. Gearing, S.Y. Kim, A.J. Kingsman, and S.M. Kingsman.** 1987. Nucleic Acids Res. **15**:10267-10284.
41. **Haugen, T.H., T.P. Cripe, G.D. Ginder, M. Karin, and L.P. Turek.** 1987. EMBO J. **6**:145-152.
42. **Hirochika, H., T.R. Broker, and L.T. Chow.** 1987. J. Virol. **61**:2599-2606.
43. **Howley, P.M., E.T. Schenborn, E. Lund, J.C. Byrne, and J.E. Dahlberg.** 1985. Mol. Cell. Biol. **5**:3310-3315.
44. **Jin, X.W., L.M. Cowsert, W.P. Pilacinski, and A.B. Jenson.** 1989. J. Gen. Virol. **70**:1133-1140.
45. **Komly, C.A., F. Breitburd, O. Croissant, and R.E. Streeck.** 1986. J. Virol. **60**:813-816.
46. **Lambert, P.F., B.A. Spalholz, and P.M. Howley.** 1987. Cell **50**:69-78.

47. **Li, R., J. Knight, G. Bream, A. Stenlund, and M. Botchan.** 1989. Genes Dev. **3**:510-526.
48. **Lusky, M., L. Berg, H. Weiher, and M. Botchan.** 1983. Mol. Cell. Biol. **3**:1108-1122.
49. **Lusky, M., and M.R. Botchan.** 1984. Cell **36**:391-401.
50. **Lusky, M., and M.R. Botchan.** 1985. J. Virol. **53**:955-965.
51. **Lusky, M., and M.R. Botchan.** 1986. Proc. Natl. Acad. Sci. USA **83**:3609-3613.
52. **Matsukura, T., T. Kanda, A. Furuno, H. Yoshikawa, T. Kawana, and K. Yoshiike.** 1986. J. Virol. **58**:979-982.
53. **McBride, A.A., R. Schlegel, and P.M. Howley.** 1988. EMBO J. **7**:533-539.
54. **Moskaluk, C., and D. Bastia.** 1987. Proc. Natl. Acad. Sci. USA **84**:1215-1218.
55. **Münger, K., W.C. Phelps, V. Bubb, P.M. Howley, and R. Schlegel.** 1989. J. Virol. ,in press.
56. **Neary, K., and D. DiMaio.** 1989. J. Virol. **63**:259-266.
57. **Patel, K.R., K.T. Smith, and M.S. Campo.** 1987. J. Gen. Virol. **68**:2117-2128.
58. **Phelps, W.C., and P.M. Howley.** 1987. J. Virol. **61**:1630-1638.
59. **Phelps, W.C., C.L. Yee, K. Münger, and P.M. Howley.** 1988. Cell **53**:539-547.
60. **Rabson, M.S., C. Yee, Y.C. Yang, and P.M. Howley.** 1986. J. Virol. **60**:626-634.
61. **Roberts, J.M., and H. Weintraub.** 1986. Cell **46**:741-752.
62. **Roesl, F., W. Waldeck, H. Zentgraf, and G. Sauer.** 1986. J. Virol. **58**:500-507.
63. **Sarver, N., M.S. Rabson, Y.C. Yang, J.C. Byrne, and P.M. Howley.** 1984. J. Virol. **52**:377-388.
64. **Schiller, J.T., W.C. Vass, and D.R. Lowy.** 1984. Proc. Natl. Acad. Sci. USA **81**:7880-7884.
65. **Schiller, J.T., W.C. Vass, K.H. Vousden, and D.R. Lowy.** 1986. J. Virol. **57**:1-6.
66. **Schneider-Gadicke, A., and E. Schwarz.** 1986. EMBO J. **5**:2285-2292.
67. **Schwarz, E., M. Dürst, C. Demankowski, O. Lattermann, R. Zech, E. Wolfsperger, S. Suhai, and H. zur Hausen.** 1983. EMBO J. **2**:2341-2348.
68. **Seedorf, K., G. Krämmer, M. Dürst, S. Suhai, and W.G. Rowekamp.** 1985. Virology **145**:181-185.
69. **Smotkin, D., and F.O. Wettstein.** 1986. Proc. Natl. Acad. Sci. USA **83**:4680-4684.
70. **Spalholz, B.A., P.F. Lambert, C.L. Yee, and P.M. Howley.** 1987. J. Virol. **61**:2128-2137.
71. **Spalholz, B.A., Y.C. Yang, and P.M. Howley.** 1985. Cell **42**:183-191.
72. **Stenlund, A., G.L. Bream, and M.R. Botchan.** 1987. Science **236**:1666-1671.
73. **Stenlund, A., J. Zabielski, H. Ahola, J. Moreno-Lopez, and U. Pettersson.** 1985. J. Mol. Biol. **182**:541-554.
74. **Storey, A., D. Pim, A. Murray, K. Osborn, L. Banks, and L. Crawford.** 1988. EMBO J. **7**:1815-1820.
75. **Swift, F.V., K. Bhat, H.B. Younghusband, and H. Hamada.** 1987. EMBO J. **6**:1339-1344.
76. **Thierry, F., J.M. Heard, K. Dartmann, and M. Yaniv.** 1987. J. Virol. **61**:134-142.
77. **Thorner, L., N. Bucay, J. Choe, and M. Botchan.** 1988. J. Virol. **62**:2474-2482.
78. **Waldeck, W., F. Roesl, and H. Zentgraf.** 1984. EMBO J. **3**:2173-2178.
79. **Watanabe, S., T. Kanda, and K. Yoshiike.** 1989. J. Virol. **63**:965-969.
80. **Weiher, H., and M.R. Botchan.** 1984. Nucleic Acids Res. **12**:2901-2916.
81. **Wilson, V.G.** 1989. Virus Res. **13**:1-14.
82. **Yang, Y.C., H. Okayama, and P.M. Howley.** 1985. Proc. Natl. Acad. Sci. USA **82**:1030-1034.
83. **Yang, Y.C., B.A. Spalholz, M.S. Rabson, and P.M. Howley.** 1985. Nature **318**:575-577.
84. **Zachow, K.R., R.S. Ostrow, and A.J. Faras.** 1987. Virology **158**:251-254.

VACCINIA VIRUS

Patricia L. Earl and Bernard Moss
Laboratory of Viral Diseases
National Institute of Allergy and Infectious Diseases
National Institutes of Health
Bethesda, Maryland 20892
August 1989

The genome of vaccinia virus is a double stranded DNA molecule approximately 185 kbp in length. The two DNA strands are covalently continuous with each terminus containing a hairpin of 104 incompletely base paired nucleotides. The genome contains a 10 kbp inverted terminal repeat, within which are multiple sets of small (70-125 bp) tandem repeats. The WR strain (ATCC VR-119) has been used for most laboratory studies and is the primary one represented here. (One exception is the sequence of HindIII F (54) which is from the LIPV strain).

Restriction maps abstracted from references 19, 26, 49, and 53 are shown in Figure 1. These were independently derived and are not necessarily superimposable. The HindIII map is shown at the top of Figures 2-7 which summarize present knowledge of vaccinia gene locations. Open reading frames (ORFs) are named according to the HindIII fragment in which they initiate and are numbered from left to right. The direction of translation of ORFs is indicated (L=leftward, R=rightward). To maintain the published ORF designation of the HindIII K fragment (10) the first ORF has been called K0L instead of K1L. Published ORF designations in the HindIII F fragment (54) have been reassigned to conform with the rest of the map.

Viral transcripts are shown as arrows indicating the direction of transcription. The apparent molecular weight of the in vitro translation product of each transcript is given above the arrow. RNAs synthesized early in infection (i.e. in the presence of inhibitors of DNA replication) are denoted by asterisks. RNAs synthesized late in infection are generally heterogeneous in size and have undefined 3' ends illustrated by dashed lines following the arrows. Two genes with "intermediate" promoters, I3L and I8R are labeled with an "i" above the transcript. In cases where map locations of ORFs, transcripts or mutants continue into an adjacent fragment or where the end points are unknown, a dotted line is used.

The map locations of temperature sensitive and drug resistant mutants are shown by horizontal lines. Mutants in the same complementation group are separated by commas; those mapping within the same limits but belonging to different complementation groups are separated by semicolons. In some instances the linear order of mutants within a particular region of the genome is known. The names of these mutants are separated by a dotted line and positioned below the line.

References are given in the figures as italicized numbers.

Figure 1

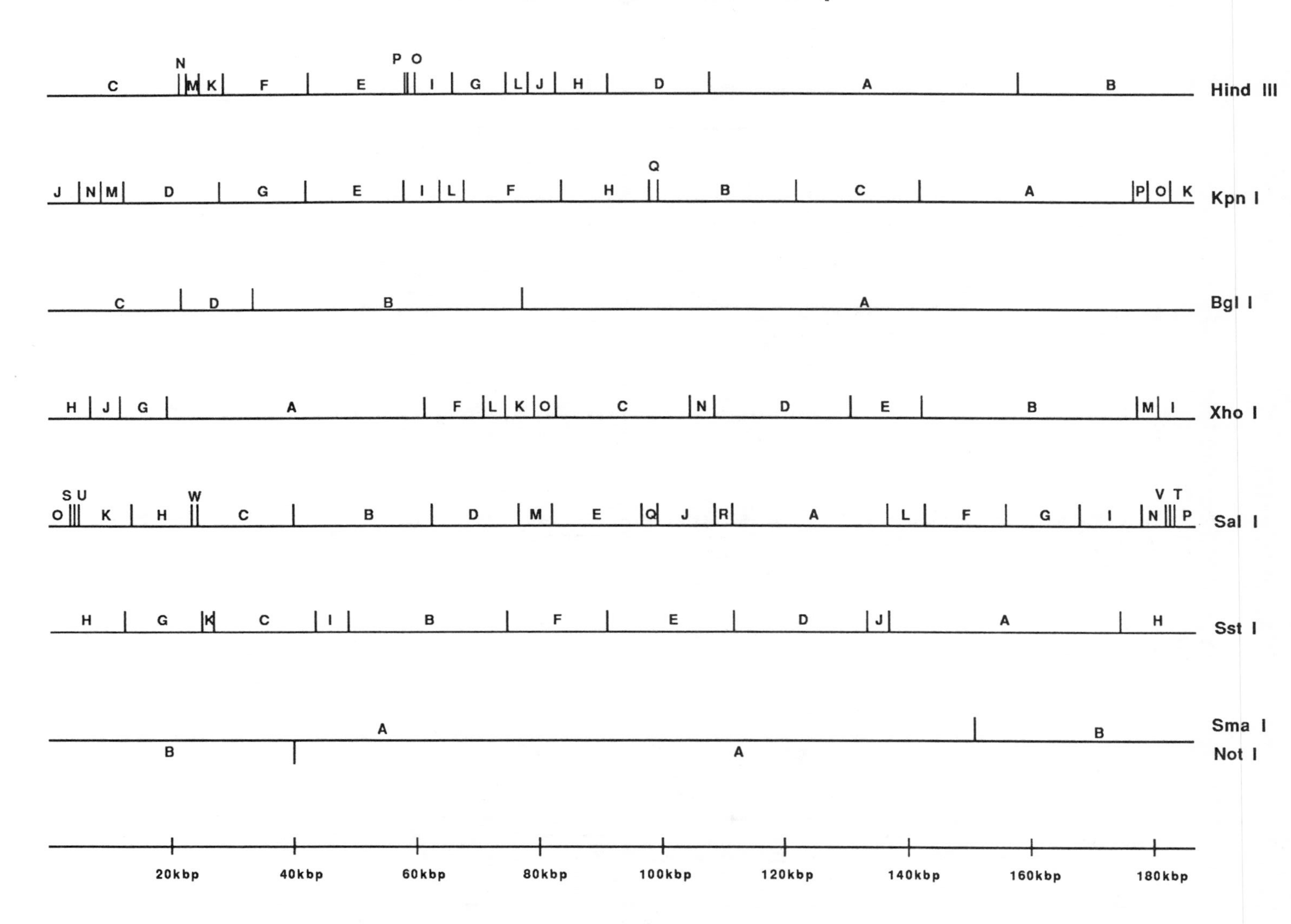
Vaccinia Virus Restriction Map
Hind III
Kpn I
Bgl I
Xho I
Sal I
Sst I
Sma I
Not I
20kbp
40kbp
60kbp
80kbp
100kbp
120kbp
140kbp
160kbp
180kbp

Figure 2

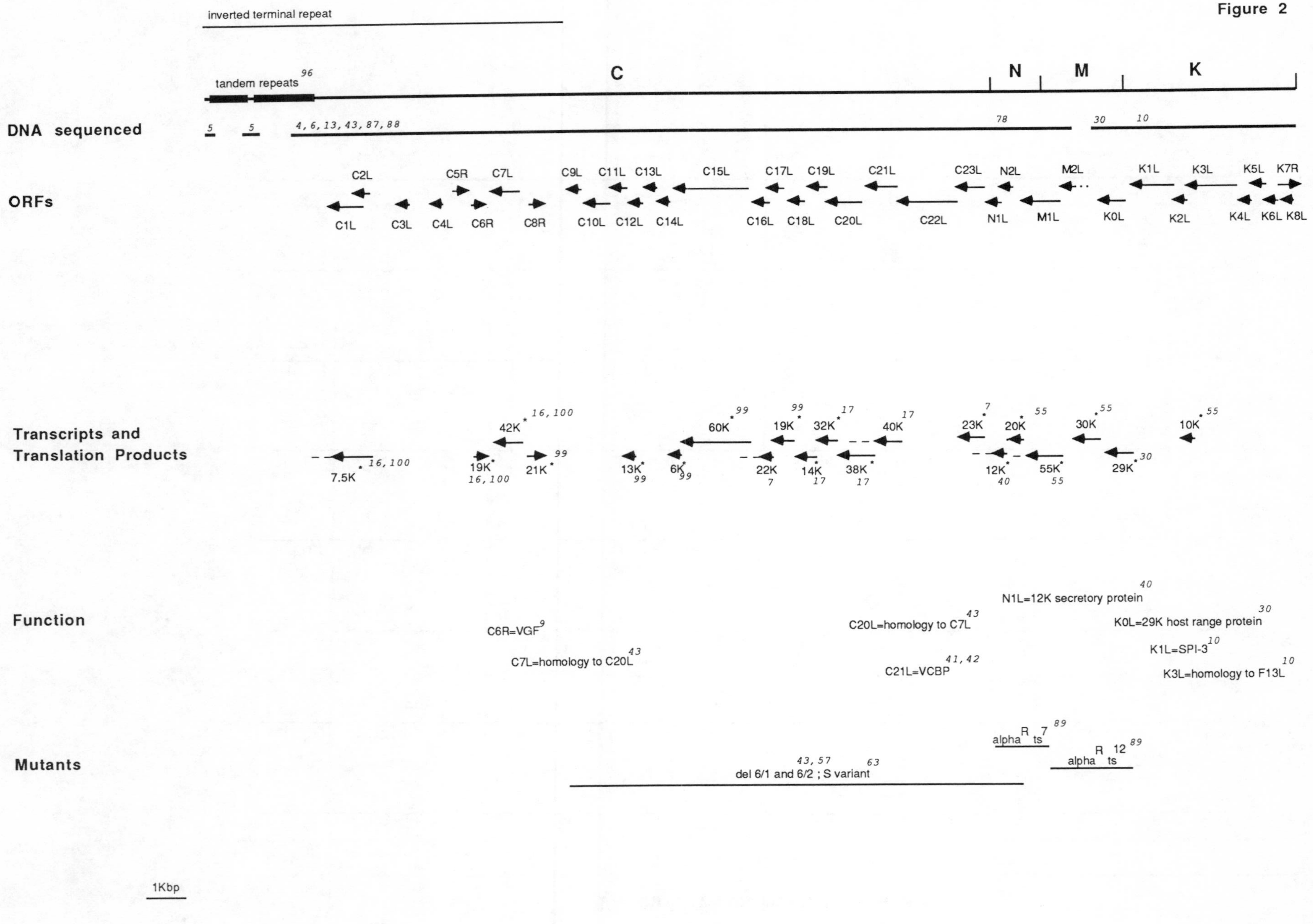

inverted terminal repeat

tandem repeats[96]

C N M K

DNA sequenced

5 5 4, 6, 13, 43, 87, 88 78 30 10

ORFs

C2L C5R C7L C9L C11L C13L C15L C17L C19L C21L C23L N2L M2L K1L K3L K5L K7R

C1L C3L C4L C6R C8R C10L C12L C14L C16L C18L C20L C22L N1L M1L K0L K2L K4L K6L K8L

Transcripts and Translation Products

7.5K[16,100] 19K[16,100] 42K[16,100] 21K[99] 13K[99] 6K[99] 60K[99] 19K[99] 22K[7] 32K[17] 14K[17] 40K[17] 38K[17] 23K[7] 12K[40] 20K[55] 55K[55] 30K[55] 29K[30] 10K[55]

Function

C6R=VGF[9]

C7L=homology to C20L[43]

C20L=homology to C7L[43]

C21L=VCBP[41,42]

N1L=12K secretory protein[40]

K0L=29K host range protein[30]

K1L=SPI-3[10]

K3L=homology to F13L[10]

Mutants

del 6/1 and 6/2[43,57] ; S variant[63]

alpha^{R} ts7[89]

alpha^{R} ts12[89]

1Kbp

Figure 3

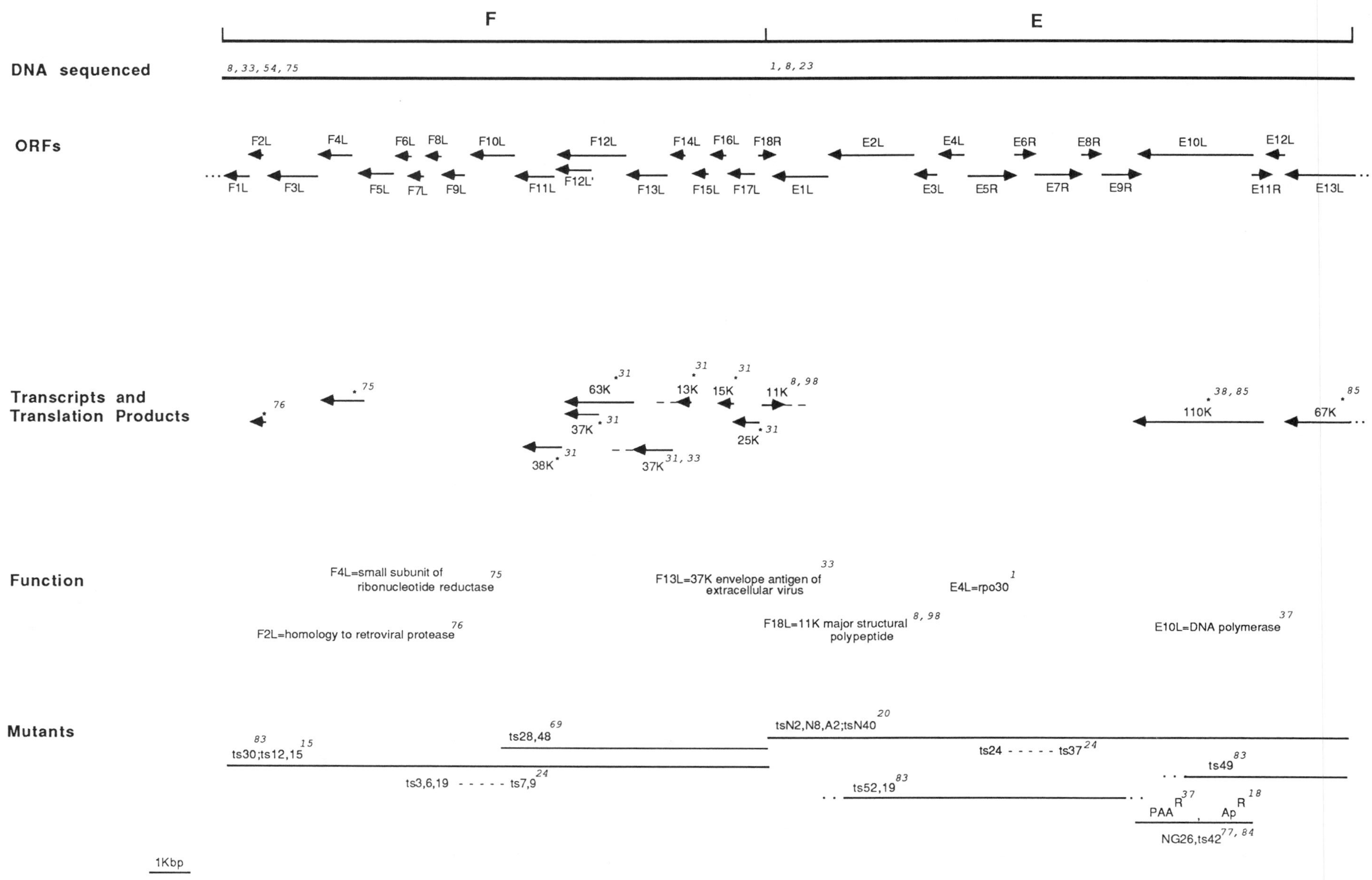

F
E
DNA sequenced
8, 33, 54, 75
1, 8, 23
ORFs
F2L F4L F6L F8L F10L F12L F14L F16L F18R E2L E4L E6R E8R E10L E12L
F1L F3L F5L F7L F9L F11L F12L' F13L F15L F17L E1L E3L E5R E7R E9R E11R E13L
Transcripts and Translation Products
76
75
63K 31
37K 31
38K 31
37K 31, 33
13K 31
15K 31
25K 31
11K 8, 98
110K 38, 85
67K 85
Function
F4L=small subunit of ribonucleotide reductase 75
F2L=homology to retroviral protease 76
F13L=37K envelope antigen of extracellular virus 33
F18L=11K major structural polypeptide 8, 98
E4L=rpo30 1
E10L=DNA polymerase 37
Mutants
ts30 83;ts12,15 15
ts28,48 69
ts3,6,19 - - - - - ts7,9 24
tsN2,N8,A2;tsN40 20
ts24 - - - - - ts37 24
ts49 83
ts52,19 83
PAA R 37, Ap R 18
NG26,ts42 77, 84
1Kbp

Figure 4

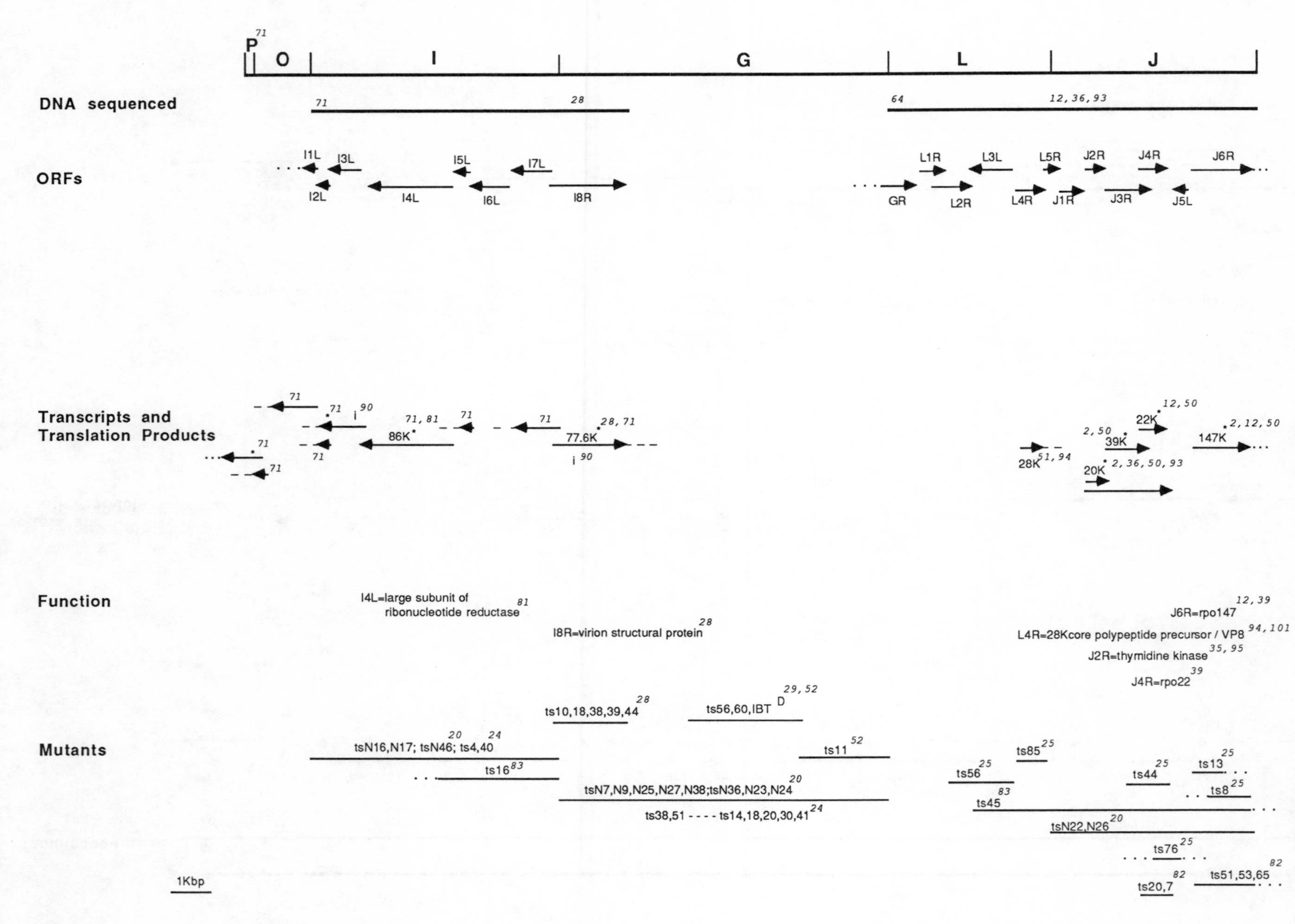

P
O
I
G
L
J
DNA sequenced
ORFs
I1L
I3L
I2L
I4L
I5L
I6L
I7L
I8R
GR
L1R
L2R
L3L
L4R
L5R
J1R
J2R
J3R
J4R
J5L
J6R
Transcripts and Translation Products
86K
77.6K
28K
20K
39K
22K
147K
Function
I4L=large subunit of ribonucleotide reductase
I8R=virion structural protein
J6R=rpo147
L4R=28Kcore polypeptide precursor / VP8
J2R=thymidine kinase
J4R=rpo22
Mutants
tsN16,N17; tsN46; ts4,40
ts16
ts10,18,38,39,44
tsN7,N9,N25,N27,N38;tsN36,N23,N24
ts38,51 - - - - ts14,18,20,30,41
ts56,60,IBT D
ts11
ts56
ts85
ts45
tsN22,N26
ts44
ts13
ts8
ts76
ts20,7
ts51,53,65
1Kbp

Figure 5

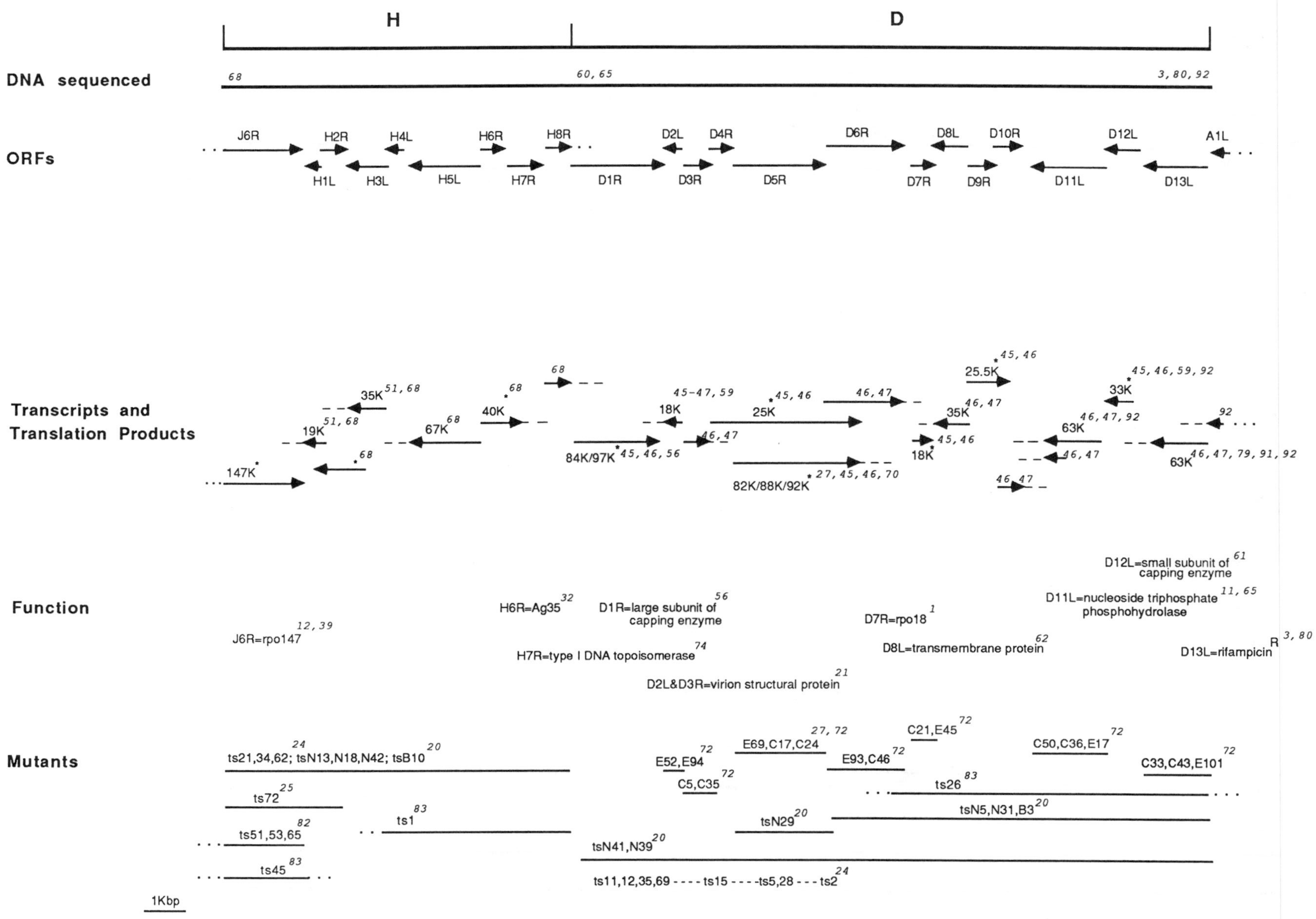

H D

DNA sequenced

68 60, 65 3, 80, 92

ORFs

J6R H2R H4L H6R H8R D2L D4R D6R D8L D10R D12L A1L

H1L H3L H5L H7R D1R D3R D5R D7R D9R D11L D13L

Transcripts and Translation Products

147K* 19K[51,68] 35K[51,68] *[68] 67K[68] 40K*[68] [68]

84K/97K*[45,46,56] 18K[45-47,59] [46,47] 25K*[45,46] 82K/88K/92K*[27,45,46,70] [46,47]

25.5K*[45,46] 35K[46,47] 18K*[45,46] [46,47]

63K[46,47,92] [46,47] 33K*[45,46,59,92] 63K[46,47,79,91,92] [92]

Function

J6R=rpo147[12,39]

H6R=Ag35[32]

H7R=type I DNA topoisomerase[74]

D1R=large subunit of capping enzyme[56]

D2L&D3R=virion structural protein[21]

D7R=rpo18[1]

D8L=transmembrane protein[62]

D12L=small subunit of capping enzyme[61]

D11L=nucleoside triphosphate phosphohydrolase[11,65]

D13L=rifampicin R[3,80]

Mutants

ts21,34,62; tsN13,N18,N42; tsB10[24,20]

ts72[25]

ts51,53,65[82]

ts45[83]

ts1[83]

E52,E94[72]

C5,C35[72]

E69,C17,C24[27,72]

tsN29[20]

tsN41,N39[20]

ts11,12,35,69 - - - - ts15 - - - -ts5,28 - - - ts2[24]

C21,E45[72]

E93,C46[72]

ts26[83]

tsN5,N31,B3[20]

C50,C36,E17[72]

C33,C43,E101[72]

1Kbp

Figure 6

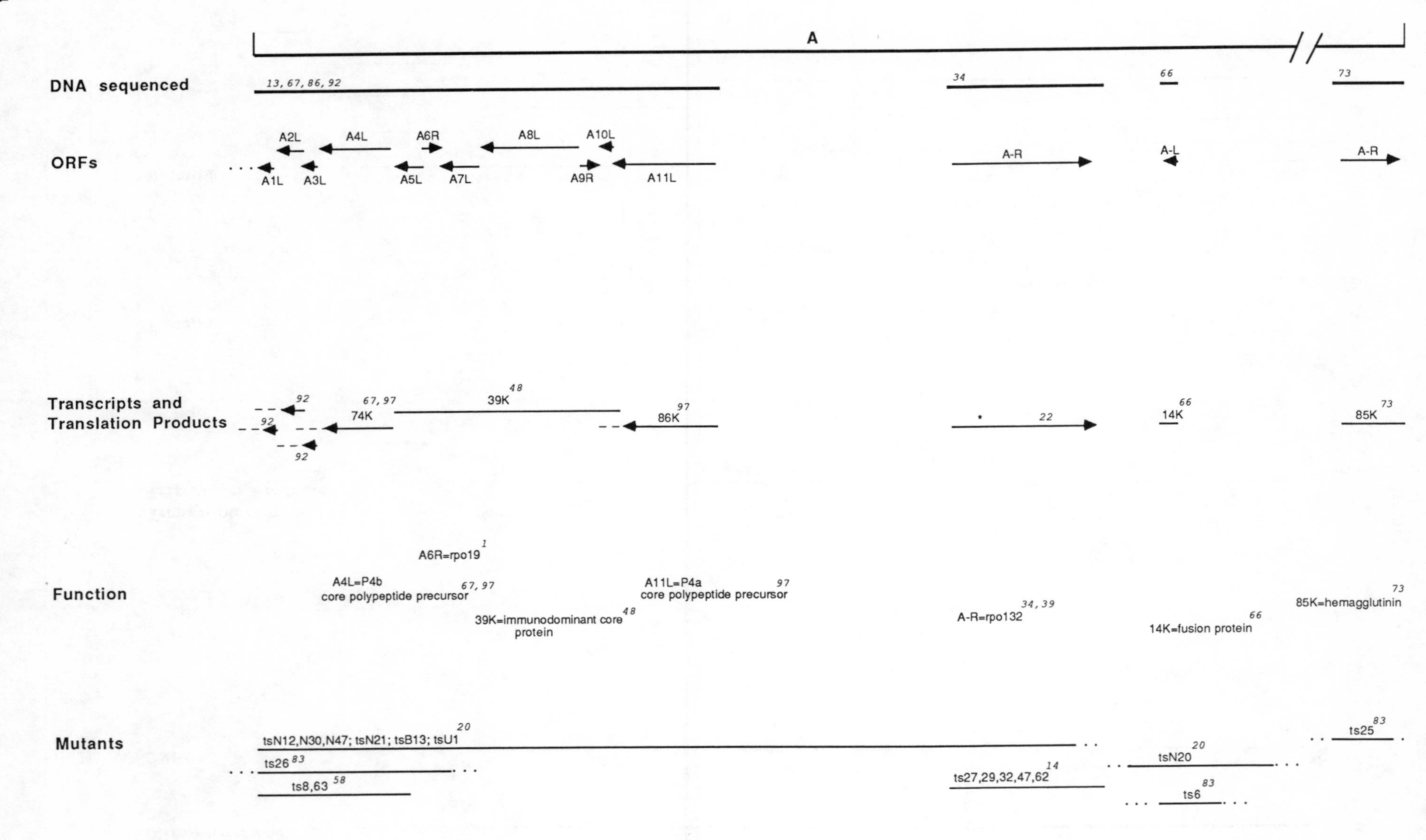
A
DNA sequenced
13, 67, 86, 92
34
66
73
ORFs
A1L
A2L
A3L
A4L
A5L
A6R
A7L
A8L
A9R
A10L
A11L
A-R
A-L
A-R
Transcripts and Translation Products
92
92
92
74K 67, 97
39K 48
86K 97
22
14K 66
85K 73
Function
A6R=rpo19 1
A4L=P4b core polypeptide precursor 67, 97
39K=immunodominant core protein 48
A11L=P4a core polypeptide precursor 97
A-R=rpo132 34, 39
14K=fusion protein 66
85K=hemagglutinin 73
Mutants
tsN12,N30,N47; tsN21; tsB13; tsU1 20
ts26 83
ts8,63 58
ts27,29,32,47,62 14
tsN20 20
ts6 83
ts25 83
1Kbp

Figure 7

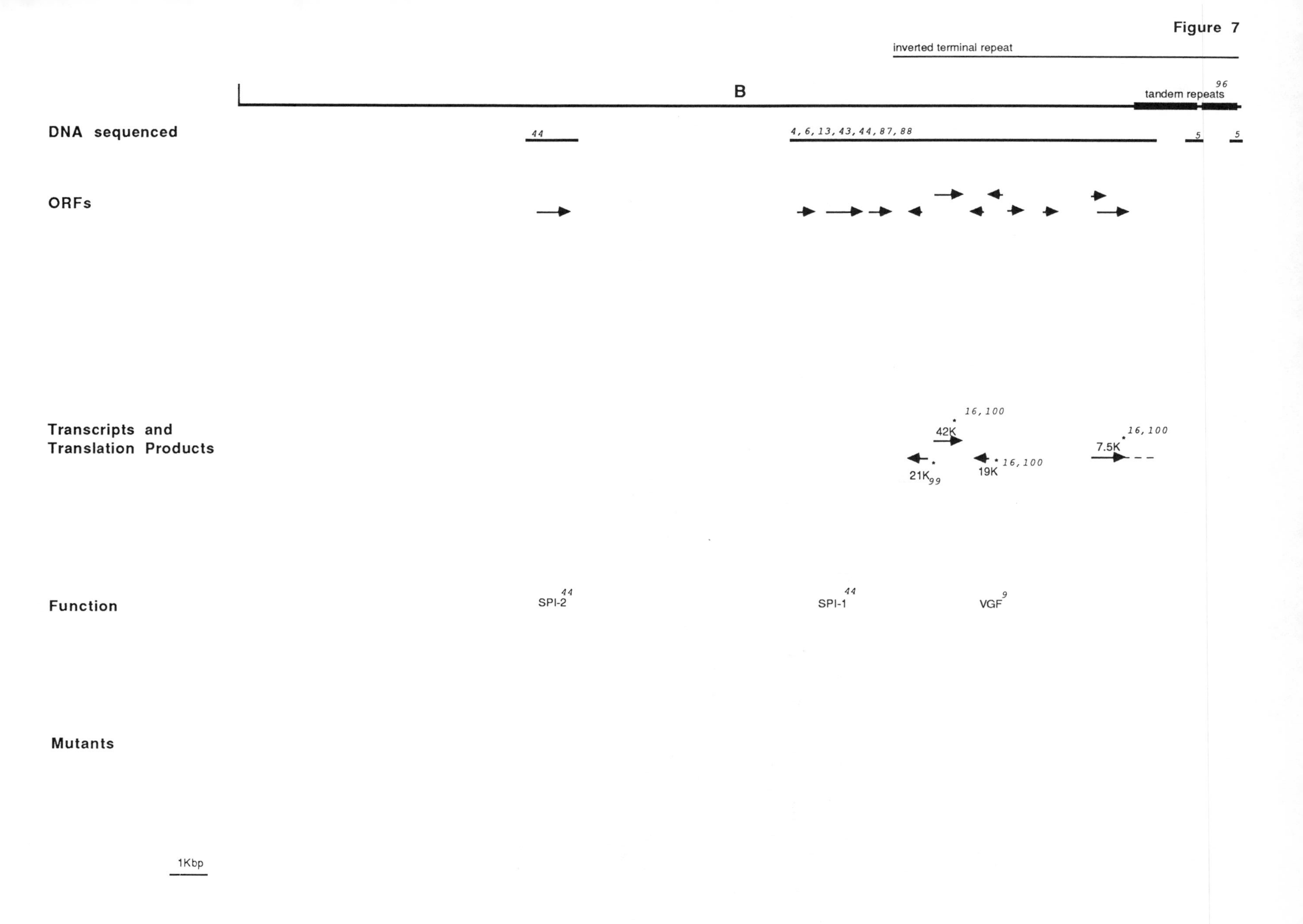
inverted terminal repeat
B
96
tandem repeats
DNA sequenced
44
4, 6, 13, 43, 44, 87, 88
5
5
ORFs
Transcripts and Translation Products
16, 100
42K
16, 100
7.5K
21K
99
19K
16, 100
Function
SPI-2
44
SPI-1
44
VGF
9
Mutants
1Kbp

Genes Encoding Proteins with Suggested Activities

Name	Activity	ORF
Vaccinia growth factor (VGF)	EGF receptor binding, cell proliferation	C6
Vaccinia complement binding protein (VCBP)	Inhibit classical complement cascade	C21
Host range protein	Replication in human cells	K0
Serine protease inhibitor 3 (SPI-3)	Protease inhibition	K1
Retroviral protease homolog	Protease?	F2
Ribonuleotide reductase, small subunit	Nucleotide metabolism	F4
RNA polymerase subunit (rpo30)	Transcription	E4
DNA polymerase	DNA replication	E10
Ribonucleotide reductase, large subunit	Nucleotide metabolism	I4
Thymidine kinase (TK)	Nucleotide metabolism	J2
RNA polymerase subunit (rpo22)	Transcription	J4
RNA polymerase subunit (rpo147)	Transcription	J6
DNA topoisomerase, type 1	DNA unwinding	H7
Capping enzyme, large subunit	mRNA modification	D1
RNA polymerase subunit (rpo18)	Transcription	D7
Nucleoside triphosphate phospho-hydrolase I (ATPase, NPH1)	Transcription?	D11
Capping enzyme, small subunit	mRNA modification	D12
RNA polymerase subunit (rpo19)	Transcription	A6
RNA polymerase subunit (rpo132)	Transcription	A?
Fusion protein	Virus entry	A?
Hemagglutinin	Virus entry	A?
Serine protease inhibitor 2 (SPI-2)	Protease inhibition	B?
Serine protease inhibitor 1 (SPI-1)	Protease inhibition	B?

REFERENCES

1. Ahn, B.-Y. et.al. Personal communication.
2. Bajszar, G. et.al. (1983) J. Virol. 45:62-72.
3. Baldick, C.J. & Moss, B. (1987) Virol. 156:138-145.
4. Baroudy, B.M. & Moss, B. (1982) Nucl. Acid Res. 10:5673-5679.
5. Baroudy, B.M. et.al. (1982) Cell 28:315-324.
6. Baroudy, B.M., Venkatesan, S. & Moss, B. Personal communication.
7. Belle Isle, H. et.al. (1981) Virol. 112:306-317.
8. Bertholet, C. et.al. (1985) Proc. Natl. Acad. Sci. 82:2096-2100.
9. Blomquist, M.C. et.al. (1984) Proc. Natl. Acad. Sci. 81:7363-7367; Reisner, A.H. (1985) Nature 313:801-803; Stroobant, P. et.al. (1985) Cell 42:383-393; Twardzik, D.R. et.al. (1985) Proc. Natl. Acad. Sci. 82:5300-5304; King, C.S. et.al. (1986) Mol. and Cell. Biol. 6:332-336; Buller, R.M. et.al. (1988) Virol.164:182-192; Buller, R.M. et.al. (1988) J.Virol. 62:866-874.
10. Boursnell, M.E.G. et.al. (1988) J. Gen. Virol. 69:2995-3003.
11. Broyles, S.S. & Moss, B. (1987) J. Virol. 61:1738-1742.
12. Broyles, S.S. & Moss, B. (1986) Proc. Natl. Acad. Sci. 83:3141-3145.
13. Cole, N.B. Personal communication.
14. Condit, R. Personal communication.
15. Condit, R.C. et.al. (1983) Virol. 128:429-443.
16. Cooper, J.A. et.al. (1981) J. Virol. 37:284-294.
17. Cooper, J.A. et.al. (1981) J. Virol. 39:733-745.
18. DeFilippes, F. J. Virol. 63:4060-4063.
19. DeFilippes, F. (1982) J. Virol. 43:136-149.
20. Drillien, R. & Spehner, D. (1983) Virol. 131:385-393.
21. Dyster, L. & Niles, E. Unpublished.
22. Earl, P.L. Unpublished.
23. Earl, P.L. et.al. (1986) Proc. Natl. Acad. Sci. 83:3659-3663.
24. Ensinger, M.J. & Rovinsky, M. (1983) J. Virol. 48:419-428.
25. Ensinger, M.J. et.al. (1985) J. Virol. 56:1027-1029.
26. Esposito, J.J. & Knight, J.C. (1985) Virol. 143:230-251.
27. Evans, E. & Traktman, P. (1987) J. Virol. 61:3152-3162.
28. Fathi, Z. & Condit, R. In preparation.
29. Fathi, Z. et.al. (1986) Virol. 155:97-105.
30. Gillard, S. et.al. (1986) Proc. Natl. Acad. Sci. 83:5573-5577.
31. Golini, F. & Kates, J.R. (1984) J. Virol. 49:459-470.
32. Gordon, J. et.al. (1988) Virol. 167:361-369.
33. Hirt, P. et.al. (1986) J. Virol. 58:757-764.
34. Holmes, M., Cole, N. & Moss, B. Personal communication.
35. Hruby, D.E. & Ball, L.A. (1982) J. Virol. 43:403-409.
36. Hruby, D.E. et.al. (1983) Proc. Natl. Acad. Sci. 80:3411-3415.
37. Jones, E.V. & Moss, B. (1984) J. Virol. 49:72-77.
38. Jones, E.V. & Moss, B. (1985) J. Virol. 53:312-315.
39. Jones, E.V. et.al. (1987) J. Virol. 61:1765-1771.
40. Kotwal, G.J. et.al. (1989) Virol. 171:579-587.
41. Kotwal, G.J. et.al. (1989) In UCLA Symposia on Molecular and Cellular Biology Vol 121, in press.
42. Kotwal, G.J. & Moss, B. (1988) Nature 335:176-178.
43. Kotwal, G.J. & Moss, B. (1988) Virol. 167:524-537.
44. Kotwal, G.J. & Moss, B. (1989) J. Virol. 63:600-606.
45. Lee-Chen, G-J. et.al. (1988) Virol. 163:64-79.
46. Lee-Chen, G-J. & Niles, E.G. (1988) Virol. 163:52-63.
47. Lee-Chen, G-J. & Niles, E.G. (1988) Virol. 163:80-92.
48. Maa, J-S. & Esteban, M. (1987) J. Virol. 61:3910-3919.

49. Mackett, M. & Archard. L.C. (1979) J. Gen. Virol. 45:683-701.
50. Mahr, A. & Roberts, B.E. (1984) J. Virol. 49:497-509.
51. Mahr, A. & Roberts, B.E. (1984) J. Virol. 49:510-520.
52. Meis, R. & Condit, R. Unpublished.
53. Merchlinsky, M. Personal communication.
54. Micryukov, N.N. et.al. (1988) Biotechnologiya 4:442-449.
55. Morgan, J.R. & Roberts, B.E. (1984) J. Virol. 51: 283-297.
56. Morgan, J.R. et.al. (1984) J. Virol. 52:206-214.
57. Moss, B. et.al. (1981) J. Virol. 40:387-395.
58. Nicholson, L. & Condit, R. Unpublished.
59. Niles, E., Unpublished.
60. Niles, E.G. et.al. (1986) Virol. 153: 96-112.
61. Niles, E.G. et.al. Submitted for publication.
62. Niles, E.G. & Seto, J. (1988) J. Virol. 62:3772-3778.
63. Panicali, D. et.al. (1981) J. Virol. 37:1000-1010.
64. Plucienniczak, A. et.al. (1985) Nucl. Acid Res. 13:985-998.
65. Rodriguez, J.F. et.al. (1986) Proc. Natl. Acad. Sci. 83:9566-9570.
66. Rodriguez, J.F. & Esteban, M. (1987) J. Virol. 61:3550-3554.
67. Rosel, J. & Moss, B. (1985) J. Virol. 56:830-838.
68. Rosel, J.L. et.al. (1985) J. Virol. 60:436-449.
69. Roseman, N. Personal communication.
70. Roseman, N.A. & Hruby, D.E. (1987) J. Virol. 61:1398-1406.
71. Schmitt, J.F.C. & Stunnenberg, H.G. (1988) J. Virol. 62:1889-1897.
72. Seto, J. et.al. (1987) Virol. 160:110-119.
73. Shida, H. (1986) Virol. 150:451-462.
74. Shuman, S. & Moss, B. (1987) Proc. Natl. Acad. Sci. 84:7478-7482.
75. Slabaugh, M. et.al. (1988) J.Virol. 62:519-527.
76. Slabaugh, M.B. & Roseman, N.A. (1989) Proc. Natl. Acad. Sci. 86:4152-4155.
77. Sridhar, P. & Condit, R.C. (1983) Virol. 128:444-457.
78. Tamin, A. et.al. (1988) Virol. 165,141-150.
79. Tartaglia, J. et.al. (1986) Virol. 150:45-54.
80. Tartaglia, J. & Paoletti, E. (1985) Virol. 147:394-404.
81. Tengelsen, L.A. et.al. (1988) Virol. 164:121-131.
82. Thompson, C.L. et.al. (1989) J. Virol. 63:705-713.
83. Thompson, C.L. & Condit. R.C. (1986) Virol. 150:10-20.
84. Traktman, P. et.al. (1989) J.Virol. 63:841-846.
85. Traktman, P. et.al. (1984) J. Virol. 49:125-131.
86. Van Meir, E. & Wittek, R. (1988) Arch. Virol. 102:19-27.
87. Venkatesan, S. et.al. (1982) J. Virol. 44:637-646.
88. Venkatesan, S. et.al. (1981) Cell 125:805-813.
89. Villareal, E.C. & Hruby, D.E. (1986) J. Virol. 57:65-70.
90. Vos, J.C. & Stunnenberg, H.G. (1988) EMBO J. 7:3487-3492.
91. Weinrich, S.L. et.al. (1985) J. Virol. 55:450-457.
92. Weinrich, S.L. & Hruby. D.E. (1986) Nucl. Acid Res. 14:3003-3016.
93. Weir, J.P. & Moss, B. (1983) J. Virol. 46:530-537.
94. Weir, J.P. & Moss, B. (1984) J. Virol. 51:662-669.
95. Weir, J.P. et.al. (1982) Proc. Natl. Acad. Sci. 79:1210-1214.
96. Wittek, R. & Moss, B. (1980) Cell 21:277-284.
97. Wittek, R. et.al. (1984) Nucl. Acid Res. 12:4835-4848.
98. Wittek, R. et.al. (1984) J. Virol. 49:371-378.
99. Wittek, R. et.al. (1981) J. Virol. 39:722-732.
100. Wittek. R. et.al. (1980) Cell 21:487-493.
101. Yang, W-P. et.al. (1988) Virol. 167:585-590.

The Human and Simian Immunodeficiency Viruses, HIV-I, HIV-II, and SIV_{mac}

Drs. Steven F. Josephs and Flossie Wong-Staal

Laboratory of Tumor Cell Biology, National Cancer Institute, Bethesda, Maryland 20892

The human retroviruses HIV-1 and HIV-2 are the etiological agents of the acquired immunodeficiency syndrome (AIDS). These viruses and one group of simian retroviruses, SIV_{mac}, are closely related. The following maps and information have been compiled from sequences and annotations as published (38).

Genes:

gag -viral core proteins
pol -polymerase:protease-reverse transcriptase/ribonuclease H-endonuclease/integrase
env -envelope
tat -transcriptional transactivator (formerly tat-3, TA)
rev -regulator of virus expression (formerly art/trs)
vif -virion infectivity factor (formerly sor,A,P',Q)
nef -negative regulatory factor (formerly 3'orf,B,E',F)
vpr -viral protein R (formerly R)
vpu -viral protein U (specific to HIV-1)
vpx -viral protein X (specific to HIV-2, SIV_{mac})

Abbreviations:

AP-1 -activator protein 1 binding sequence
CTF/NF-1 -CCAAT binding transcription factor/nuclear factor 1
Fos/Jun -cellular fos and jun oncogene encoded proteins
IL-2 -interleukin 2 gene (sequence similarity)
IL-2R -interleukin 2 receptor gene (sequence similarity)
LBP-1 -leader binding protein 1
LTR -long terminal repeat sequences
NFAT-1 -nuclear factor of activated T cells
NFkB -nuclear factor kappa B
NRE -negative regulatory element
sp1 -nuclear factor sp1 binding sites
TAR -transactivation response element
TAX -transcriptional activator (Human T-lymphotropic virus I)
UBP-1 -untranslated region binding protein 1
UBP-2 -untranslated region binding protein 2

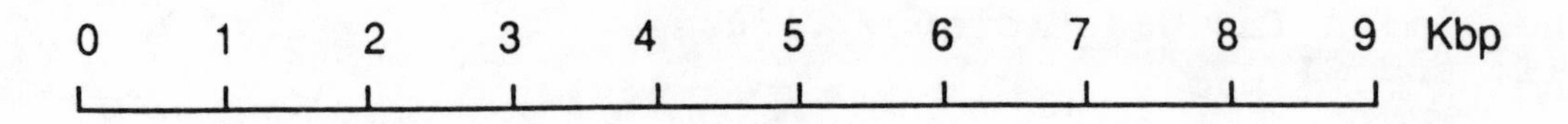

0
1
2
3
4
5
6
7
8
9
Kbp

HIV-1

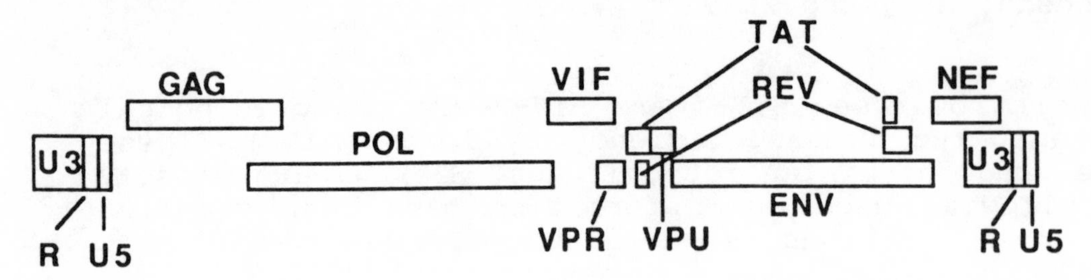

TAT
GAG
VIF
REV
NEF
POL
U3
U3
ENV
VPR
VPU
R
U5
R
U5

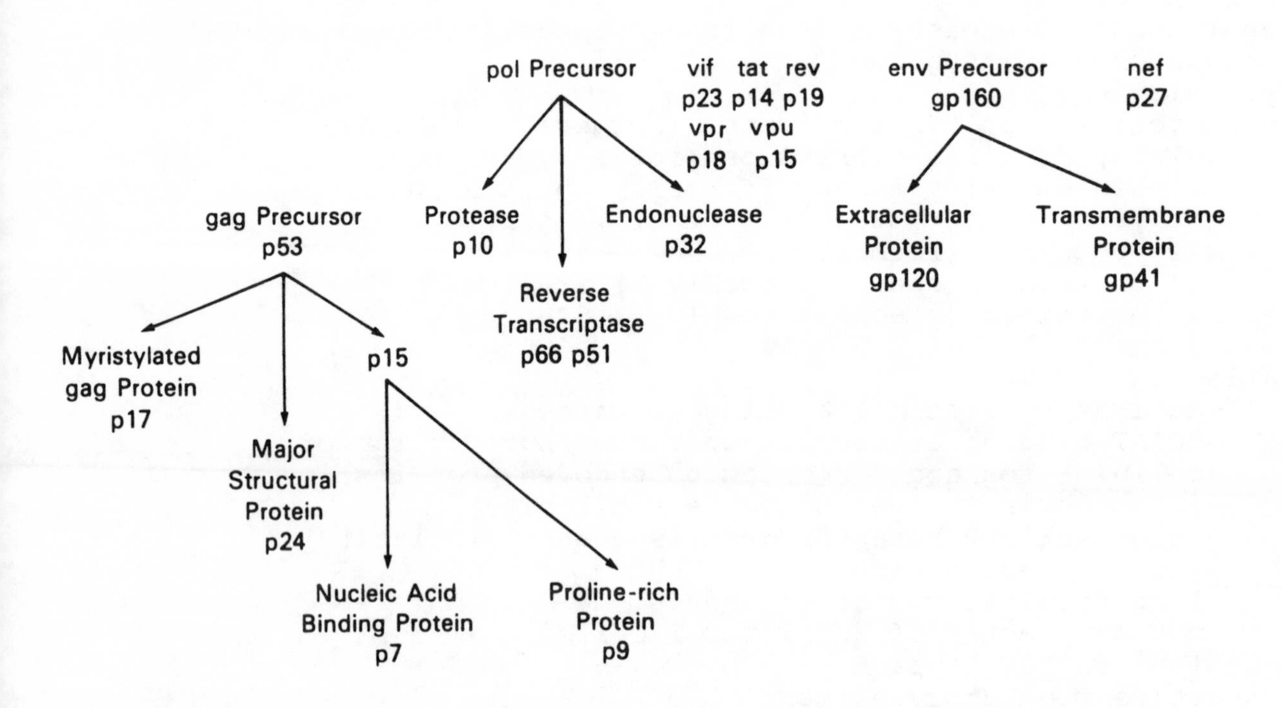

pol Precursor
vif p23
tat p14
rev p19
vpr p18
vpu p15
env Precursor gp160
nef p27
gag Precursor p53
Protease p10
Endonuclease p32
Extracellular Protein gp120
Transmembrane Protein gp41
Reverse Transcriptase p66 p51
Myristylated gag Protein p17
p15
Major Structural Protein p24
Nucleic Acid Binding Protein p7
Proline-rich Protein p9

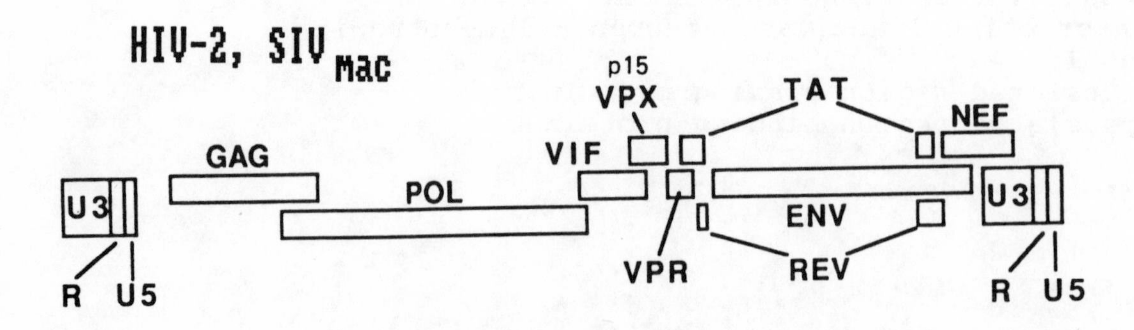

HIV-2, SIV mac
p15
VPX
TAT
NEF
VIF
GAG
POL
U3
U3
ENV
VPR
REV
R
U5
R
U5

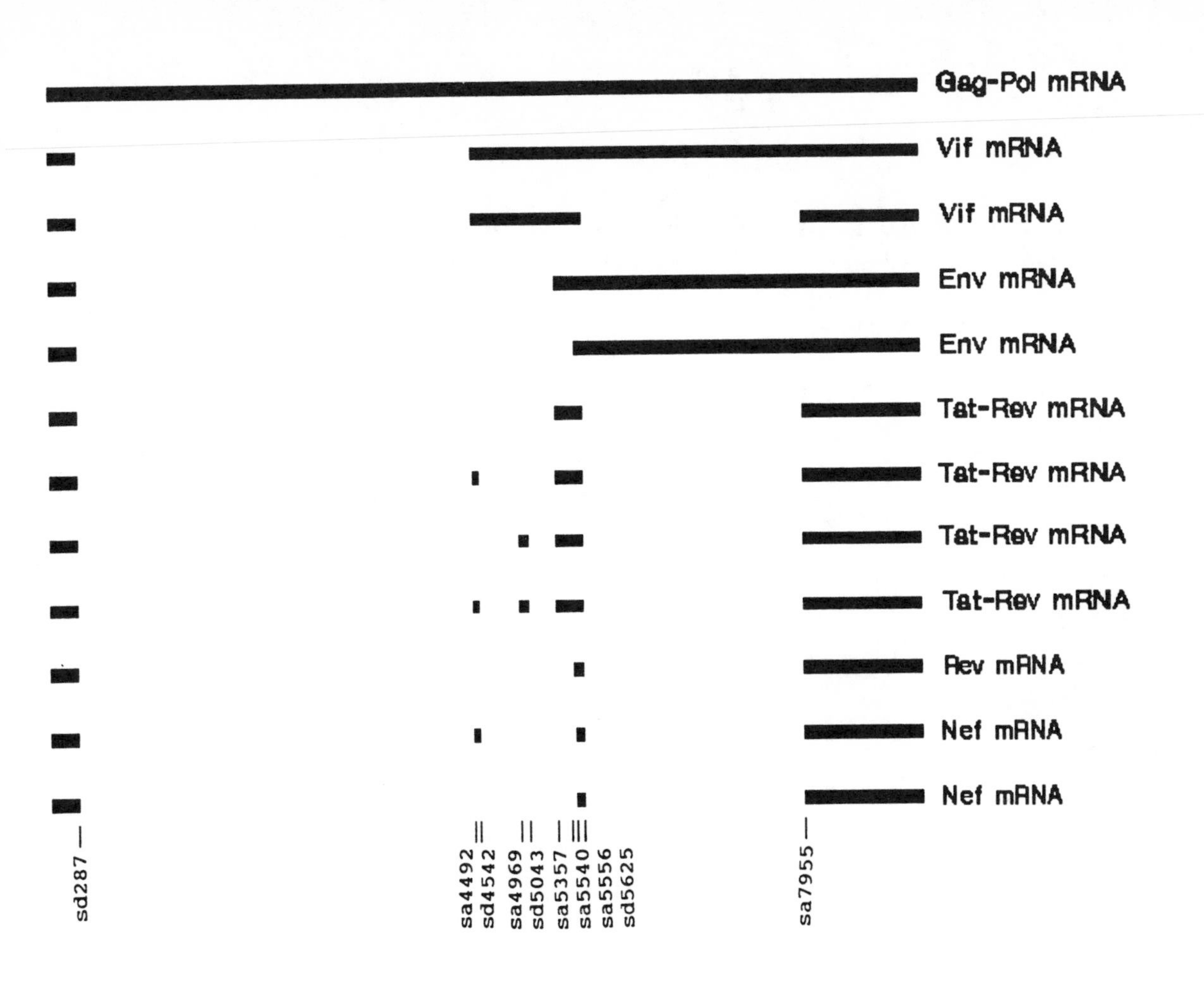
Gag-Pol mRNA
Vif mRNA
Vif mRNA
Env mRNA
Env mRNA
Tat-Rev mRNA
Tat-Rev mRNA
Tat-Rev mRNA
Tat-Rev mRNA
Rev mRNA
Nef mRNA
Nef mRNA
sd287
sa4492
sd4542
sa4969
sd5043
sa5357
sa5540
sa5556
sd5625
sa7955

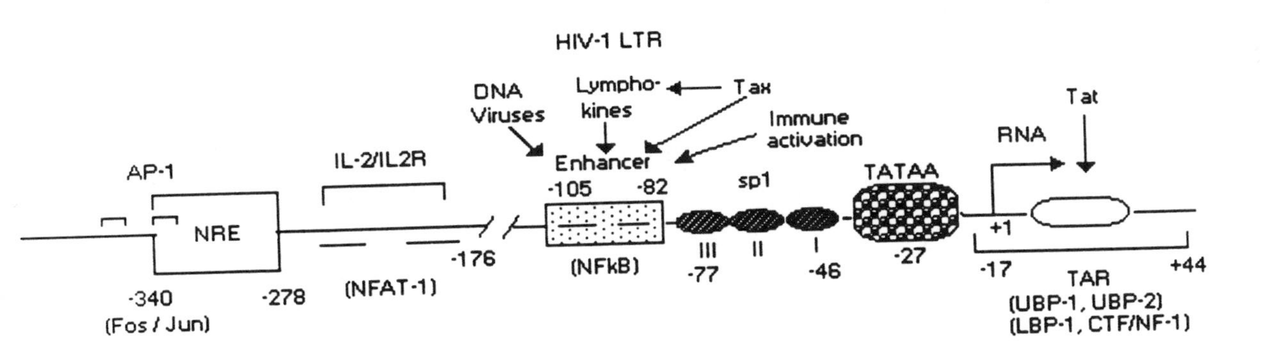
HIV-1 LTR
DNA Viruses
Lympho-kines
Tax
Immune activation
Tat
RNA
AP-1
IL-2/IL2R
Enhancer
-105
-82
sp1
TATAA
NRE
-176
(NFkB)
III
II
I
-77
-46
-27
+1
-17
+44
TAR
(UBP-1, UBP-2)
(LBP-1, CTF/NF-1)
-340
-278
(Fos / Jun)
(NFAT-1)

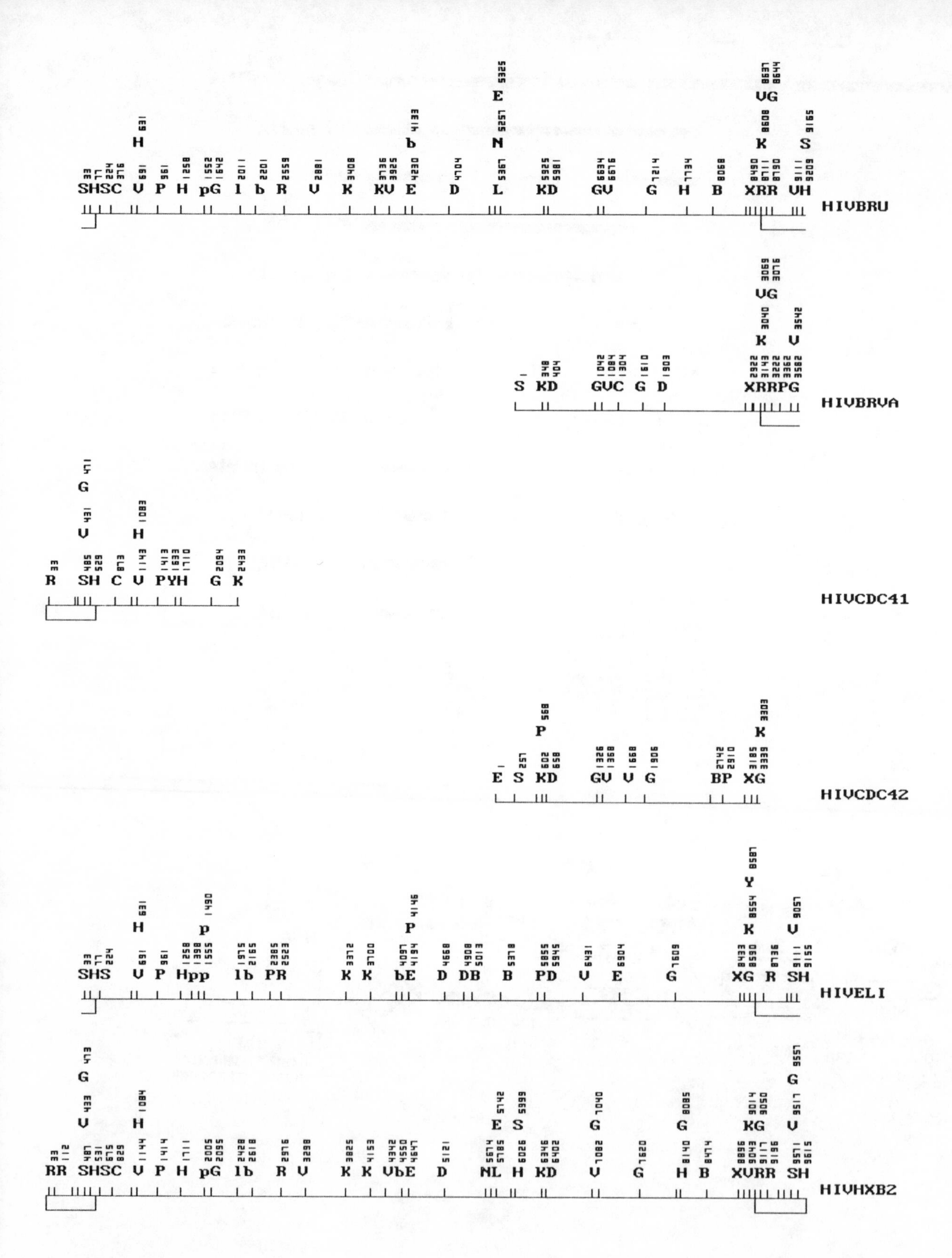
HIVBRU
HIVBRVA
HIVCDC41
HIVCDC42
HIVELI
HIVHXB2

H 599
SHGC 1 45 193 344; V 599; P 929; H 1226; pG 1520 1610; E 1874
HIVJH31

P 332
KD 366 422; RVVC 903 1054 1141 1361; H 2196; L 2551
HIVJH32

KD 152 208; V 721; G 885; DY 1743 1907; P 2451
HIVJY1

P 4175; S 9165
MH 27 75; V 705; B 3353; b 4126; G 7653; M 9160
SY 32 152; V 723; P 993; p 1584; I 2004; b 2194; R 2552; P 3011; K 3401; KVYE 3729 3918 4005 4223; D 4697; B 5407; KD 5518 5574; VG 6472 6630; Y 7658; K 8607; VH 9112 9208
HIVMAL

G 472; G 9577
V 432; H 1082; E 5758; H 6989; V 9537
R 112; SHSC 486 530 674 827; V 1142; P 1421; H 1718; pG 2012 2102; I 2444; bPR 2634 2854 2992; V 3314; K 3841; KVb 4169 4358 4566; H 4962; D 5137; NL 5690 5800; S 6014; KD 6358 6414; RG 6874 7070; G 7641; SDH 7788 7931 8158; L 8513; VKGR 8881 9035 9071 9217; SH 9551 9556 9635
HIVMN

G 473; V 9034; G 9548
V 433; E 5743; S 5995; G 7031; K 9005; V 9508
R 112; SHG 487 531 679; V 1145; P 1415; H 1712; pG 2006 2096; I 2429; bPR 2619 2839 2977; V 3299; K 3826; K 4154; Vb 4343 4551; D 5122; L 5785; H 6026; KD 6343 6399; V 7073; G 7611; H 8131; B 8465; AXGRR 8648 8887 9041 9108 9187; SH 9552 9606
HIVNL43

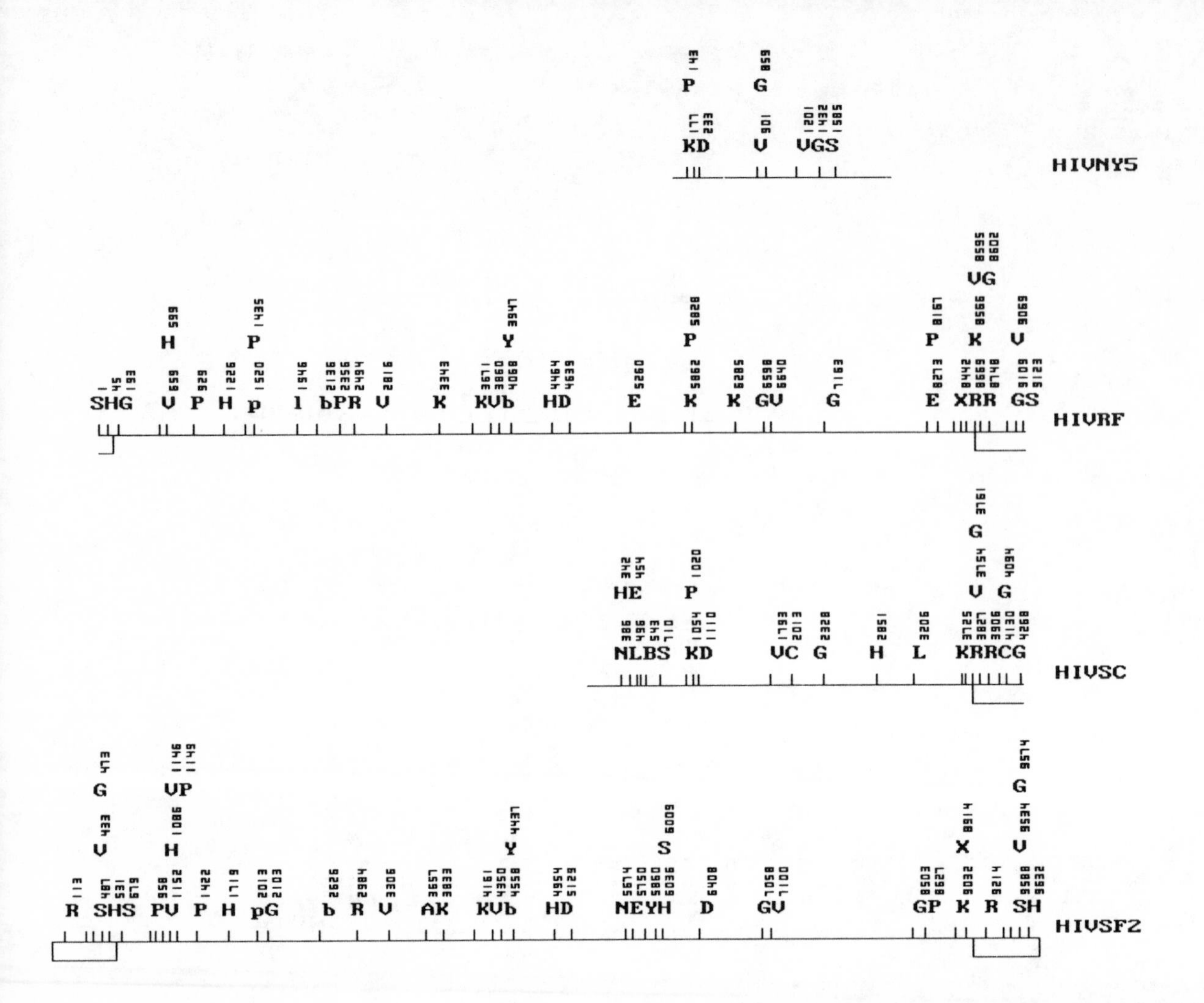
HIVNY5
HIVRF
HIVSC
HIVSF2

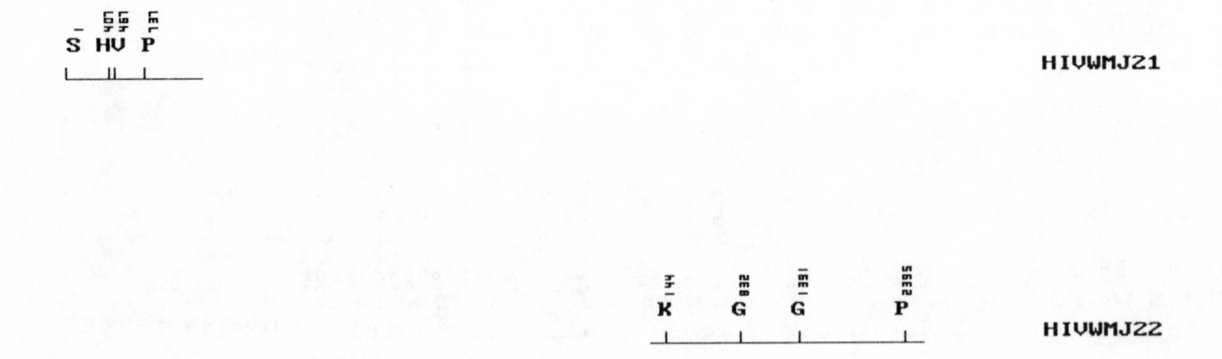
HIVWMJ21
HIVWMJ22

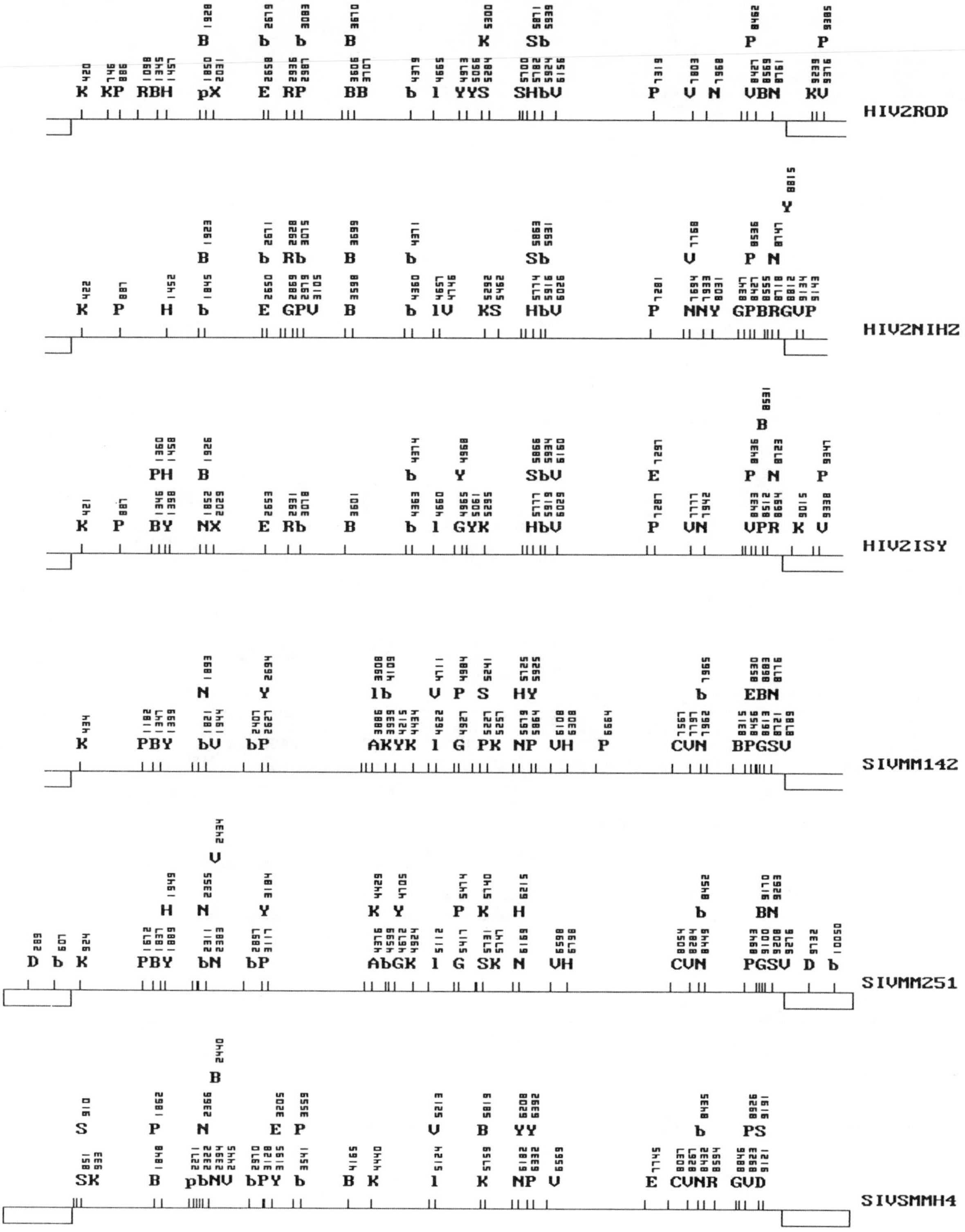

B b b B K Sb P P

K KP RBH pX E RP BB b I YYS SHbV P V N VBN KV

HIV2ROD

Y

B b Rb B b Sb V P N

K P H b E GPV B b IV KS HbV P NNY GPBRGVP

HIV2NIH2

B

PH B b Y SbV E P N P

K P BY NX E Rb B b I GYK HbV P VN VPR K V

HIV2ISY

N Y Ib V P S HY b EBN

K PBY bV bP AKYK I G PK NP VH P CVN BPGSV

SIVMM142

V

H N Y K Y P K H b BN

D b K PBY bN bP AbGK I G SK N VH CVN PGSV D b

SIVMM251

B

S P N E P V B YY b PS

SK B pbNV bPY b B K I K NP V E CVNR GVD

SIVSMMH4

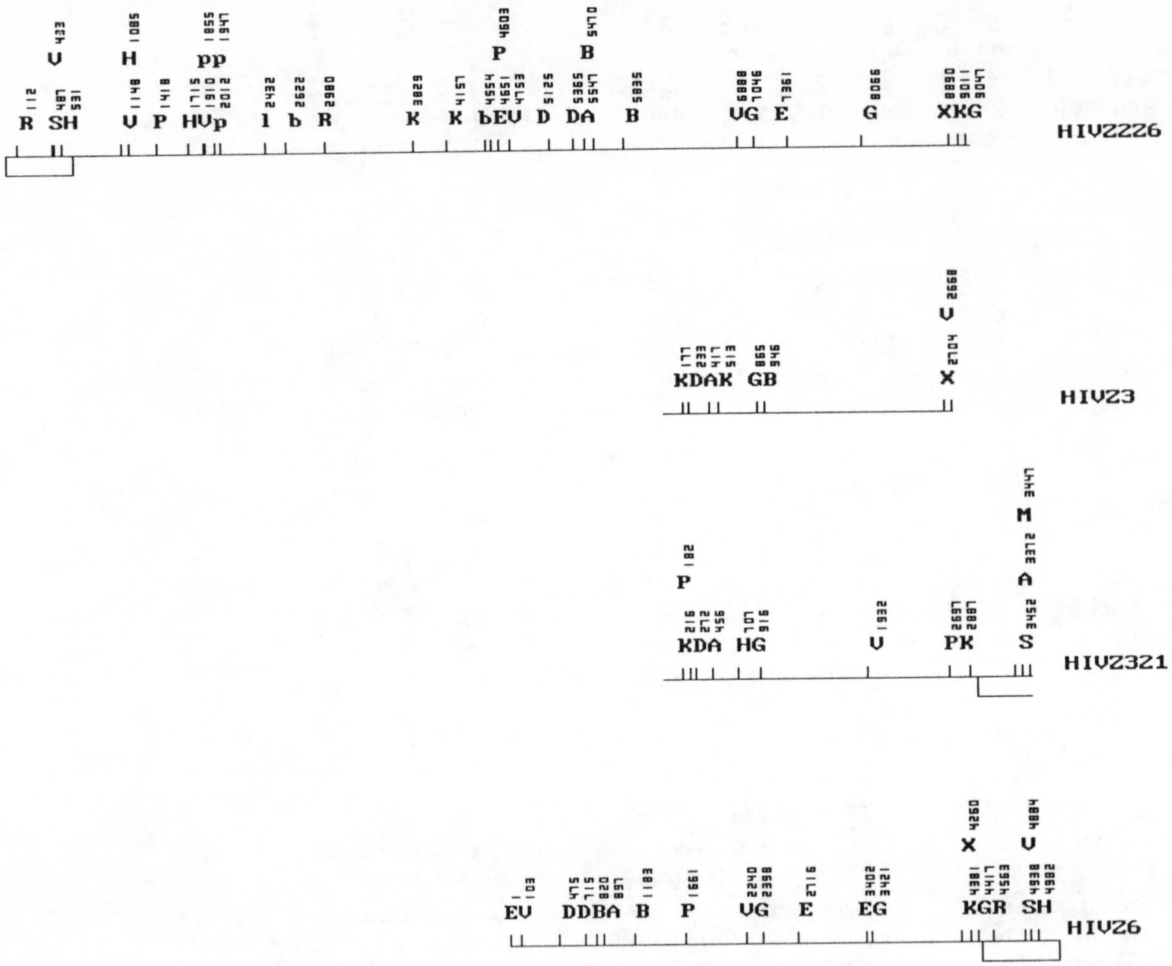

HIV2226

HIV23

HIV2321

HIV26

Restriction Enzyme Sites			
A - Hpa I	E - Eco RI	L - Sal I	R - Eco RV
b - Bal I	G - Bgl II	M - Sma I	S - Sac I
B - Bam HI	H - Hind III	N - Nco I	V - Pvu II
C - Cla I	K - Kpn I	p - Apa I	X - Xho I
D - Nde I	l - Bcl I	P - Pst I	Y - Xba I

HIVMN (9738)
2000
4000
6000
8000
10000
HIVHXB2 (9718)
HIVNL43 (9709)
HIVSF2 (9737)
HIVZ2Z6 (9081)
HIVCDC41 (2438)
HIVMAL (9229)
HIVBRU (9229)
HIVELI (9176)
HIVRF (9128)
HIVJH31 (1897)
HIVWMJ21 (1278)
HIVMN (9738)
HIVZ6 (5159)
HIVSC (4273)
HIVCDC42 (3373)
HIVJH32 (2903)
HIVBRVA (3600)
HIVZ321 (3457)
HIVZ3 (2709)
HIVNY5 (2125)
HIVJY1 (2653)
HIVWMJ22 (2568)

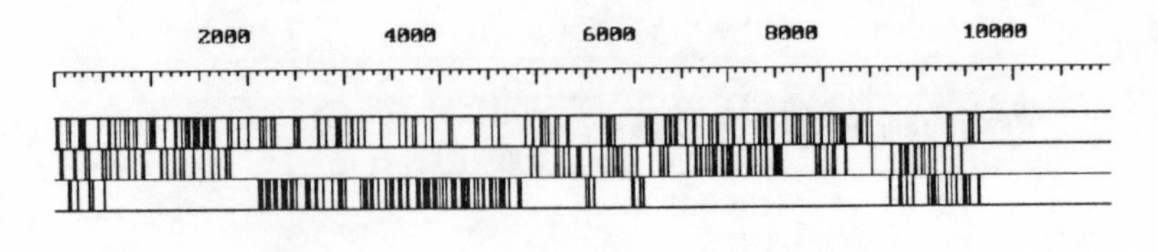

HIV2ISY (9636)

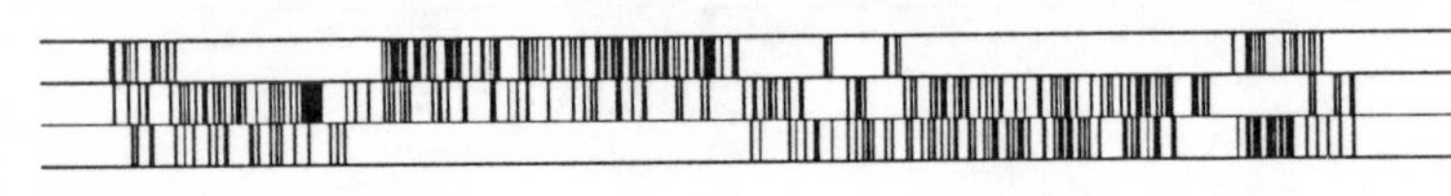

SIVMM142 (9646)

SIVMM251 (10277)

SIVSMMH4 (10241)

References:HIV-1
1. Adachi, A., et. al. 1986. J. Virol. 59:284-291.
2. Alizon, M., et al. 1986 Cell 46:63-74.
3. Allan, J., et al. 1985. Science 228:1091-1093.
4. Allan, J., et al. 1985. Science 230:820-823.
5. Anand, R., et al. 1989. Virology 168:79-89.
6. Arya, S. K., et al. 1985. Science 229:69-73.
7. Arya, S.K. and Gallo, R.C. 1986. Proc. Nat. Acad. Sci. (USA) 83:2209-2213.
8. Barin, F., et al. 1985. Science 228:1094-1096.
9. Benn, S., et al. 1985. Science 230:949-951.
10. Cullen B.R., et al. 1986. Cell 46:973-982.
11. Crowl, R., et al. 1985. Cell 41:979-986.
12. Desai, S.M., et. al. 1986. Proc. Natl. Acad. Sci. (USA) 83:8380-8384.
13. Ensoli, B., et al. 1989. EMBO J. 8:
14. Feinberg, M., et al. 1986. Cell 46:807-817.
15. Feng, S. and Holland E.C. 1988. Nature 334:165-167.
16. Fisher, A.G., et al. 1986. Nature 320:367-373.
17. Frankel, A.D., et al. 1988. Science 240:70-73.
18. Franza, B.R. Jr., et al. 1987. Nature 330:391-395.
19. Franza, B.R. Jr., et al. 1988. Science 239:1150-1153.
20. Garcia, J.A., et al. 1988. EMBO J. 7:3143-3147.
21. Garcia, J.A., et al. 1989. EMBO J. 8:765-778.
22. Green M. and Lowenstein P.M. 1988. Cell 55:1179-1188.
23. Gurgo, C., et al. 1988. Virology 164:531-536.
24. Hahn, B., et al. 1984. Nature 312:166-169.
25. Hahn, B., et al. 1986. Science 232:1548-1553.
26. Hauber, J. and Cullen, B.R. 1988. J. Virol. 62:673-679.
27. Hauber, J., et al. 1987. Proc. Nat. Acad. Sci. (USA)84:6364-6368.
28. Jakobovits A., et al. 1988. Mol. Cell Biol. 8:2555-2561.
29. Jeang K.-T. et al. 1988. Proc. Nat. Acad. Sci. (USA) 85:8291-8292.
30. Jones, K.A., et al. 1988. Genes Dev. 2:1101-1114.
31. Kan, N., et al. 1986. Science 231:1553-1555.
32. Kao, S.-Y., et al. 1987. Nature 330:489-493.
33. Knight D.M., et al. 1987. Science 237:837-840.
34. Kuppuswamy M., et al. 1989. Nucleic Acids Res. 17:3551-3561.
35. Lee, T.-H., et al. 1986. Science 231:1546-1549.
36. Malim, M.H., et al. 1989. Nature 338:254-257.
37. Muessing, M.A., et al. 1985. Nature 313:450-458.
38. Myers G., et al. (Eds.) 1989. Human Retroviruses and AIDS. Los Alamos National Laboratory, Los Alamos.
39. Nabel, G. and Baltimore, D. 1987. Nature 326:711-713.
40. Okamoto, T. and Wong-Staal, F. 1986. Cell 47:29-35.
41. Patarca, R. and Haseltine, W.A. 1987. Aids Res. Retroviruses 3:1-2
42. Rabson, A.B., et al. 1985. Science 229:1388-1340.
43. Ratner, L., et al. 1985. Nature 313:277-284.
44. Ratner, L., et al. 1985. Nucleic Acids Res. 13:8219-8229.
45. Ratner, L., et al. 1987. AIDS Res. Hum. Retroviruses 3:57-59.
46. Rice, A.P. and Mathews, M.B. 1988. Nature 332:551-553.
47. Rosen, C., et al. 1985. Cell 41:813-823.
48. Rosen, C., et al. 1986. Nature 319:555-559.
49. Ruben, S., et al. 1989. J. Virol. 73:1-8.
50. Sadaie, M.R., et al. 1988. Science 239:910-913.
51. Sadaie, M.R., et al. 1988. Proc. Nat. Acad. Sci. (USA) 85:9224-9228.
52. Selby, M.J., et al. 1989. Genes Dev. 3:547-558.
53. Sanchez-Pescador, R., et al. 1985. Science 227:484-492.
54. Shaw, G., et al. 1984. Science 227:1165-1171.
55. Siekevitz, M., et al. 1987. Science 328:1575-1578.
56. Sodroski, J., et al. 1985. Science 229:74-77.
57. Sodroski, J., et al. 1986. Nature 321:412-417.
58. Srinivasan, A., et al. 1987. Gene 52:71-82.
59. Starcich, B., et al. 1985. Science 227:538-540.
60. Starcich, B., et al. 1986. Cell 45:637-648.
61. Tong-Starksen, S.E., et. al. 1987. Proc. Natl. Acad. Sci. (USA) 84:6845-6849.
62. Toohey, M.G. and Jones, K. 1989. Genes Dev. 3:265-282.
63. Veronese, F.D., et al. 1986. Science 229:1402-1405.
64. Wain-Hobson, S., et al. 1985. Cell 40:9-17.
65. Willey, L., et al. 1986. Proc. Nat. Acad. Sci. (USA) 83:5038-5042.
66. Wong-Staal, F., et al. 1985. Science 229:759-762.
67. Wong-Staal, F. and Gallo, R.C. 1985. Nature 317:395-403.
68. Wong-Staal, F., et al. 1987. AIDS Res. Hum. Retroviruses 3:33-39
69. Wright, C., et al. 1986. Science 324:988-992.
70. Wu, F.K., et al. 1988. EMBO J. 7:2117-2129.
71. Yourno, J., et al. 1988. AIDS Res. Hum. Retroviruses 4:165-173.
HIV-2, SIV_{mac}
72. Chakrabarti, L., et al. 1987. Nature 328:543-547.
73. Clavel, F., et al. 1986. Nature 324:691-695
74. Daniel, M.D., et al. 1988. J. Virol. 62:4123-4128.
75. Fukasawa, M. et al. 1988. Nature 333:457-461.
76. Franchini, G. et al. 1987. Nature 328:539-543.
77. Franchini, G. et al. 1987. AIDS Res. Hum. Retroviruses 3:11-17.
78. Franchini, G. et al. 1988. AIDS Res. Hum. Retroviruses 4:243-250
79. Franchini, G. et al. 1989. Proc. Nat. Acad. Sci. (USA) 86:2433-2437.
80. Henderson, L.E. et al. 1988. Science 24:199-201.
81. Guo, H.-G., et al. 1987. AIDS Res. Hum. Retroviruses 3:177-185
82. Guyader, M., et al. 1987. Nature 326:662-669.
83. Hirsch, V., et. al. 1987. Cell 49:307-319.
84. Hirsch, V., et al. 1989. Nature 339:389-392.
85. Kanki, P.J., et al. Science 232: 238-240.
86. Zagury, J.F., et al. 1988. Proc. Natl. Acad. Sci. (USA) 85:5941-5945.
Acknowledgement: We thank Edward Wu and Richard Guilfoyle, Pan Data Systems, Rockville, MD, for computer graphics and program design.

October, 1986

Drs. Lee Ratner[1] and Flossie Wong-Staal[2]
[1]Departments of Medicine and Microbiology & Immunology, Washington University, St. Louis, Missouri 63110
[2]Laboratory of Tumor Cell Biology, National Cancer Institute, Bethesda, Maryland 20892

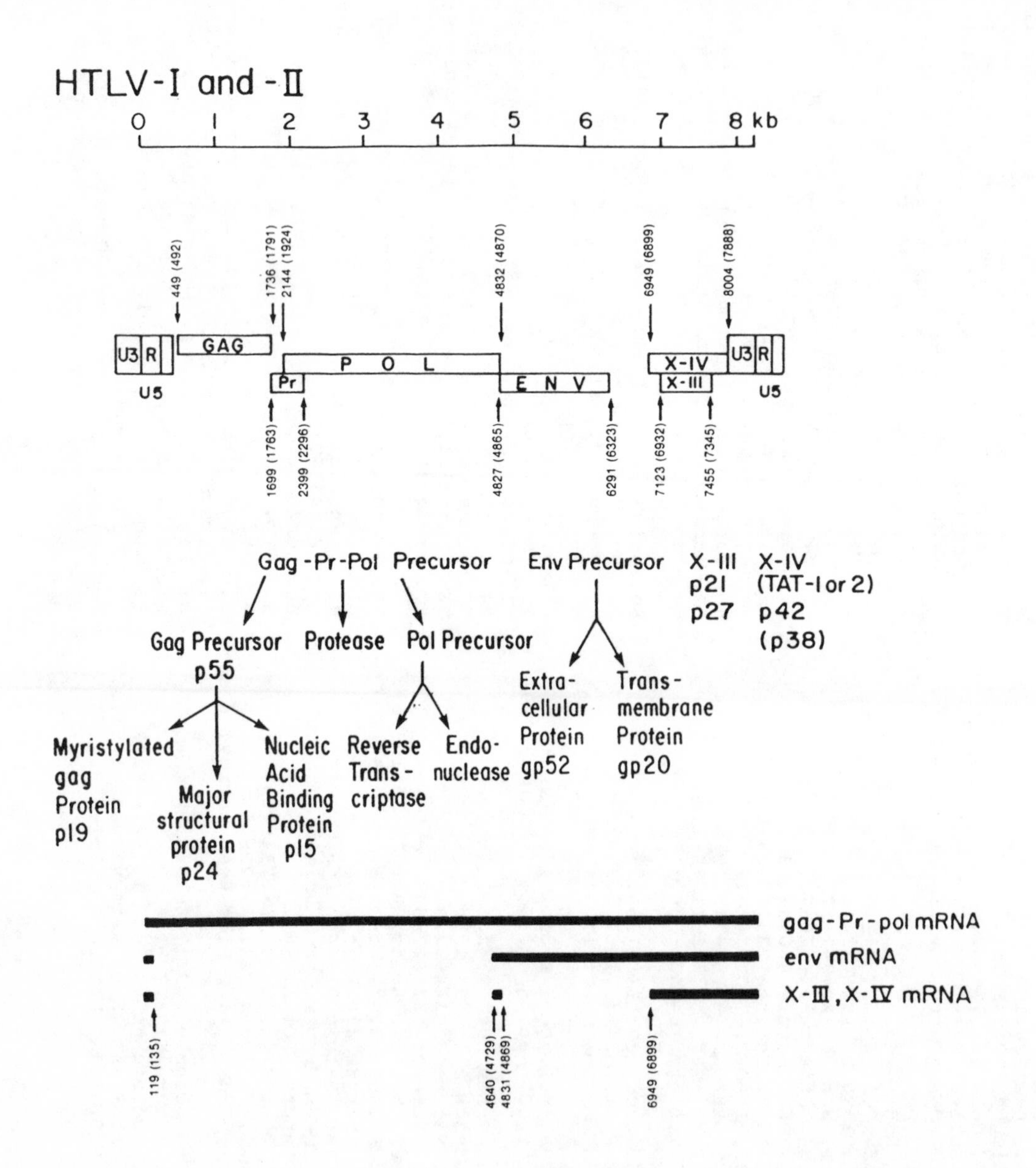

LTRs:

	Size of LTR Regions				tRNA Primer Binding Site	Location of Polyadenylation Signal
	LTR	U3	R	U5		
HTLV-I	759	354	229	176	Pro	U3
HTLV-II	767	316	248	203	Pro	U3

HTLV-I_{ATK}

HTLV-I_{Ch}

HTLV-I_{MT2}

HTLV-Ib_{EI}

HTLV-II_{MO}

HTLV-II_{JP}

HTLV-II_{NRA}

REFERENCES:

HTLV-I:
1. Felber, B.K., et al. 1985. Science 229:275-279.
2. Gallo, R.C. 1984. Cancer Surv. 3:113-159.
3. Hahn, B., et al. 1984. Int. J. Cancer 34:613-618.
4. Josephs, S.F., et al. 1984. Virol. 139:340-345.
5. Kiyokawa, T., et al. 1985. Proc. Nat. Acad. Sci. (USA) 82:8359-8363.
6. Lee, T.-H., et al. 1984. Proc. Nat. Acad. Sci. (USA) 81:3856-3860.
7. Ratner, L., et al. 1985. J. Virol. 54:781-790.
8. Rosen, C., et al. 1985. Proc. Nat. Acad. Sci. (USA) 82:6502-6506.
9. Seiki, M., et al. 1983. Proc. Nat. Acad. Sci. (USA) 80:3618-3622.
10. Seiki, M., et al. 1985. Science 228:1532-1534.
11. Shaw, G.M., et al. 1984. Proc. Nat. Acad. Sci. (USA) 81:4544-4548.
12. Shimotohno, K., et al. 1985. Proc. Nat. Acad. Sci. (USA) 82:302-306.
13. Slamon, D., et al. 1985. Science 228:1427-1430.
14. Sodroski, J., et al. 1984. Science 225:381-384.
15. Sodroski, J., et al. 1985. Science 228:1430-1434.
16. Wachsman, W., et al. 1984. Science 226:177-179.
17. Wachsman, W., et al. 1985. Science 228:1534-1537.

HTLV-II:
18. Chen, I.S.Y., et al. 1983. Nature 305:502-505.
19. Devare, S., et al. 1985. Virol. 142:206-210.
20. Gelmann, E.P., et al. 1984. Proc. Nat. Acad. Sci. (USA) 81:993-997.
21. Hahn, B., et al. In: Gottlieb, M.J. and Groopman, J.E. (eds). Acquired Immune Deficiency Syndrome pp. 59-72, Alan R. Liss, Inc. New York.
22. Haseltine, W., et al. 1984. Science 225:419-421.
23. Kalyanaraman, V.S., et al. 1982. Science 318:571-573.
24. Kalyanaraman, V.S., et al. 1985. EMBO J. 4:1455-1460.
25. Rosenblatt, J.D., et al. 1986. New Engl. J. Med. 315:372-377.
26. Schultz, A. and Oroszolan, S. 1984. Virol. 133:431-437.
27. Shimotohno, K., et al. 1984. Proc. Nat. Acad. Sci. (USA) 81:1079-1083.
28. Shimotohno, K., et al. 1984. Proc. Nat. Acad. Sci. (USA) 81:6657-6661.
29. Shimotohno, K., et al. 1985. Proc. Nat. Acad. Sci. (USA) 82:3101-3105.
30. Sodroski, J., et al. 1984. Science 225:421-424.
31. Sodroski, J., et al. 1984. Proc. Nat. Acad. Sci. (USA) 81:4617-4621.

THE NON-HUMAN LENTIVIRUSES

Thierry Huet and Simon Wain-Hobson
Laboratoire de Biologie et Immunologie Moléculaire des Rétroviruses
25, Rue de Dr. Roux, Institut Pasteur, 75724 Paris cedex 15

The lentiviruses represent a non-oncogenic subgroup of the retrovirus group (Weiss et al., 1985). They include the human immunodeficiency viruses type 1 (Barré-Sinoussi et al., 1983) and type 2 (Clavel et al., 1986). The molecular genetics of these viruses are dealt with by Wong-Staal and collegues in this volume.

Lentiviruses, like RNA viruses, are genetically highly variable. Thus there are an enormous number of strains. The SIVagm group of viruses appear to be particularly divergent. Due to the extensive internal genetic variation within an isolate no single clone is representative of a virus (Goodenow et al., 1989). In addition in vitro cultivation provides a strong selective pressure (Meyerhans et al., 1989). For these reasons this chapter provides a qualitative overview of lentiviral genetics. For those working with molecular clones there is no substitute to having the relevent sequences in a personnal computer file.

References

Barré-Sinoussi et al., Science **220**, 868-871 (1983).
Benveniste et al., J. Virol. **60**, 483-490 (1986).
Blum et al., Virology **142**, 270-277 (1985).
Braun et al., J. Virol. **61**, 4046-4054 (1987).
Chakrabarti et al., Nature **328**, 543-547 (1987).
Chiu et al., Nature **317**, 366-368 (1985).
Clavel et al., Science **233**, 343-346 (1986).
Colombini et al., PNAS **86**, 4813-4817 (1989).
Crawford et al., Science **207**, 997-999 (1980).
Daniel et al., Science **228**, 1201-1204 (1985).
Doolittle, Nature **339**, 338-339 (1989).
Doolittle et al., Cur. Top. Microbiol. Immunol. In press (1989)
Dorn & Derse, J. Virol. **62**, 3522-3526 (1988).
Essex and Kanki, Nature **331**, 621-622 (1988).
Franchini et al., Nature **328**, 539-543 (1987).
Fukasawa et al., Nature **333**, 457-461 (1988).
Fultz et al., PNAS **83**, 5286-5290 (1986).
Gonda et al., Nature **330**, 388-391 (1987).
Goodenow et al., J. AIDS In press (1989).
Harris et al., Virology **113**, 573-583 (1981).
Hess et al., J. Virol. **60**, 385-393 (1986).
Hirsch et al., Cell **49**, 307-319 (1987).
Hirsch et al., J. Med. Primatol. in press (1989a).
Hirsch et al., Nature **339**, 389-392 (1989b).
Kanki et al., Science **230**, 951-954 (1985).
Kanki et al., Science **232**, 238-243 (1986).
Kawakami et al., Virology **158**, 300-312 (1987).
Kestler et al., Nature **331**, 619-622 (1988).
Lowenstine et al., Int. J. Cancer **38**, 563-574 (1986).
Mazarin et al., J.Virol. **62**, 4813-4814 (1988).
Meyerhans et al., Cell In press (1989).
Molineaux et al., Gene **23**, 137-144 (1983).
Murphy-Corb et al., Nature **321**, 435-437 (1986).
Myers et al., HIV Sequence Database, Los Alamos, (1989).
Ohta et al., Int. J. Cancer **41**, 115-122 (1988).
Olmsted et al., PNAS **86**, 2448-2452 (1989).
Pederson et al., Science **235**, 790-793 (1987).
Pyper et al., J. Virol. **58**, 665-670 (1986).
Querat et al., J. Virol **52**, 672-679 (1984).
Rushlow et al., Virology **155**, 309-321 (1986).
Sherman et al., **62**, 120-126 (1988).
Sonigo et al., Cell **42**, 369-382 (1985).
Stephens et al., Science **231**, 589-594 (1986).
Tsujimoto et al., J.Virol. **62**, 4044-4050 (1988).
van der Maaten et al., J. Natl. Can. Inst. **49**, 1649-1657 (1972).
Vigne et al., Virology **161**, 218-227 (1987).
Yaniv et al., Virology **145**, 340-345 (1985).
Weiss et al., RNA Tumour Viruses, Cold Spring Harbor Labs, (1985).

Simian Immunodeficiency Virus (SIV)

SIVmac First isolated from rhesus monkey, *Macaca mulatta* (Daniel et al., 1985). Only found in captive macaques, including *M. nemestrina* (Benveniste et al., 1986) and *M. fascicularis* (Kestler et al., 1988). Pathogenic in Rhesus monkeys.
Molecular cloning and sequencing: strain 142 (Chakrabarti et al., 1987). Other isolates cloned in Kestler et al., 1988.

Previously described isolates of STLV3-agm (Kanki et al., 1985), and HTLV4 (Kanki et al., 1986) turned out to be contaminants of SIVmac strain 251 (Kestler et al., 1988; Essex and Kanki, 1988). Thus STLV3agm clones and sequences are those of SIVmac strain 251 (Franchini et al., 1987 and Hirsch et al., 1987).

Genetic organization:

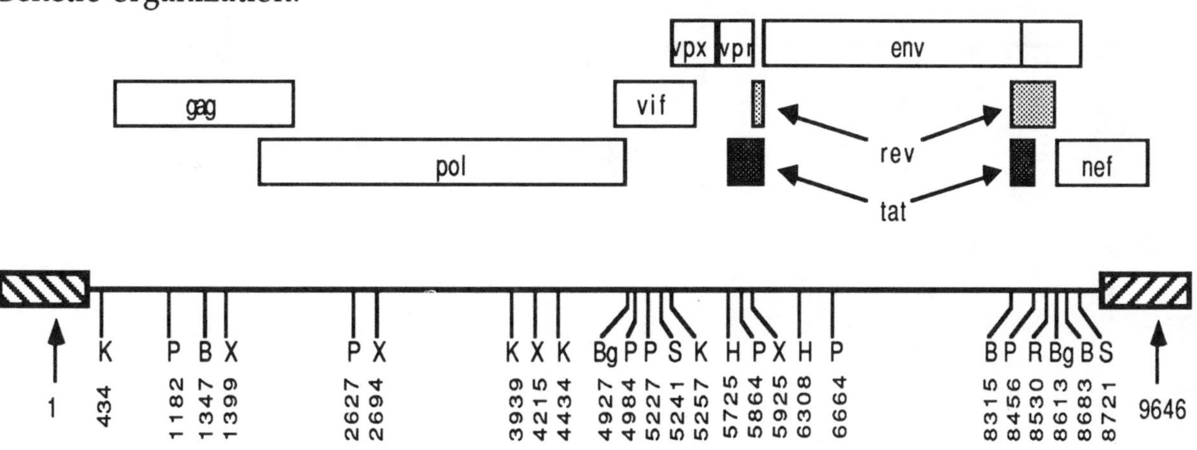

LTR and ORF coordinates

U3 : 514 bp
R : 176 bp
U5 : 140 bp

gag : 551 - 2068
pol : 1726 - 4893
vif : 4826 - 5467
vpx : 5298 - 5633
vpr : 5637 - 5939
tat : 5788 - 6083 / 8301 - 8394
rev : 6014 - 6083 /8301 - 8551
env : 6090 - 8735
nef : 8572 - 9357

Remarks: A number of clones encode an in phase stop codon in the 3' part of env coincident with the splice acceptor site for tat and rev. Clones are infectious. Same stop as in HIV2 ROD.
Splice pattern for tat, rev and nef gene products (Colombini et al., 1989). An unusual intron of 145bp in the R/U5 region is described.

SIVsm Isolated from sooty mangabey, *Cercocebus atys* (Murphy-Corb et al., 1986; Fultz et al., 1986; Lowenstine et al., 1986). Only isolated from captive mangabeys for the moment. Apparently not pathogenic in mangabeys. Molecular cloning: strain Delta/F236 (Hirsch et 1989a). Sequencing: Hirsch et al., 1989b.

Genetic organization:

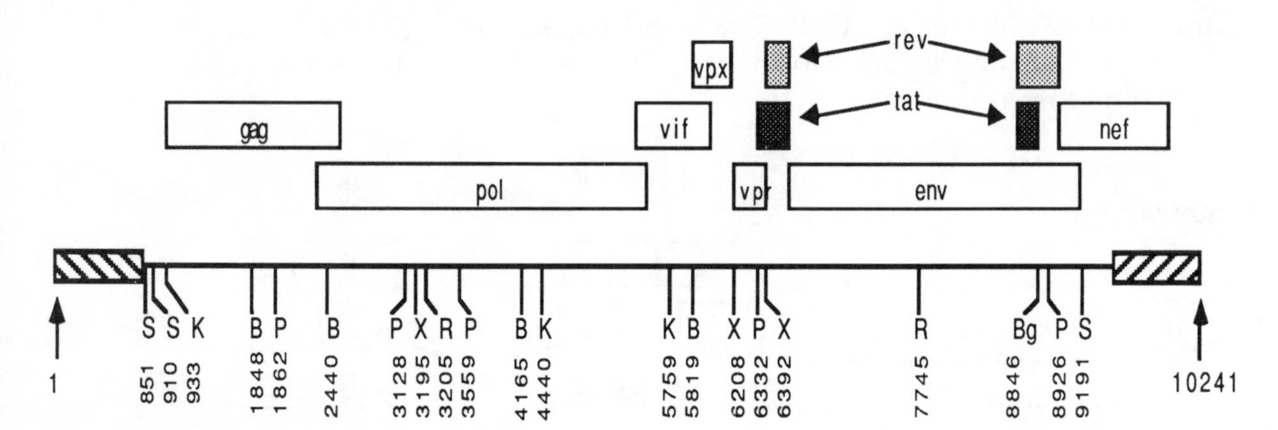

LTR and ORF coordinates

U3 : 514 bp	gag : 1049 - 2569
R : 176 bp	pol : 2338 - 5395
U5 : 125 bp	vif : 5328 - 5969
	vpx : 5800 - 6135
	vpr : 6140 - 6406
	tat : 6291 - 6544 / 8768 - 8864
	rev : 6481 - 6544 / 8768 - 9006
	env : 6551 - 9205
	nef : 9042 - 9968

Remarks: No in phase stop codon in the env gene sequence. Clone SMH4 encodes an insertion of an A at position 4529 in the integrase portion of the pol orf which is not present in another clone SMH3. SMH4 encodes an extended nef open reading frame.

SIVagm First authentic isolates from wild caught African green monkeys, *Cercopithicus aethops*, (Ohta et al., 1988). Apparently not pathogenic in African green monkeys.
Molecular cloning and sequencing: Fukasawa et al., 1988.

Genetic organization:

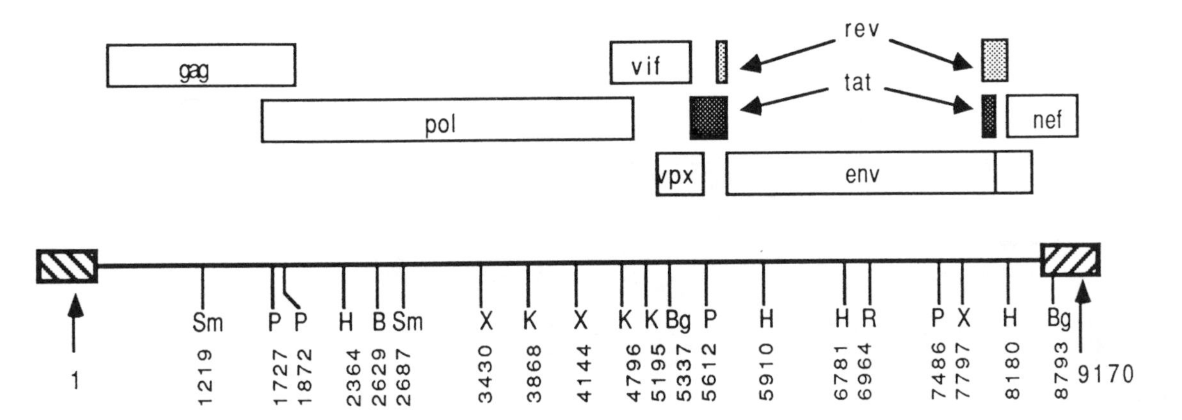

LTR and ORF coordinates

U3 : 507 bp
R : 117 bp
U5 : 102 bp

gag : 432 - 1988
pol : 1634 - 4816
vif : 4755 - 5459
vpx : 5236 - 5592
tat : 5546 - 5763 / 7984 - 8065
rev : 5703 - 5763 / 7984 - 8174
env : 5770 - 8364
nef : 8201 - 8887

Remarks: In phase stop codon in 3' of env is not in the same location as in SIVmac. Not all strains encode stop codons. Genome does not encode a vpr gene, vpx and tat overlap directly.

SIVmnd Isolated from wild caught mandrills, *Papio (Mandrillus) sphinx*, (Tsujimoto et al., 1988). Apparently not pathogenic in mandrills.
Molecular cloning: Tsujimoto et al., 1988.
Sequencing: Hayami, unpublished data.

Feline Immunodeficiency Virus (FIV)

FIV Formerly feline T-lymphotropic virus (Pederson et al., 1987). Widespread in cats. Pathogenic.
Molecular cloning: Olmsted et al., 1989. Infectious clone.
Sequencing: LTR (Olmsted et al., 1989).

Remarks: 9.4kb genome showing weak cross-hybridization to the Visna, CAEV and SIVsmm genomes. Small LTR (355bp) is characteristic of the ungulate LTRs. PBS is complementary to tRNAlys3.

Ungulate Lentiviruses

Visna Visna virus. Referred to as maedi virus, ovine maedi/visna virus (OMVV). Sometimes refered to as progressive pneumonia virus or zwoegerziekte virus. Pathogenic in sheep, its natural host.
Provirus: 9.6kb linear DNA unintegrated form, with a single stranded region in the center of the genome (at 4.75 +/- 0.25kb; Harris et al., 1981; Blum et al., 1985).
Molecular cloning: strain 1514 (Harris et al., 1981).
Sequencing: Sonigo et al., 1985; Braun et al., 1987.

Genetic organization:

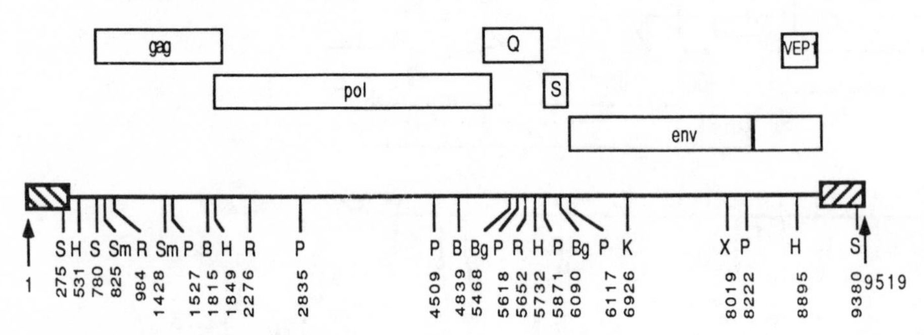

LTR and ORF coordinates

U3 : 254 bp
R : 97 bp
U5 : 63 bp

gag : 743 - 2068
pol : 1942 - 5259
Q : 5219 - 5908
S : 5911 - 6192
env : 6210 - 9158

Remarks: In phase stop codon up from the membrane spanning region in env indicates that the clone is defective. mRNA splicing pattern (Vigne et al., 1987; Mazarin et al., 1988). VEP1 (viral early protein) is encoded by two exons. The amino terminal part of the protein comes from an exon carrying the first 45 residues of the env protein sequence. This is spliced to the orf marked VEP1producing an orf of 167 residues. S gene probably encodes the transactivator protein.

OPAV Ovine Pulmonary Adenomatosis Virus. Isolated from tumourous tissue samples (Querat et al., 1984). Probably not oncogenic; a variant of visna virus.

CAEV Caprine Arthritis Encephalitis Virus. Pathogenic virus of goats (has been called goat leukoencephalopathy virus, GLV; Crawford et al., 1980).
Provirus: Integrated and unintegrated linears (Yaniv et al., 1985; Molineaux et al., 1983).
Molecular cloning: Yaniv et al., 1985.
Sequencing: 625 bases of *pol* (Chiu et al., 1985); LTR (Hess et al., 1986).

Remarks: Hybridizes to visna DNA (Pyper et al., 1986).

BIV Bovine Immunodeficiency Virus. First isolated in 1972 (van der Maaten et al., 1972). Pathogenic in cows.
Molecular cloning: Gonda et al., 1987.
Sequencing: partial pol sequence, Gonda et al., 1987.

EIAV Equine Infectious Anemia Virus. Pathogenic in horses.
Provirus: Integrated and unintegrated 8.2kb forms.
Molecular cloning and sequencing: Kawakami et al., 1987.
Other sequence data: Stephens et al., 1986; Rushlow et al., 1986.

Genetic organization:

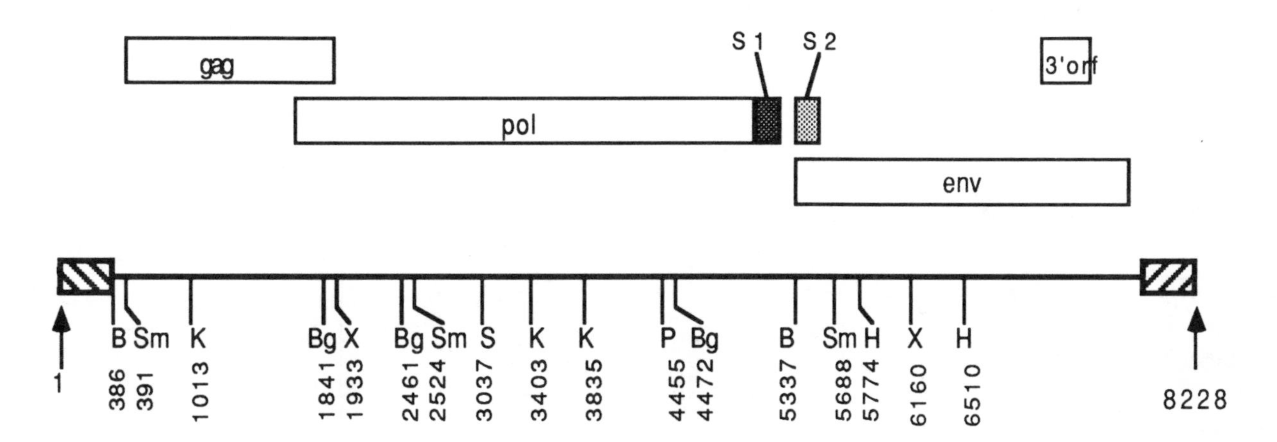

LTR and ORF coordinates

U3 : 207 bp	gag : 465 - 1922
R : 79 bp	pol : 1682 - 5119
U5 : 35 bp	S1 : 5123 - 5272
	env : 5312 - 7887
	3'orf : 7233 - 7637

Remarks: The LTR gag and pol genes are characteristic of the ungulate lentiviruses. However the env sequence is so unlike any of the lentiviral env proteins as to suggest that the env sequences are not of lentiviral origin. The transactivator gene product is encoded either by the S1 or S2 orfs (Sherman et al., 1988; Dorn & Derse 1988).

Legend to restriction sites:

B: BamHI Bg: BglII R: EcoRI H: HindIII K: KpnI P: PstI S: SacI Sl: SalI Sm: SmaI
X: XbaI Xh: XhoI

Retroviral phylogeny and interrelationships have been extensively described by Doolittle (Doolittle, 1989; Doolittle et al., 1989) and Myers (1989).

March, 1984

Animal Retrovirus Restriction Maps

Compiled by John M. Coffin
Dept. Molecular Biology and Microbiology
Tufts University School of Medicine
Boston, Massachusetts 02111

Derived from:

Molecular Biology of Tumor Viruses:
RNA Tumor Viruses
Second Edition
R. Weiss, N. Teich, H. Varmus,
and J. Coffin (eds.)
Cold Spring Harbor Laboratory
Volume II
1984

Sequence data supplied
by C. Van Beveren

AKV

AKR endogenous murine leukemia virus, mouse

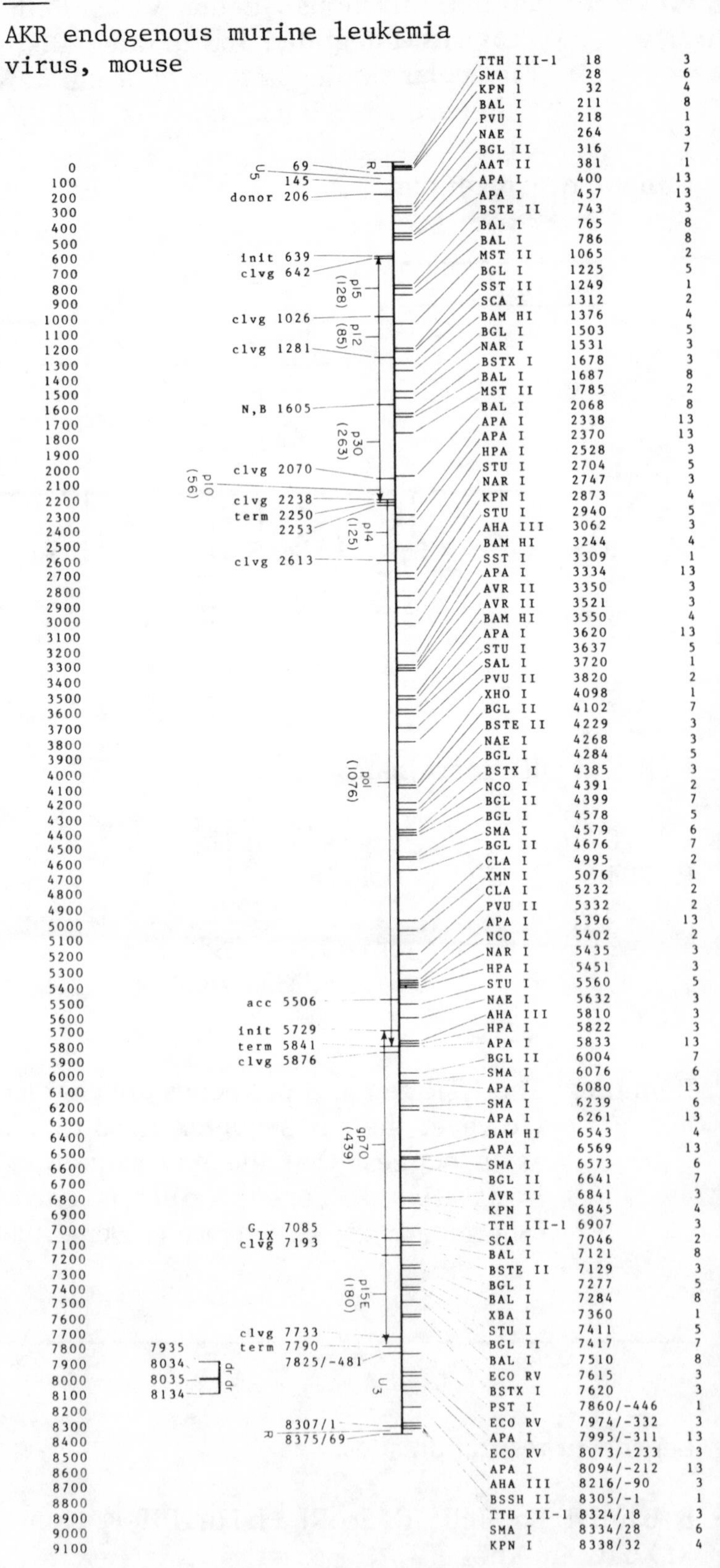

W. Herr (pers. comm.) and NAR 10:6931, 1982

Mo-MLV

Moloney murine leukemia virus, mouse

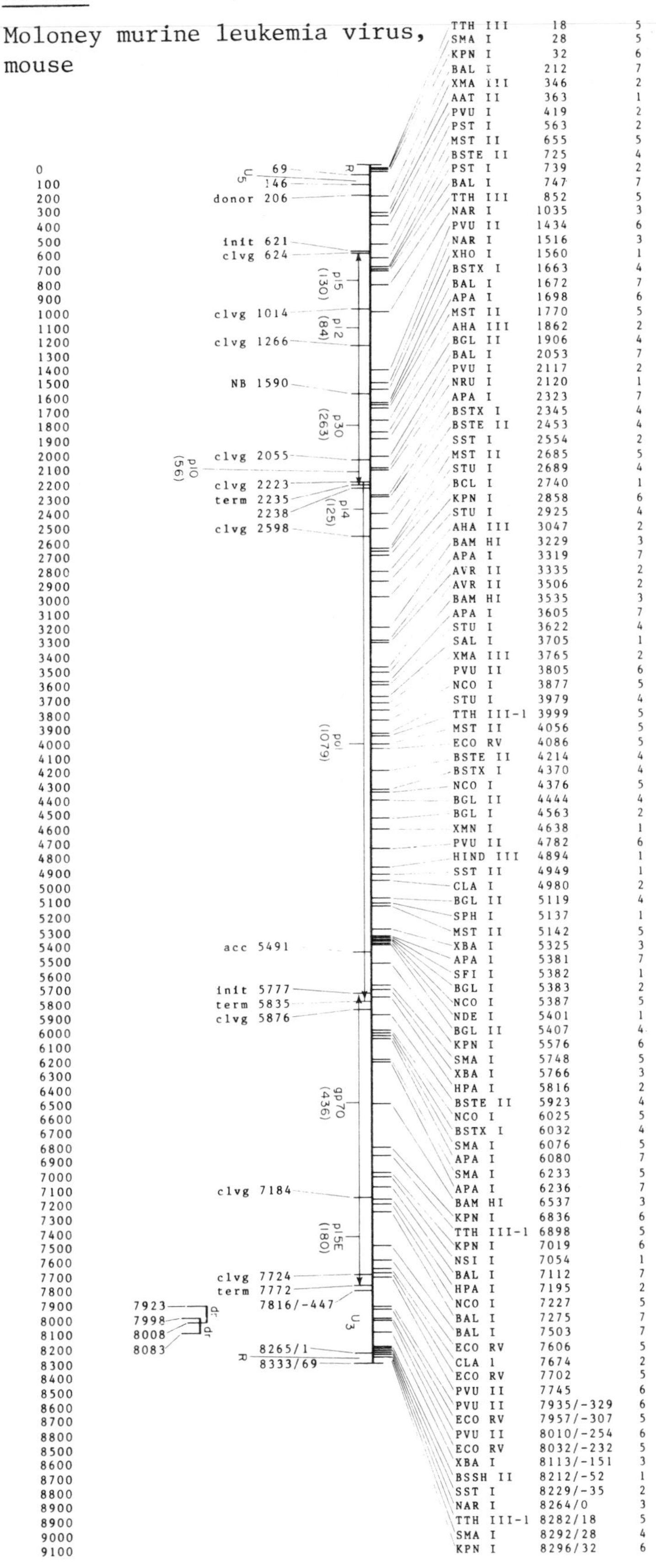

R
U5
69
146
donor 206
init 621
clvg 624
p15 (130)
clvg 1014
p12 (84)
clvg 1266
NB 1590
p30 (263)
clvg 2055
p10 (56)
clvg 2223
term 2235
2238
p14 (125)
clvg 2598
pol (1079)
acc 5491
init 5777
term 5835
clvg 5876
gp70 (436)
clvg 7184
p15E (180)
clvg 7724
term 7772
7816/-447
U3
7923
7998
8008
8083
dr
dr
8265/1
8333/69
R
TTH III 18 5
SMA I 28 5
KPN I 32 6
BAL I 212 7
XMA III 346 2
AAT II 363 1
PVU I 419 2
PST I 563 2
MST II 655 5
BSTE II 725 4
PST I 739 2
BAL I 747 7
TTH III 852 5
NAR I 1035 3
PVU II 1434 6
NAR I 1516 3
XHO I 1560 1
BSTX I 1663 4
BAL I 1672 7
APA I 1698 6
MST II 1770 5
AHA III 1862 2
BGL II 1906 4
BAL I 2053 7
PVU I 2117 2
NRU I 2120 1
APA I 2323 7
BSTX I 2345 4
BSTE II 2453 4
SST I 2554 2
MST II 2685 5
STU I 2689 4
BCL I 2740 1
KPN I 2858 6
STU I 2925 4
AHA III 3047 2
BAM HI 3229 3
APA I 3319 7
AVR II 3335 2
AVR II 3506 2
BAM HI 3535 3
APA I 3605 7
STU I 3622 4
SAL I 3705 1
XMA III 3765 2
PVU II 3805 6
NCO I 3877 5
STU I 3979 4
TTH III-1 3999 5
MST II 4056 5
ECO RV 4086 5
BSTE II 4214 4
BSTX I 4370 4
NCO I 4376 5
BGL II 4444 4
BGL I 4563 2
XMN I 4638 1
PVU II 4782 6
HIND III 4894 1
SST II 4949 1
CLA I 4980 2
BGL II 5119 4
SPH I 5137 1
MST II 5142 5
XBA I 5325 3
APA I 5381 7
SFI I 5382 1
BGL I 5383 2
NCO I 5387 5
NDE I 5401 1
BGL II 5407 4
KPN I 5576 6
SMA I 5748 5
XBA I 5766 3
HPA I 5816 2
BSTE II 5923 4
NCO I 6025 5
BSTX I 6032 4
SMA I 6076 5
APA I 6080 7
SMA I 6233 5
APA I 6236 7
BAM HI 6537 3
KPN I 6836 6
TTH III-1 6898 5
KPN I 7019 6
NSI I 7054 1
BAL I 7112 7
HPA I 7195 2
NCO I 7227 5
BAL I 7275 7
BAL I 7503 7
ECO RV 7606 5
CLA I 7674 2
ECO RV 7702 5
PVU II 7745 6
PVU II 7935/-329 6
ECO RV 7957/-307 5
PVU II 8010/-254 6
ECO RV 8032/-232 5
XBA I 8113/-151 3
BSSH II 8212/-52 1
SST I 8229/-35 2
NAR I 8264/0 3
TTH III-1 8282/18 5
SMA I 8292/28 4
KPN I 8296/32 6

Notes: Shinnick et al., Nature 293:543-548, 1981. No sites: ASU II, ECO RI, MLU I, MST I, NAE I, NOT I. Clone: pMLV-1. LTR: 594 bp. 0-point: 8264.

Pr-RSV

Prague Rous sarcoma virus (src), chicken

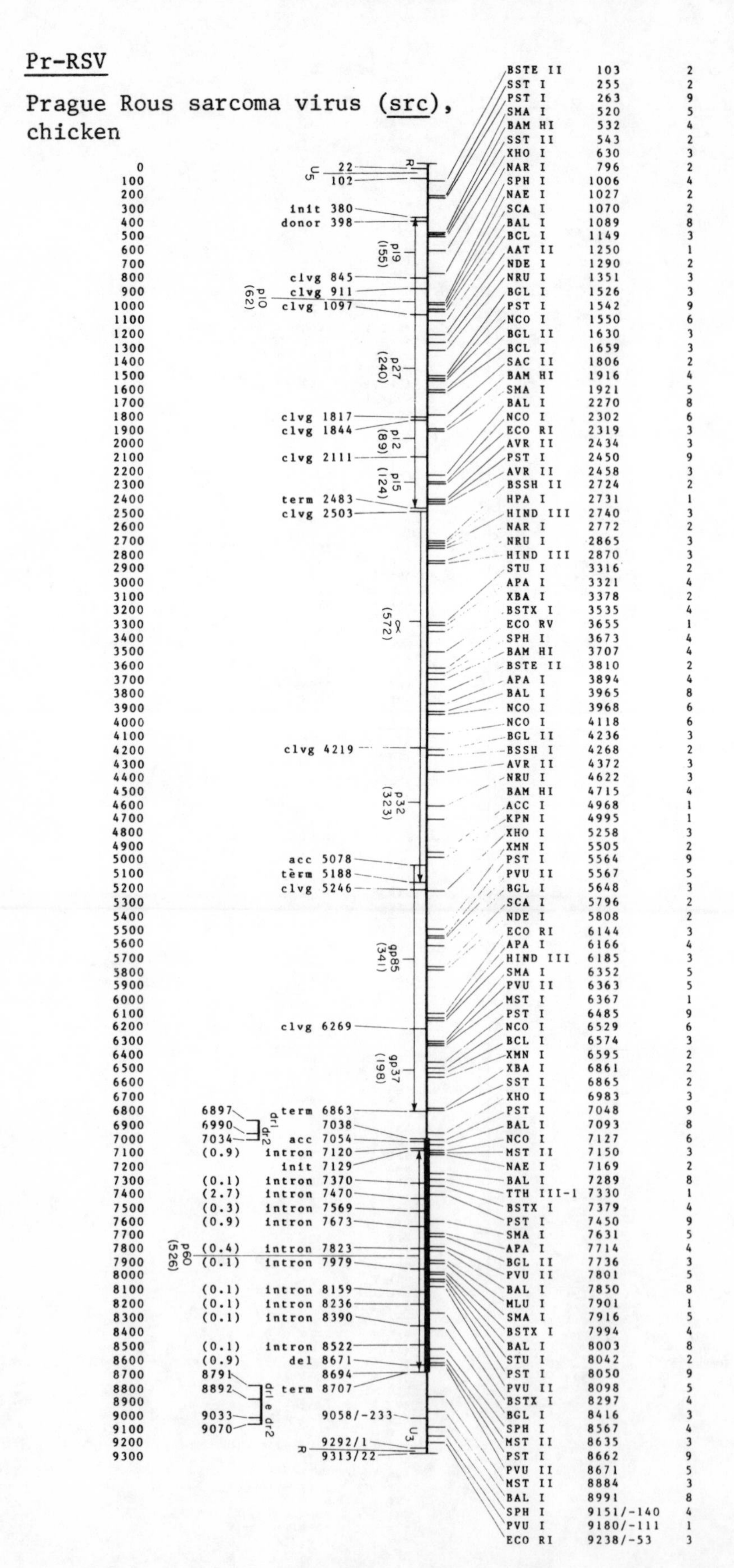

Notes: Schwartz et al., Cell 32:853-869, 1983. Introns from Takeya and Hanafusa, Cell 32:881-890, 1983. Lower case d marks deletion in v-src relative to c-src. LTR: 335 bp; 0-point: 9291. 3' terminus of src uncertain. No sites: AHA III, CLA I, NOT I, NSI I, SAL I, SFI I, XMA III.

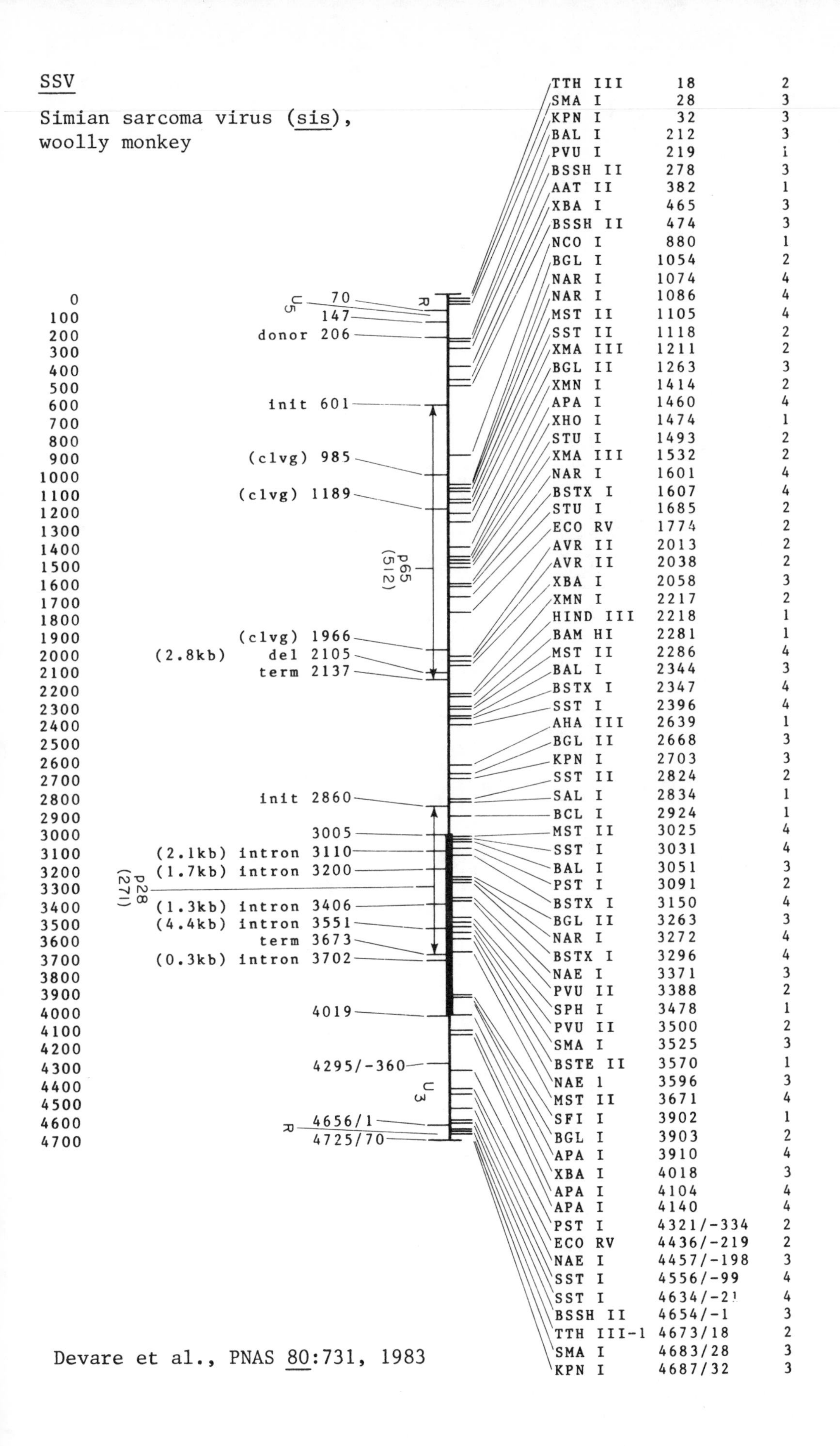
SSV
Simian sarcoma virus (sis),
woolly monkey
0
100
200
300
400
500
600
700
800
900
1000
1100
1200
1300
1400
1500
1600
1700
1800
1900
2000
2100
2200
2300
2400
2500
2600
2700
2800
2900
3000
3100
3200
3300
3400
3500
3600
3700
3800
3900
4000
4100
4200
4300
4400
4500
4600
4700
U5
R
70
147
donor 206
init 601
(clvg) 985
(clvg) 1189
p65
(512)
(clvg) 1966
(2.8kb) del 2105
term 2137
init 2860
3005
(2.1kb) intron 3110
(1.7kb) intron 3200
p28
(271)
(1.3kb) intron 3406
(4.4kb) intron 3551
term 3673
(0.3kb) intron 3702
4019
4295/-360
U3
R
4656/1
4725/70
TTH III 18 2
SMA I 28 3
KPN I 32 3
BAL I 212 3
PVU I 219 1
BSSH II 278 3
AAT II 382 1
XBA I 465 3
BSSH II 474 3
NCO I 880 1
BGL I 1054 2
NAR I 1074 4
NAR I 1086 4
MST II 1105 4
SST II 1118 2
XMA III 1211 2
BGL II 1263 3
XMN I 1414 2
APA I 1460 4
XHO I 1474 1
STU I 1493 2
XMA III 1532 2
NAR I 1601 4
BSTX I 1607 4
STU I 1685 2
ECO RV 1774 2
AVR II 2013 2
AVR II 2038 2
XBA I 2058 3
XMN I 2217 2
HIND III 2218 1
BAM HI 2281 1
MST II 2286 4
BAL I 2344 3
BSTX I 2347 4
SST I 2396 4
AHA III 2639 1
BGL II 2668 3
KPN I 2703 3
SST II 2824 2
SAL I 2834 1
BCL I 2924 1
MST II 3025 4
SST I 3031 4
BAL I 3051 3
PST I 3091 2
BSTX I 3150 4
BGL II 3263 3
NAR I 3272 4
BSTX I 3296 4
NAE I 3371 3
PVU II 3388 2
SPH I 3478 1
PVU II 3500 2
SMA I 3525 3
BSTE II 3570 1
NAE 1 3596 3
MST II 3671 4
SFI I 3902 1
BGL I 3903 2
APA I 3910 4
XBA I 4018 3
APA I 4104 4
APA I 4140 4
PST I 4321/-334 2
ECO RV 4436/-219 2
NAE I 4457/-198 3
SST I 4556/-99 4
SST I 4634/-21 4
BSSH II 4654/-1 3
TTH III-1 4673/18 2
SMA I 4683/28 3
KPN I 4687/32 3
Devare et al., PNAS 80:731, 1983

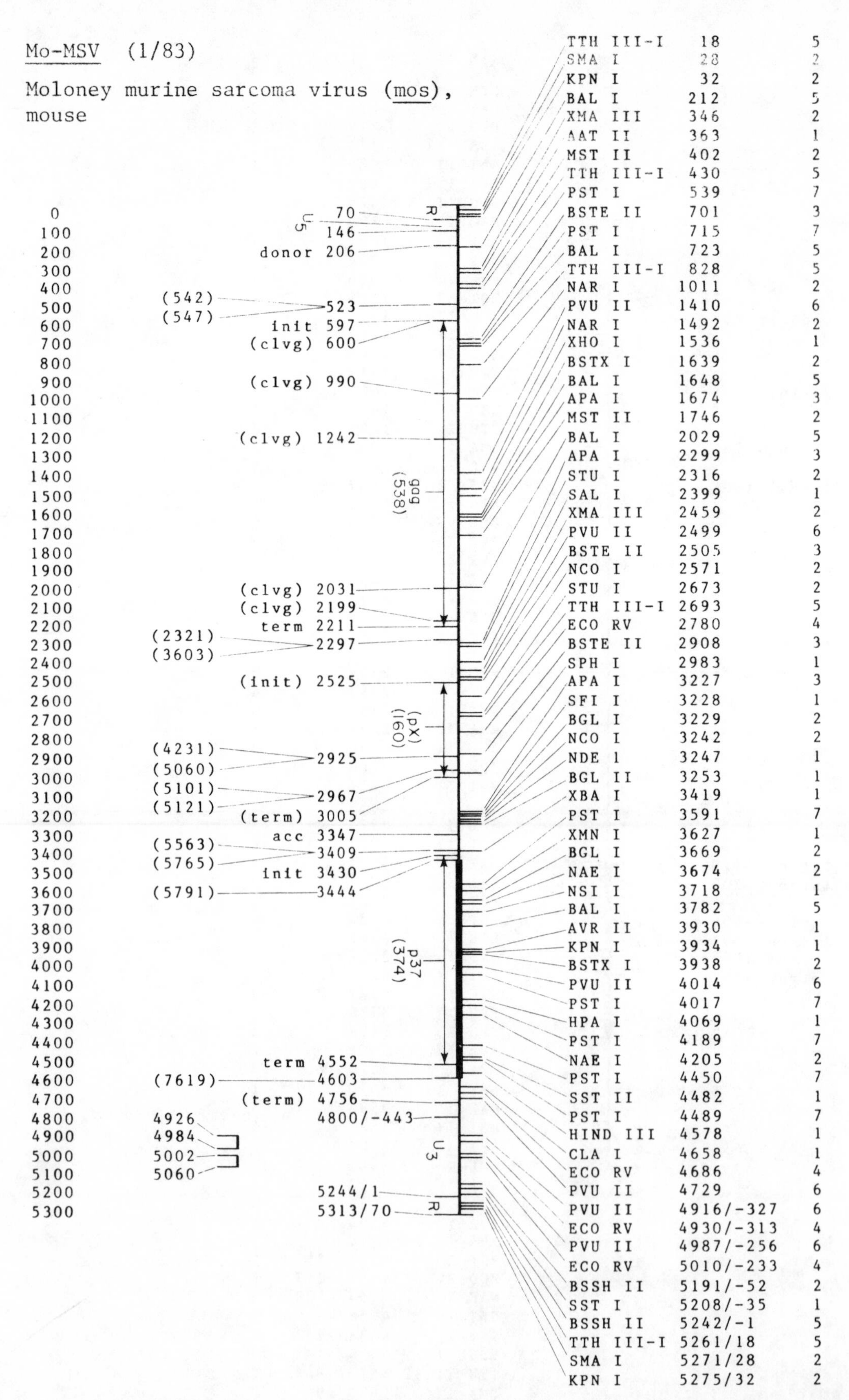
Mo-MSV (1/83)
Moloney murine sarcoma virus (mos), mouse
0
100
200
300
400
500
600
700
800
900
1000
1100
1200
1300
1400
1500
1600
1700
1800
1900
2000
2100
2200
2300
2400
2500
2600
2700
2800
2900
3000
3100
3200
3300
3400
3500
3600
3700
3800
3900
4000
4100
4200
4300
4400
4500
4600
4700
4800
4900
5000
5100
5200
5300
R
U5
70
146
donor 206
(542)
(547)
523
init 597
(clvg) 600
(clvg) 990
(clvg) 1242
gag (538)
(clvg) 2031
(clvg) 2199
term 2211
(2321)
(3603)
2297
(init) 2525
(pX) (160)
(4231)
(5060)
2925
(5101)
(5121)
2967
(term) 3005
acc 3347
(5563)
(5765)
3409
init 3430
(5791)
3444
p37 (374)
term 4552
(7619)
4603
(term) 4756
4800/-443
U3
4926
4984
5002
5060
5244/1
5313/70
R
TTH III-I 18 5
SMA I 28 2
KPN I 32 2
BAL I 212 5
XMA III 346 2
AAT II 363 1
MST II 402 2
TTH III-I 430 5
PST I 539 7
BSTE II 701 3
PST I 715 7
BAL I 723 5
TTH III-I 828 5
NAR I 1011 2
PVU II 1410 6
NAR I 1492 2
XHO I 1536 1
BSTX I 1639 2
BAL I 1648 5
APA I 1674 3
MST II 1746 2
BAL I 2029 5
APA I 2299 3
STU I 2316 2
SAL I 2399 1
XMA III 2459 2
PVU II 2499 6
BSTE II 2505 3
NCO I 2571 2
STU I 2673 2
TTH III-I 2693 5
ECO RV 2780 4
BSTE II 2908 3
SPH I 2983 1
APA I 3227 3
SFI I 3228 1
BGL I 3229 2
NCO I 3242 2
NDE I 3247 1
BGL II 3253 1
XBA I 3419 1
PST I 3591 7
XMN I 3627 1
BGL I 3669 2
NAE I 3674 2
NSI I 3718 1
BAL I 3782 5
AVR II 3930 1
KPN I 3934 1
BSTX I 3938 2
PVU II 4014 6
PST I 4017 7
HPA I 4069 1
PST I 4189 7
NAE I 4205 2
PST I 4450 7
SST II 4482 1
PST I 4489 7
HIND III 4578 1
CLA I 4658 1
ECO RV 4686 4
PVU II 4729 6
PVU II 4916/-327 6
ECO RV 4930/-313 4
PVU II 4987/-256 6
ECO RV 5010/-233 4
BSSH II 5191/-52 2
SST I 5208/-35 1
BSSH II 5242/-1 5
TTH III-I 5261/18 5
SMA I 5271/28 2
KPN I 5275/32 2

Notes: Van Beveren et al., Cell 27:97-108, 1981. Add 445 to get VB numbers. LTR: 590. 0-point: 5243. No sites: AHA III, ASU II, BAM HI, BCL I, ECO RI, MLU I, MST I, NOT I, NRU I, SCA I, PVU I.

Ab-MLV

Abelson murine leukemia virus (abl), mouse

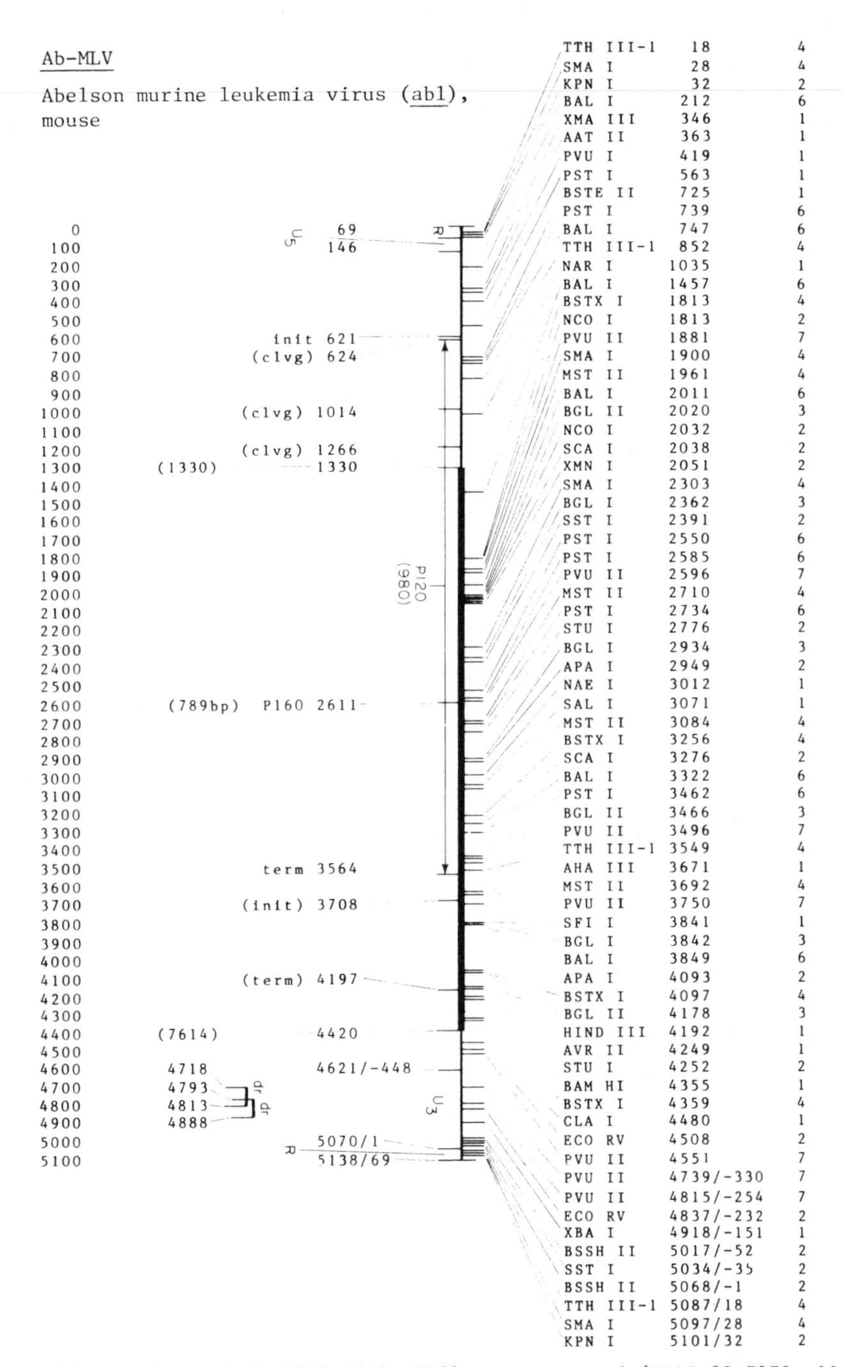

Notes: Reddy et al. PNAS 80:3623-3627, 1983, as corrected (PNAS 80:7372, 1983). Clone: PAM. Add 446 to ge Reddy et al. numbers. No sites: ASU II, BCL I, ECO RI, HPA I, MLU I, MST I, SST II, SPH I, XHO I.

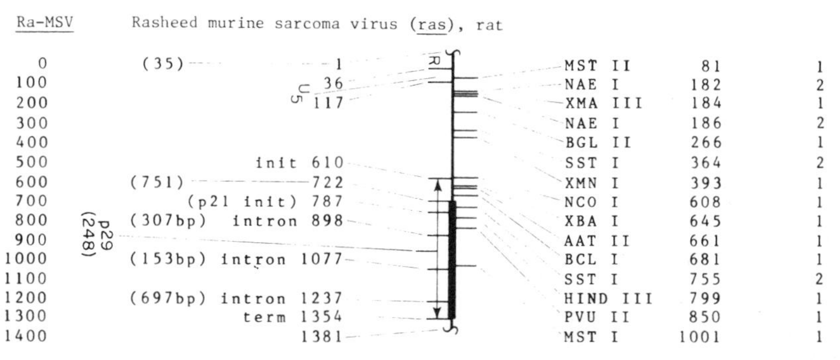

Rasheed et al., Science 221:155-157, 1983

Fr-SFFV$_p$

Friend spleen focus-forming virus, mouse

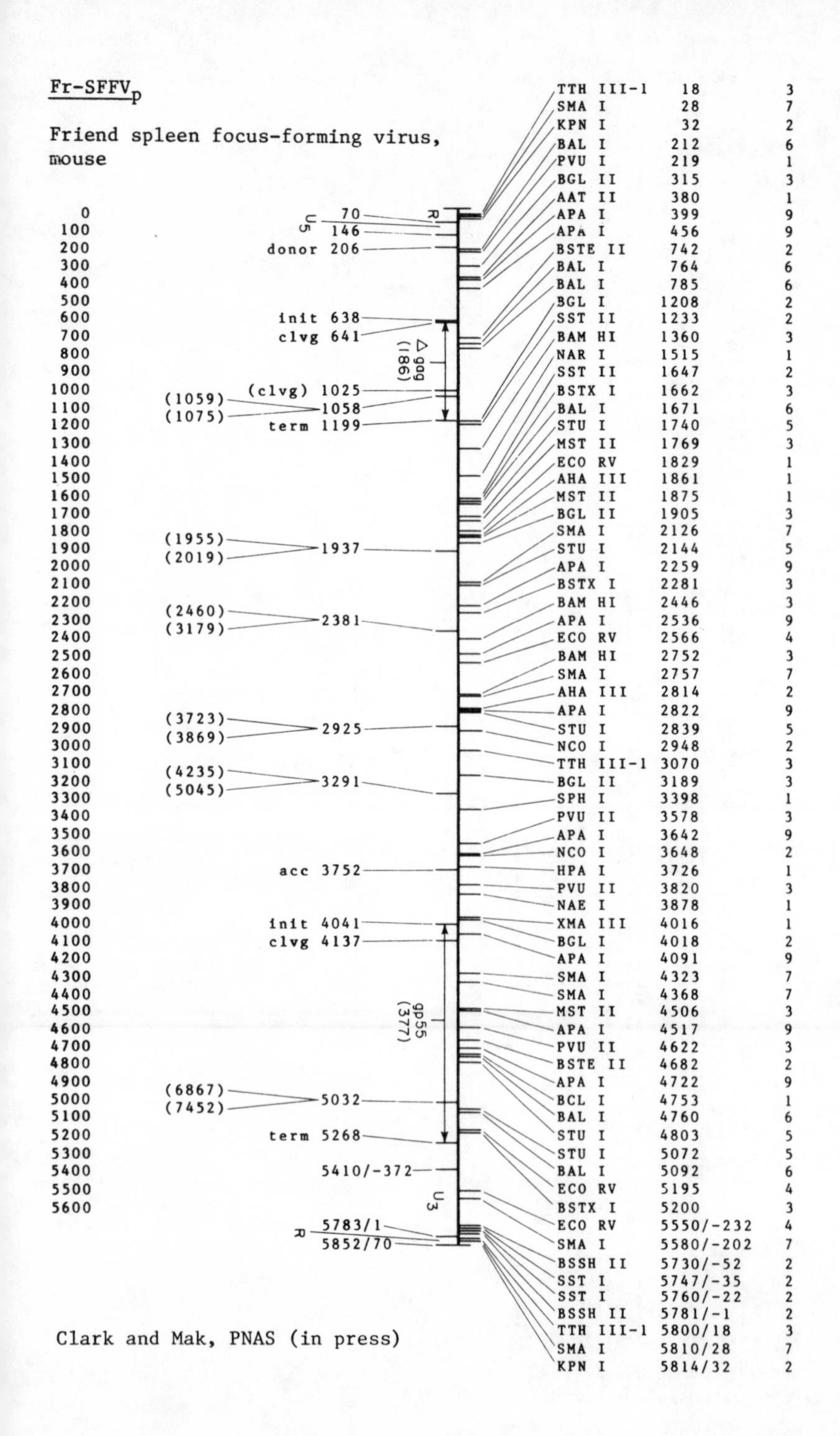

0
100
200
300
400
500
600
700
800
900
1000
1100
1200
1300
1400
1500
1600
1700
1800
1900
2000
2100
2200
2300
2400
2500
2600
2700
2800
2900
3000
3100
3200
3300
3400
3500
3600
3700
3800
3900
4000
4100
4200
4300
4400
4500
4600
4700
4800
4900
5000
5100
5200
5300
5400
5500
5600
R
U5
70
146
donor 206
init 638
clvg 641
Δ gag (186)
(clvg) 1025
(1059)
(1075)
1058
term 1199
(1955)
(2019)
1937
(2460)
(3179)
2381
(3723)
(3869)
2925
(4235)
(5045)
3291
acc 3752
init 4041
clvg 4137
gp55 (377)
(6867)
(7452)
5032
term 5268
5410/-372
U3
R
5783/1
5852/70
TTH III-1 18 3
SMA I 28 7
KPN I 32 2
BAL I 212 6
PVU I 219 1
BGL II 315 3
AAT II 380 1
APA I 399 9
APA I 456 9
BSTE II 742 2
BAL I 764 6
BAL I 785 6
BGL I 1208 2
SST II 1233 2
BAM HI 1360 3
NAR I 1515 1
SST II 1647 2
BSTX I 1662 3
BAL I 1671 6
STU I 1740 5
MST II 1769 3
ECO RV 1829 1
AHA III 1861 1
MST II 1875 1
BGL II 1905 3
SMA I 2126 7
STU I 2144 5
APA I 2259 9
BSTX I 2281 3
BAM HI 2446 3
APA I 2536 9
ECO RV 2566 4
BAM HI 2752 3
SMA I 2757 7
AHA III 2814 2
APA I 2822 9
STU I 2839 5
NCO I 2948 2
TTH III-1 3070 3
BGL II 3189 3
SPH I 3398 1
PVU II 3578 3
APA I 3642 9
NCO I 3648 2
HPA I 3726 1
PVU II 3820 3
NAE I 3878 1
XMA III 4016 1
BGL I 4018 2
APA I 4091 9
SMA I 4323 7
SMA I 4368 7
MST II 4506 3
APA I 4517 9
PVU II 4622 3
BSTE II 4682 2
APA I 4722 9
BCL I 4753 1
BAL I 4760 6
STU I 4803 5
STU I 5072 5
BAL I 5092 6
ECO RV 5195 4
BSTX I 5200 3
ECO RV 5550/-232 4
SMA I 5580/-202 7
BSSH II 5730/-52 2
SST I 5747/-35 2
SST I 5760/-22 2
BSSH II 5781/-1 2
TTH III-1 5800/18 3
SMA I 5810/28 7
KPN I 5814/32 2

Clark and Mak, PNAS (in press)

SK770 (ski), chicken

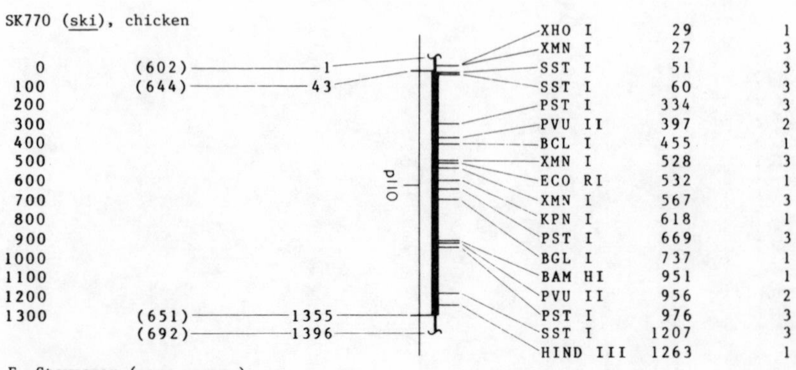

0
100
200
300
400
500
600
700
800
900
1000
1100
1200
1300
(602)
(644)
1
43
p110
(651)
(692)
1355
1396
XHO I 29 1
XMN I 27 3
SST I 51 3
SST I 60 3
PST I 334 3
PVU II 397 2
BCL I 455 1
XMN I 528 3
ECO RI 532 1
XMN I 567 3
KPN I 618 1
PST I 669 3
BGL I 737 1
BAM HI 951 1
PVU II 956 2
PST I 976 3
SST I 1207 3
HIND III 1263 1

E. Stavnezer (pers. comm.)

FeSV

M-Feline sarcoma virus (fms), cat

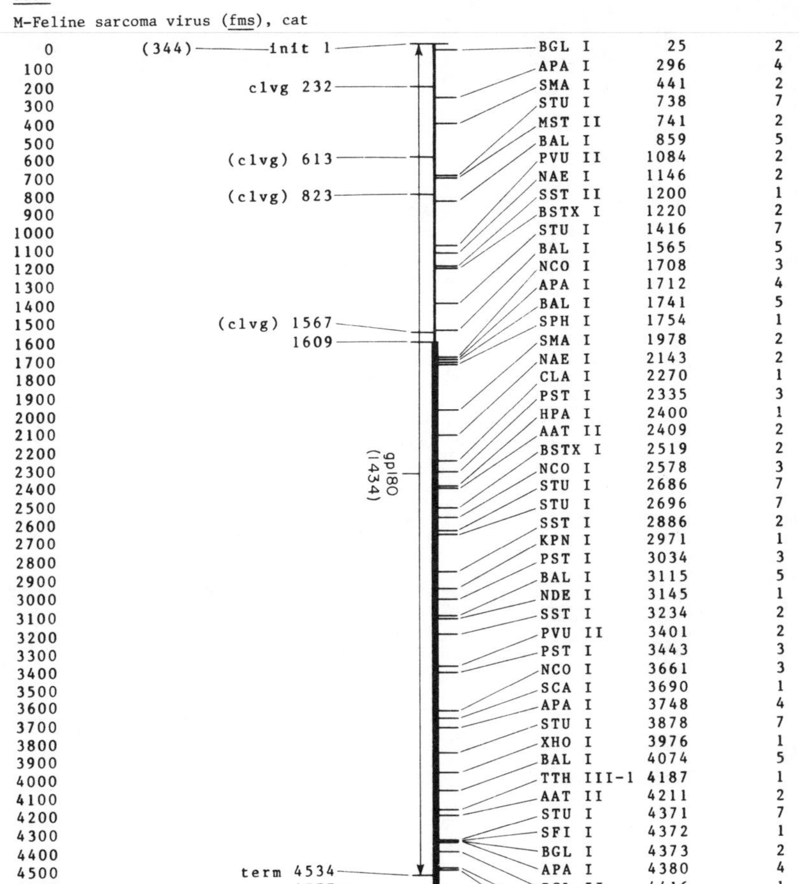

Hampe et al. (submitted)

MSV-3611

Murine sarcoma virus 3611 (raf), mouse

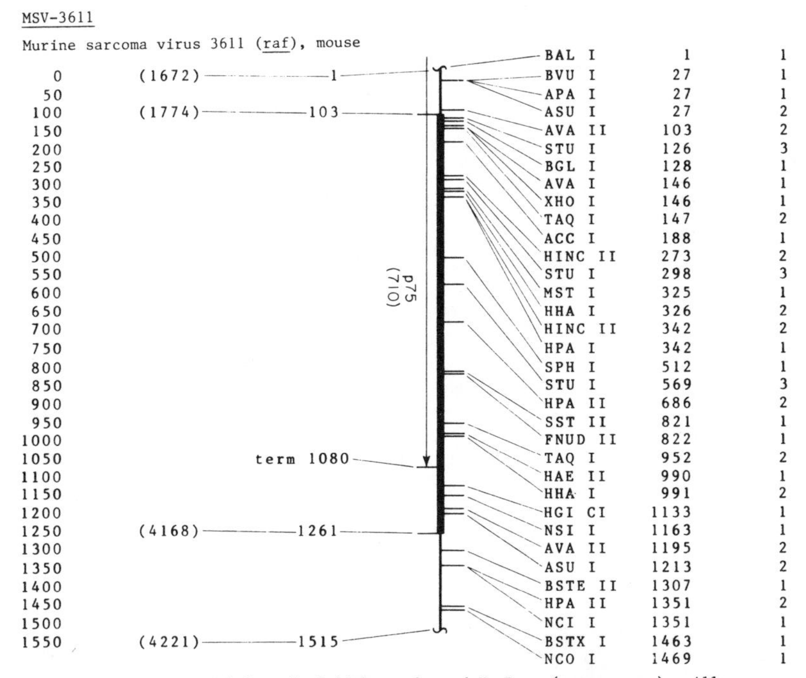

Notes: Mark, G., A. Schultz, M. Goldsborough, and U. Rapp (pers. comm.). All enzymes with 3 or fewer sites. No sites: AAT II, ACY I, AHA III, ASU II, AVR II, BAM HI, BCL I, BGL II, BSSH II, CLA I, ECO RI, ECO RV, HGA I, HLIA I, GIAD I, HIND III, KPN I, BLU I, MST II, NAE I, NAR I, NDE I, NOT I, NRU I, PST I, PVU I, PVU II, SST I, SAL I, SCA I, SFI I, SMA I, THH III-I, XBA I, XMA III, XMN I.

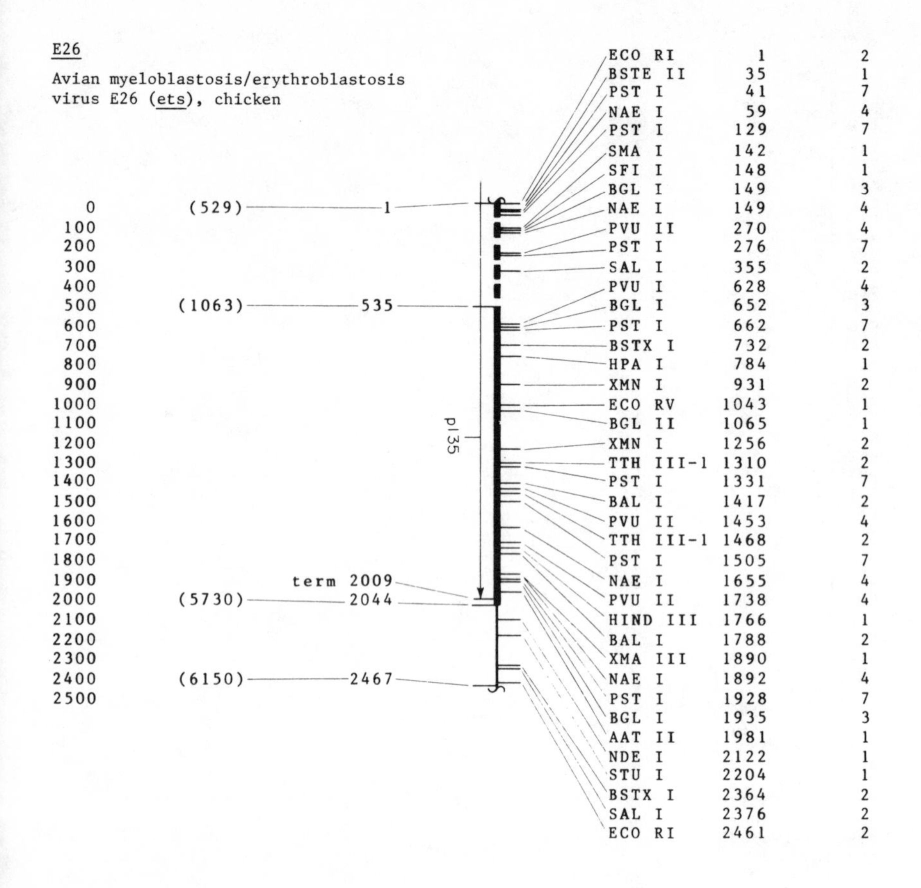
E26
Avian myeloblastosis/erythroblastosis virus E26 (ets), chicken
0
100
200
300
400
500
600
700
800
900
1000
1100
1200
1300
1400
1500
1600
1700
1800
1900
2000
2100
2200
2300
2400
2500
(529) 1
(1063) 535
p135
term 2009
(5730) 2044
(6150) 2467
ECO RI 1 2
BSTE II 35 1
PST I 41 7
NAE I 59 4
PST I 129 7
SMA I 142 1
SFI I 148 1
BGL I 149 3
NAE I 149 4
PVU II 270 4
PST I 276 7
SAL I 355 2
PVU I 628 4
BGL I 652 3
PST I 662 7
BSTX I 732 2
HPA I 784 1
XMN I 931 2
ECO RV 1043 1
BGL II 1065 1
XMN I 1256 2
TTH III-1 1310 2
PST I 1331 7
BAL I 1417 2
PVU II 1453 4
TTH III-1 1468 2
PST I 1505 7
NAE I 1655 4
PVU II 1738 4
HIND III 1766 1
BAL I 1788 2
XMA III 1890 1
NAE I 1892 4
PST I 1928 7
BGL I 1935 3
AAT II 1981 1
NDE I 2122 1
STU I 2204 1
BSTX I 2364 2
SAL I 2376 2
ECO RI 2461 2

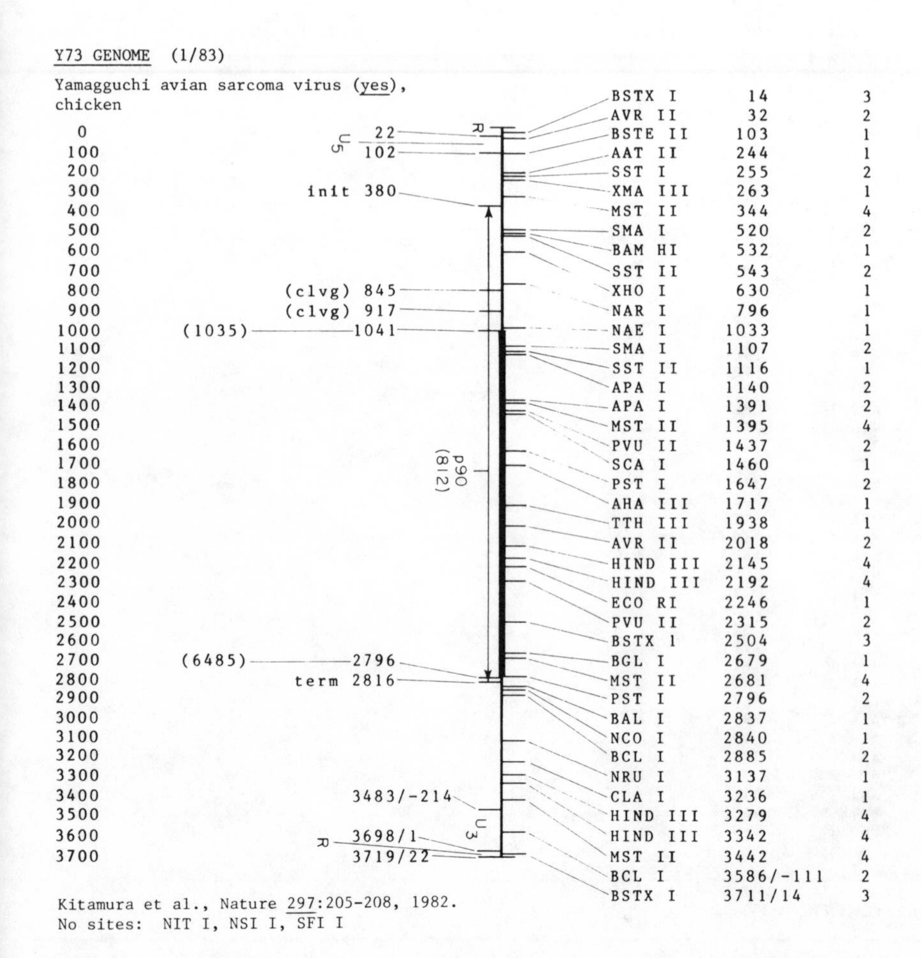
Y73 GENOME (1/83)
Yamagguchi avian sarcoma virus (yes), chicken
0
100
200
300
400
500
600
700
800
900
1000
1100
1200
1300
1400
1500
1600
1700
1800
1900
2000
2100
2200
2300
2400
2500
2600
2700
2800
2900
3000
3100
3200
3300
3400
3500
3600
3700
R
U5
22
102
init 380
(clvg) 845
(clvg) 917
(1035) 1041
p90 (812)
(6485) 2796
term 2816
3483/-214
U3
R
3698/1
3719/22
BSTX I 14 3
AVR II 32 2
BSTE II 103 1
AAT II 244 1
SST I 255 2
XMA III 263 1
MST II 344 4
SMA I 520 2
BAM HI 532 1
SST II 543 2
XHO I 630 1
NAR I 796 1
NAE I 1033 1
SMA I 1107 2
SST II 1116 1
APA I 1140 2
APA I 1391 2
MST II 1395 4
PVU II 1437 2
SCA I 1460 1
PST I 1647 2
AHA III 1717 1
TTH III 1938 1
AVR II 2018 2
HIND III 2145 4
HIND III 2192 4
ECO RI 2246 1
PVU II 2315 2
BSTX I 2504 3
BGL I 2679 1
MST II 2681 4
PST I 2796 2
BAL I 2837 1
NCO I 2840 1
BCL I 2885 2
NRU I 3137 1
CLA I 3236 1
HIND III 3279 4
HIND III 3342 4
MST II 3442 4
BCL I 3586/-111 2
BSTX I 3711/14 3
Kitamura et al., Nature 297:205-208, 1982.
No sites: NIT I, NSI I, SFI I

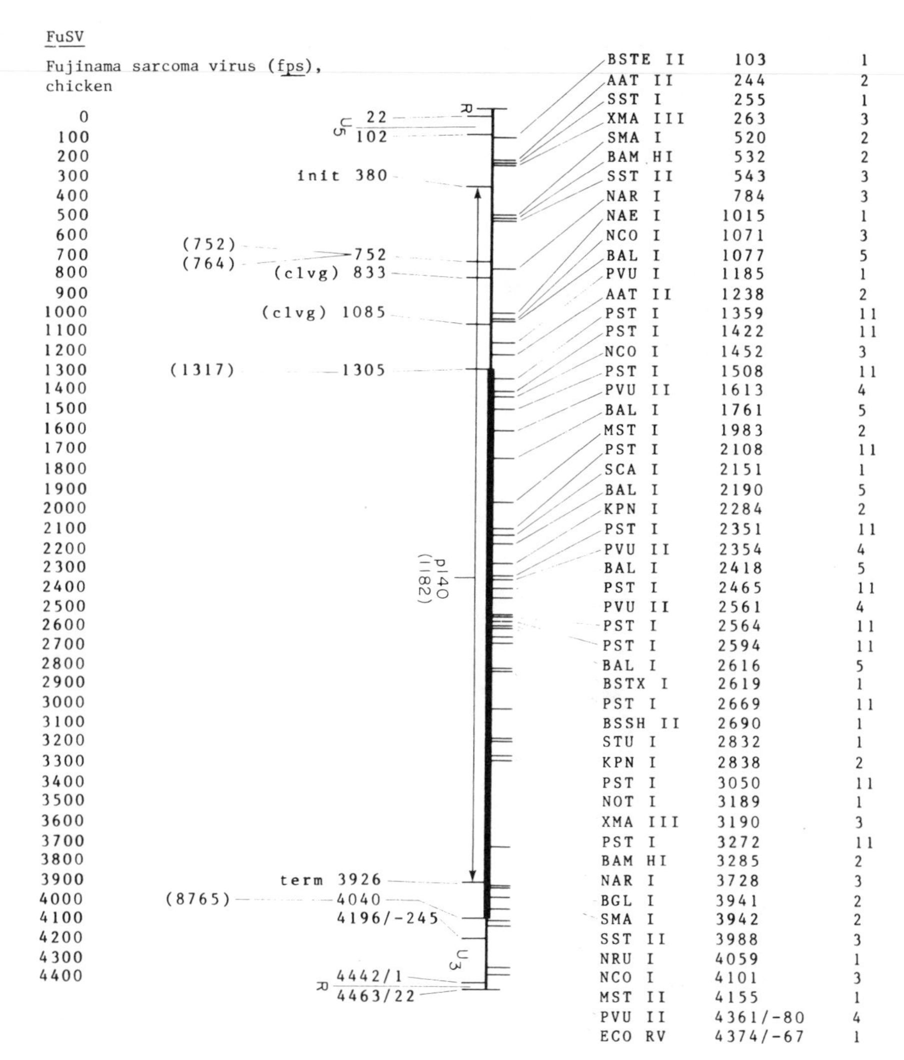

Notes: Shibuya and Hanafusa, Cell 30:787-795, 1982. 3' fps-virus junction is in a region which diverges between strains and is terefore approximate. LTR: 347 bp. 0-point: 4441. No sites: ACC I, AHA III, ASU II, AVR II, BGL II, CLA I, ECO RI, HINC II, HPA I, MLU I, NDE I, NSI I, SAL I, SFI I, SPH I, TTH III, XBA I, XHO I, XMN I.

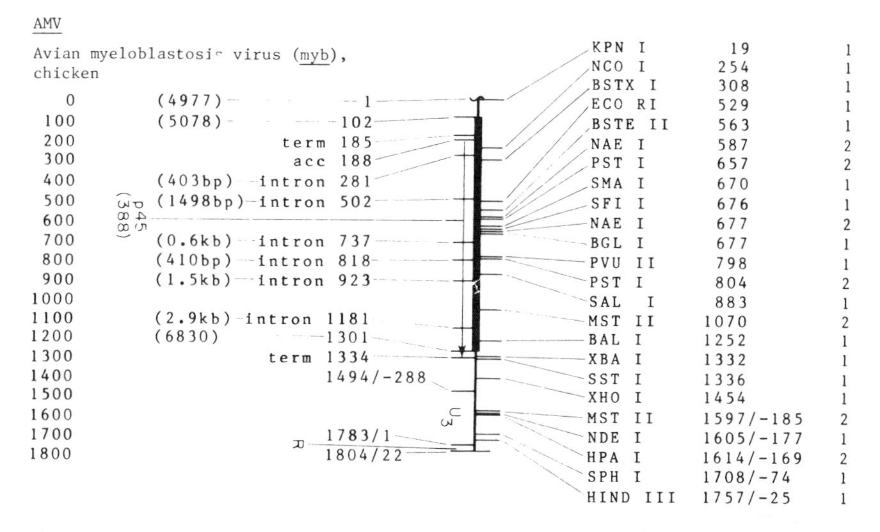

Notes: From Klempnauer et al., Cell 31:453-463. Clone pVM2. No sites: AAT II, ACY I, AHA III, APA I, ASU II, AVR II, BAM HI, BCL, BGL II, BSSH II, CLA I, ECO RV, HGI DI, MLU I, MST I, NAR I, NOT I, NSI I, PVU I, SCA I, SFI I, SST II, STU I, TTH III, XMA III, XMN I.

AEV

Avian erythroblastosis virus (erbB), chicken

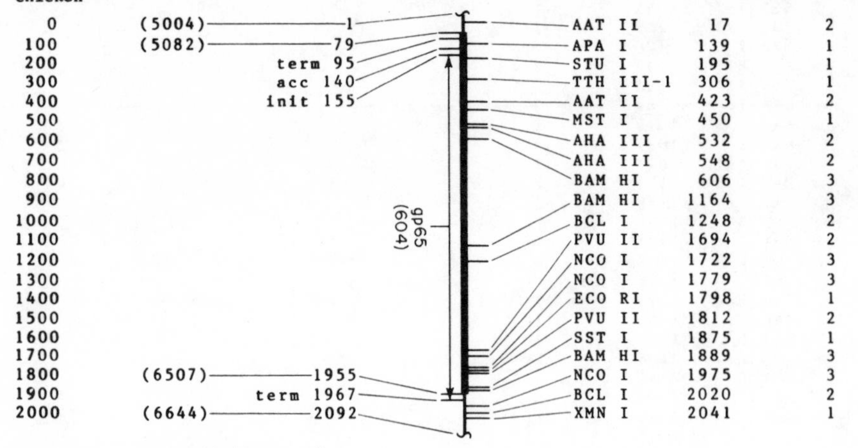

Yamamoto et al., Cell (in press).
From AEV-H; X-ref to Pr-C.

MC29

Myelocytomatosis virus (myc), chicken

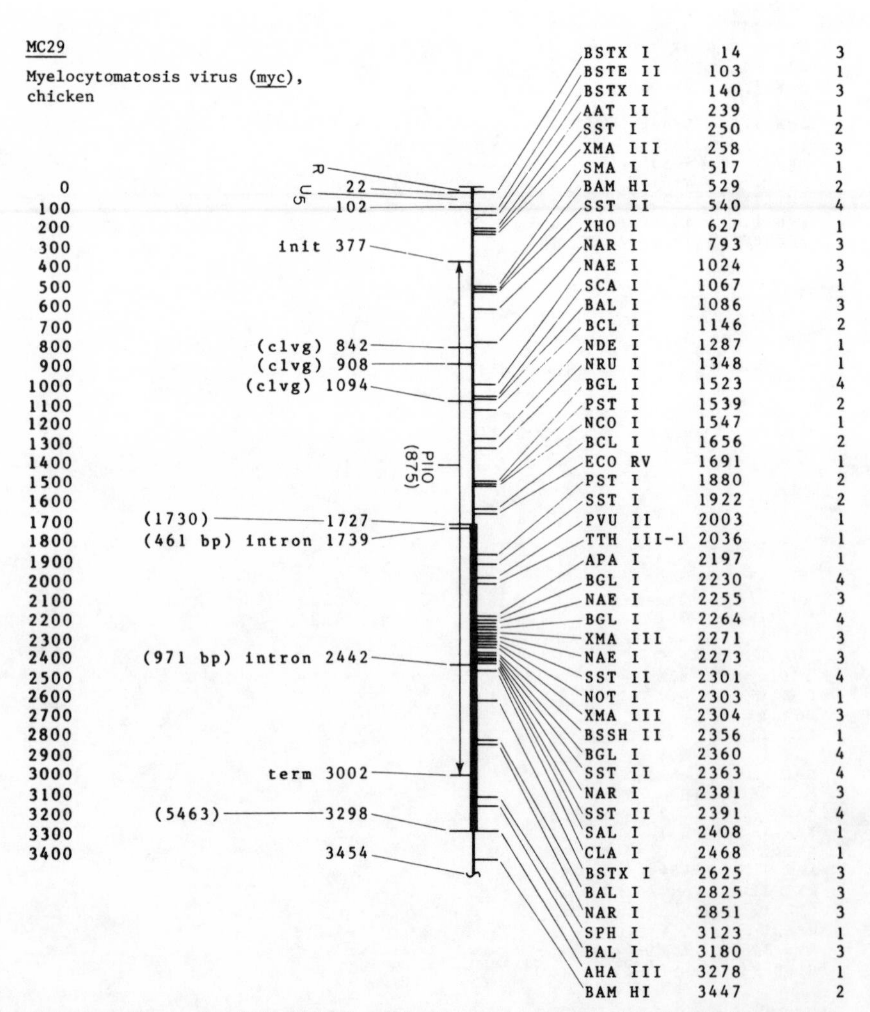

Notes: Reddy et al., PNAS 80:2500, 1983. No sites: AAT II, ASU II, AVR III, BAM HI, BCL I, BGL II, BSTE II, ECO RI, HIND III, HPA I, KPN I, MLU I, MST I, MST II, NCO I, NDE I, NRU I, NSI I, PVU I, SCA I, SFI I, SMA I, STU I, XBA I, XHO I. Introns from Watson et al., PNAS 80:2146-2150, 1983.

GA-FeSV

Gardner-Arnstein feline sarcoma virus (fes), cat

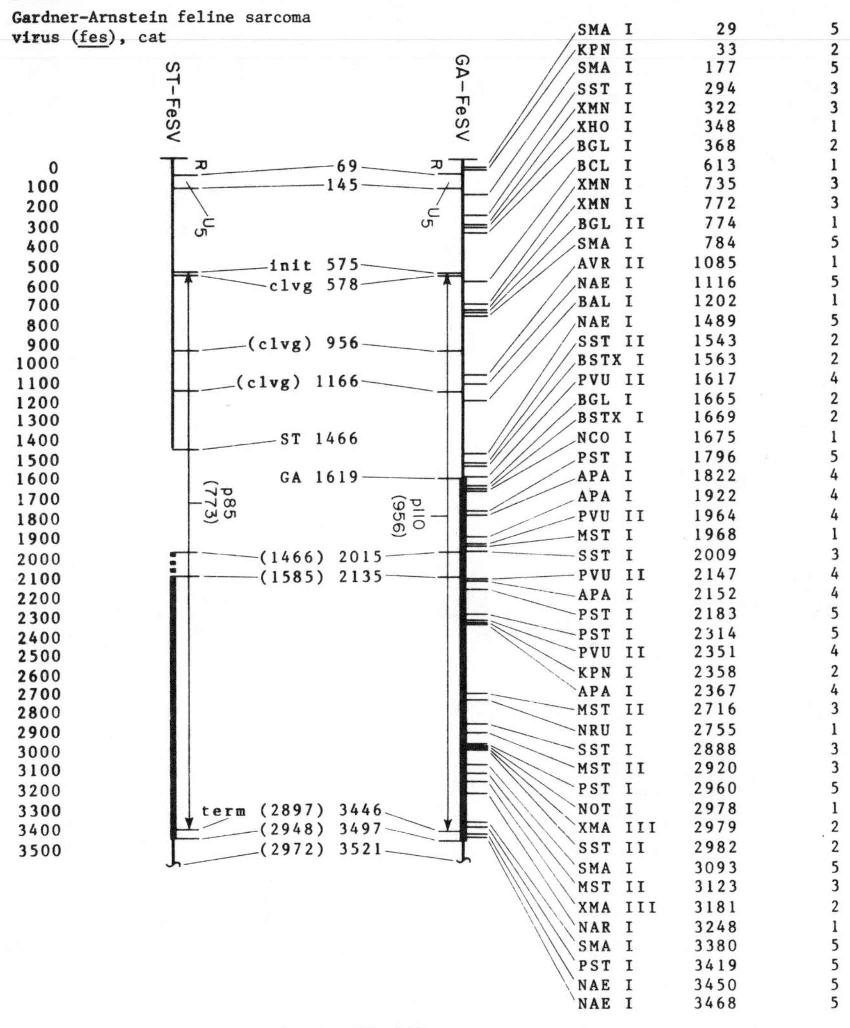

Notes: Hampe et al., Cell 30:775-785, 1982. No sites: AAT II, AHA III, ASU II, BAM HI BSSH II, BSTE II, CLA I, ECO RI, ECO RV, HIND III, HPA I, MLU I, NDE I, NSI I, PVU I, SCA I, SFI I, SPH I, STU I, TTH III, XBA I, XHO I.

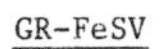

Gardner-Rasheed feline sarcoma virus (fgr), cat

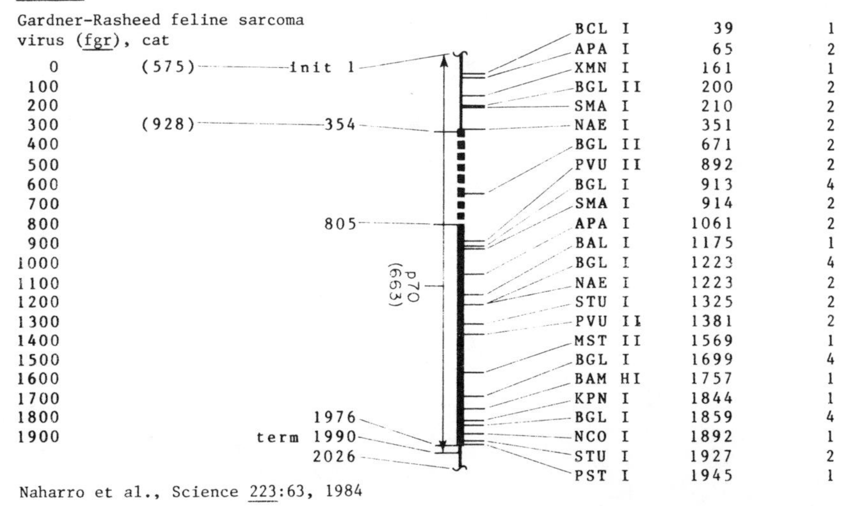

Naharro et al., Science 223:63, 1984

FBJ-OV

Finkel-Biskis-Jinkins osterosarcoma virus (fos), mouse

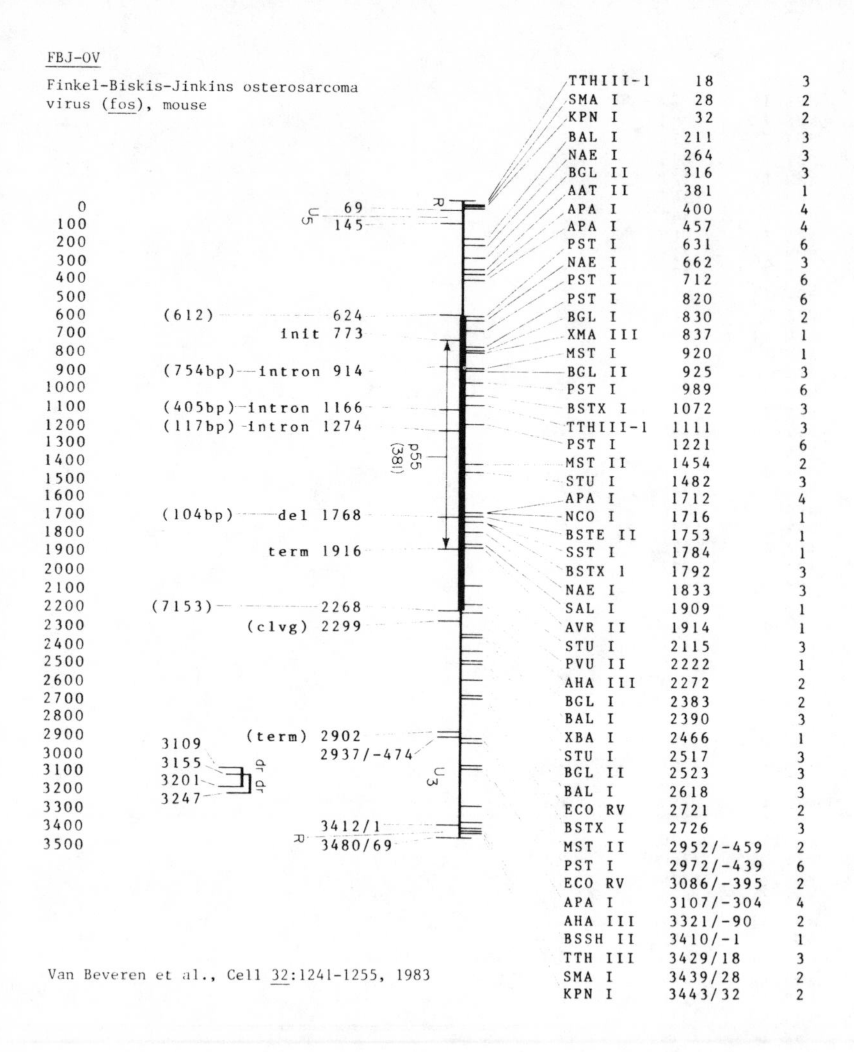

TTHIII-1	18	3
SMA I	28	2
KPN I	32	2
BAL I	211	3
NAE I	264	3
BGL II	316	3
AAT II	381	1
APA I	400	4
APA I	457	4
PST I	631	6
NAE I	662	3
PST I	712	6
PST I	820	6
BGL I	830	2
XMA III	837	1
MST I	920	1
BGL II	925	3
PST I	989	6
BSTX I	1072	3
TTHIII-1	1111	3
PST I	1221	6
MST II	1454	2
STU I	1482	3
APA I	1712	4
NCO I	1716	1
BSTE II	1753	1
SST I	1784	1
BSTX I	1792	3
NAE I	1833	3
SAL I	1909	1
AVR II	1914	1
STU I	2115	3
PVU II	2222	1
AHA III	2272	2
BGL I	2383	2
BAL I	2390	3
XBA I	2466	1
STU I	2517	3
BGL II	2523	3
BAL I	2618	3
ECO RV	2721	2
BSTX I	2726	3
MST II	2952/-459	2
PST I	2972/-439	6
ECO RV	3086/-395	2
APA I	3107/-304	4
AHA III	3321/-90	2
BSSH II	3410/-1	1
TTH III	3429/18	3
SMA I	3439/28	2
KPN I	3443/32	2

Van Beveren et al., Cell 32:1241-1255, 1983

FBR GENOME

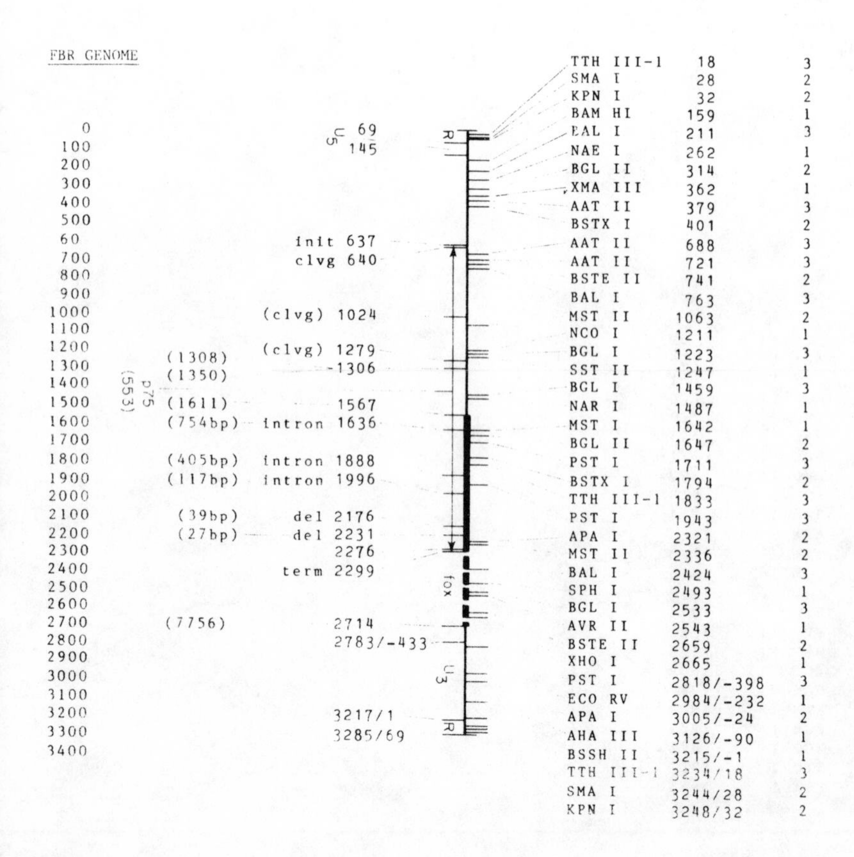

TTH III-1	18	3
SMA I	28	2
KPN I	32	2
BAM HI	159	1
EAL I	211	3
NAE I	262	1
BGL II	314	2
XMA III	362	1
AAT II	379	3
BSTX I	401	2
AAT II	688	3
AAT II	721	3
BSTE II	741	2
BAL I	763	3
MST II	1063	2
NCO I	1211	1
BGL I	1223	3
SST II	1247	1
BGL I	1459	3
NAR I	1487	1
MST I	1642	1
BGL II	1647	2
PST I	1711	3
BSTX I	1794	2
TTH III-1	1833	3
PST I	1943	3
APA I	2321	2
MST II	2336	2
BAL I	2424	3
SPH I	2493	1
BGL I	2533	3
AVR II	2543	1
BSTE II	2659	2
XHO I	2665	1
PST I	2818/-398	3
ECO RV	2984/-232	1
APA I	3005/-24	2
AHA III	3126/-90	1
BSSH II	3215/-1	1
TTH III-1	3234/18	3
SMA I	3244/28	2
KPN I	3248/32	2

ARV

Avian reticuloendotheliosis virus (rel), turkey

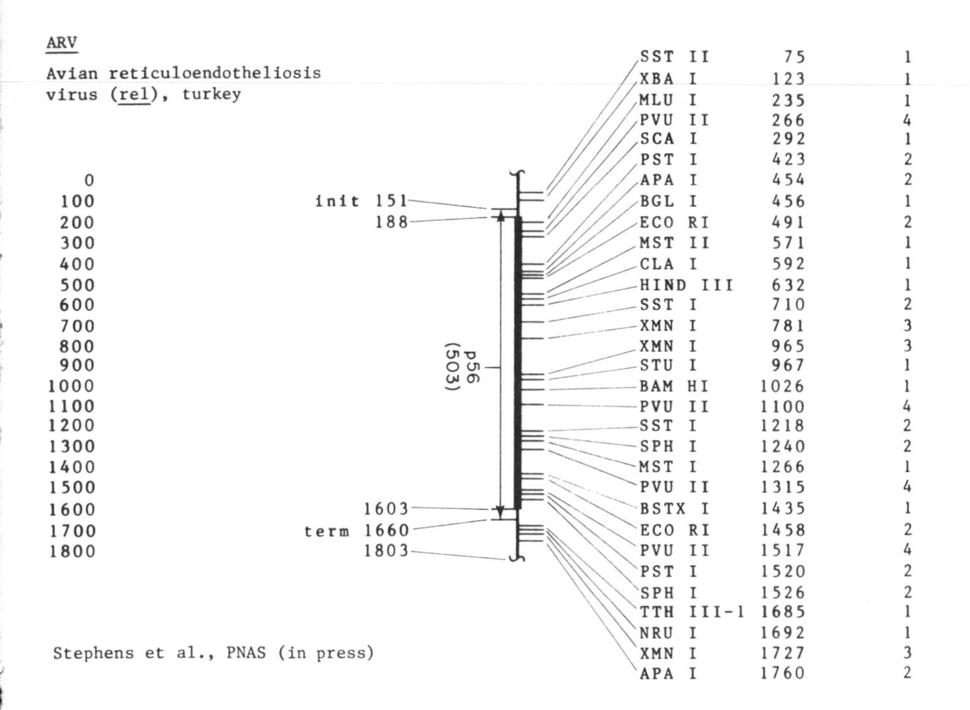

Stephens et al., PNAS (in press)

BaEV

Baboon endogenous virus

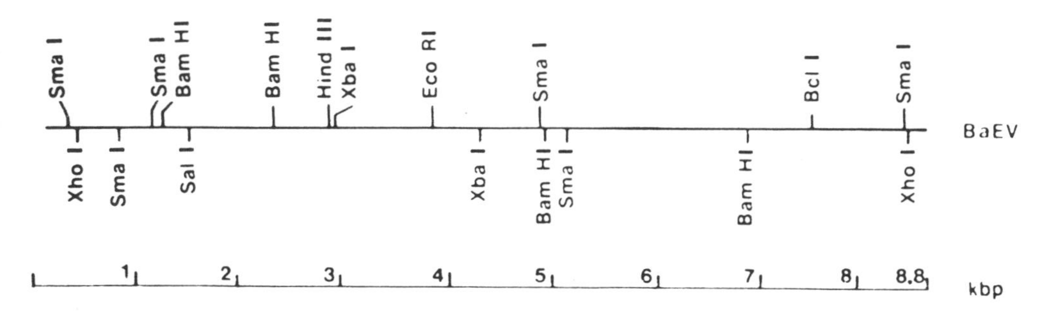

Battula, N. and G.J. Todaro, J. Virol. 36:709, 1980.
Cohen et al., J. Virol. 34:28, 1980.

RD114

Feline endogenous virus, cat

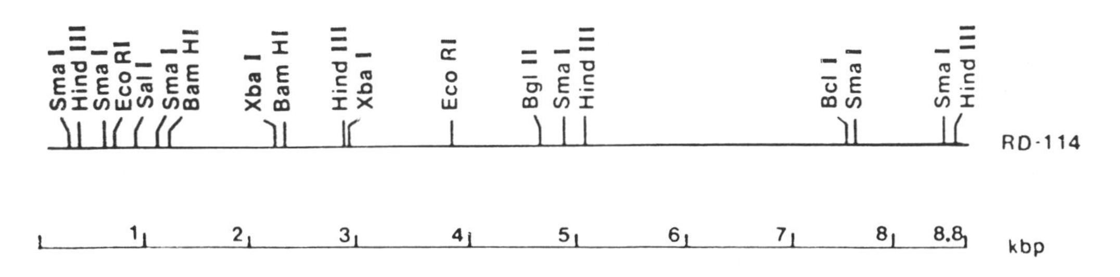

R. H. Reeves and S. J. O'Brien, J. Virol. 52:164, 1984.

INDEX

Ab-MLV (*abl*), 1.173
Acinetobacter calcoaceticus, 2.98
Adenoviruses, 1.98
Aedes (Protomacleaya) triseriatus (mosquito), 3.184
Aedes (Stegomyia) aegypti (mosquito), 3.179
AEV (*erbB*), 1.178
African green monkey (*Cercopithecus aethiops*), 4.141
AKV, 1.168
AMV (*myb*), 1.177
Anopheles albimanus, 3.190
Anopheles quadrimaculatus, 3.194
Anther smut fungus (*Ustilago violacea*), 3.94
Aotus trivirgatus (owl monkey), 4.148
Arabidopsis thaliana, 6.94
ARV (*rel*), 1.181
Aspergillus nidulans, 3.97
  mitochondrial genome, 3.109
  nuclear genes, 3.97

Baboon (*Papio papio, hamadryas, cynocephalus*), 4.140
Bacillus subtilis, 2.28
Bacteriophage, 1.3
  λ, 1.3
  P1, 1.52
  P2, 1.63
  P4, 1.70
  ϕ29, 1.74
  ϕX174, 1.79
  T4, 1.24
BaEV, 1.181
Barley (*Hordeum vulgare* L.), 6.125
BIV, 1.166
Blast fungus (*Magnaporthe grisea*), 3.83
Bos taurus (cow), 4.121
Bovine immunodeficiency virus (BIV), 1.166
BPV (bovine papillomaviruses), 1.129
Brassica oleracea (broccoli) RFLP, 6.103
Broccoli (*Brassica oleracea*) RFLP, 6.103

Caenorhabditis elegans, 3.111
CAEV-CO, 75G63 (caprine arthritis encephalitis virus), 1.166
Canis familiaris (dog), 4.105
Caprine arthritis encephalitis virus, 1.166
Capuchin monkey (*Cebus capucinus*), 4.142
Cat (*Felis catus*), 4.102
Caulobacter crescentus, 2.100
Cebus capucinus (capuchin monkey), 4.142
Cercopithecus aethiops (African green monkey), 4.141
Chicken (*Gallus gallus*), 4.168
Chimpanzee (*Pan troglodytes*), 4.135
Chinese hamster (*Cricetulus griseus*), 4.88
  CHO cells, 4.92
  nuclear genes, 4.88
Chlamydomonas reinhardtii, 2.113
  chloroplast genome, 2.122
  mitochondrial genome, 2.121
  nuclear gene loci, 2.113
Coprinus cinereus, 3.76
Corn (*Zea mays* L.), 6.39
Cow (*Bos taurus*), 4.121
Cricetulus griseus (Chinese hamster), 4.88
  CHO cells, 4.92
  nuclear genes, 4.88
CRPV (cotton tail rabbit papillomavirus), 1.129

Deermouse (*Peromyscus maniculatus*), 4.96
Dictyostelium discoideum (slime mold, cellular), 3.3
Dog (*Canis familiaris*), 4.105
DPV (deer papillomavirus), 1.129
Drosophila melanogaster (fruit fly), 3.134
  biochemical loci, 3.134
  cloned genes, 3.162
  in situ hybridization data, 3.158
Drosophila pseudoobscura (fruit fly), 3.188

E26 (*ets*), 1.176
EEPV (European elk papillomavirus), 1.129
EIAV (equine infectious anemia virus), 1.167
Epstein-Barr virus, 1.102
Equine infectious anemia virus (EIAV), 1.167
Equus caballus (horse), 4.107
Escherichia coli, 2.54

FBJ-OV (*fos*), 1.180
FBR, 1.180
Feline immunodeficiency virus (FIV), 1.165
Felis catus (cat), 4.102
FeSV (*fms*), 1.175
Fish, Poeciliid, 4.160
Fish, salmonid, 4.151
Fish, xiphophorus, 4.160
FIV, 1.165
Fr-SFFVp, 1.174
Fruit fly (*Drosophila melanogaster; Drosophila pseudoobscura*), 3.134, 3.188
FuSV (*fps*), 1.177

GA-FeSV (*fes*), 1.179
Gallus gallus (chicken), 4.168
Garden pea (*Pisum sativa*), 6.106
Gibbon (*Hylobates [Nomascus] concolor*), 4.138
Glycine max L. Merr. (soybean), 6.68
Gorilla (*gorilla gorilla*), 4.136
Gorilla gorilla (gorilla), 4.136
GR-FeSV (*fgr*), 1.179

Haemophilus influenzae Rd, 2.110
Hepatitis B virus, 1.121
Herpes simplex virus, 1.115
HIV (HIV-I, HIV-II), 1.149
Homo sapiens, 5.3
  cell surface receptors, 5.198, 5.202
  chromosome rearrangements in human neoplasia, 5.186
  DNA restriction fragment length polymorphisms (RFLPs), 5.211
  endogenous retroviral sequences, 5.198, 5.202
  fragile sites, 5.193
  genetic linkage maps, 5.134, 5.158, 5.183
  growth factors, 5.198, 5.202
  human gene map, 5.3, 5.48
  lymphokines, 5.198, 5.202
  mitochondrial chromosome, 5.132
  mitochondrial DNA, 5.246
  morbid anatomy, 5.115
  nuclear genes, 5.48
  proto-oncogenes, 5.197
Hordeum vulgare L. (barley), 6.125

Horse (*Equus caballus*), 4.107
HPV (human papillomaviruses), 1.130, 1.131
HTLV-I, -II, 1.160
Human. See *Homo sapiens*, 5.3
Human immunodeficiency virus (HIV), 1.149
Human T-lymphotropic retroviruses (HTLV-I, HTLV-II), 1.160
Hylobates (Nomascus) concolor (gibbon), 4.138

Immunodeficiency virus,
  bovine (BIV), 1.166
  feline (FIV), 1.165
  human (HIV), 1.149
  simian (SIV), 1.149, 1.163

Lactuca sativa (lettuce), 6.100
λ, 1.3
Lentiviruses, 1.162. *See also* Ungulate lentiviruses, 1.166
Leopard frog (*Rana pipiens*), 4.164
Lettuce (*Lactuca sativa*), 6.100
Lycopersicon esculentum (tomato), 6.3

Macaca mulatta (rhesus monkey), 4.139
Magnaporthe grisea (blast fungus), 3.83
Marmoset, cotton-topped (*Saquinus oedipus*), 4.147
Marsupials, 4.129
MC29 (*myc*), 1.179
Meriones unquiculatus (Mongolian gerbil), 4.101
Mesocricetus auratus (Syrian hamster), 4.100
Microcebus murinus (mouse lemur), 4.143
Mink, American (*Mustela vison*), 4.126
Mo-MLV, 1.169
Mo-MSV (*mos*), 1.172
Mongolian gerbil (*Meriones unquiculatus*), 4.101
Monkey,
  African green (*Cercopithecus aethiops*), 4.141
  capuchin (*Cebus capucinus*), 4.142
  owl (*Aotus trivirgatus*), 4.148
  rhesus (*Macaca mulatta*), 4.139
Monotremes, 4.129
Mosquito
  Aedes (*Protomacleaya*) *triseriatus*, 3.184
  Aedes (*Stegomyia*) *aegypti*, 3.179
  Anopheles albimanus, 3.190
  Anopheles quadrimaculatus, 3.194
Mouse (*Mus musculus*), 4.3
  DNA clones and probes, 4.37
  nuclear genes, 4.3
  retroviral and cancer-related genes, 4.69
  RFLPs, 4.37
Mouse lemur (*Microcebus murinus*), 4.143
MSV-3611 (*raf*), 1.175
Mus musculus (mouse), 4.3
  DNA clones and probes, 4.37
  nuclear genes, 4.3
  retroviral and cancer-related genes, 4.69
  RFLPs, 4.37
Mustela vison (American mink), 4.126

Nasonia vitripennis (parasitic wasp), 3.198
Neisseria gonorrhoeae, 2.90
Neurospora crassa, 3.9
  mitochondrial genes, 3.19
  nuclear genes, 3.9

Oncorhynchus (salmonid fish), 4.151
Orangutan (*Pongo pygmaeus*), 4.137
Oryctolagus cuniculus (rabbit), 4.114
Ovine pulmonary adenomatosis virus (OPAV), 1.166
Ovis aries (sheep), 4.118
Owl monkey (*Aotus trivirgatus*), 4.148

P1, 1.52
P2, 1.63
P4, 1.70
ϕ29, 1.74
ϕX174, 1.79
Pan troglodytes (chimpanzee), 4.135
Papillomaviruses, 1.128
  BPV-1, -2 (bovine), 1.129
  CRPV (cotton tail rabbit), 1.129
  DPV (deer), 1.129
  EEPV (European elk), 1.129
  HPV-1a, -5, -6b, -8, -11, -16, -18, -31, -33 (human), 1.130
Papio papio, *hamadryas*, *cynocephalus* (baboon), 4.140
Paramecium tetraurelia, 2.130
Parasitic wasp (*Nasonia vitripennis*), 3.198
Pea, garden (*Pisum sativum*), 6.106
Peromyscus maniculatus (deermouse), 4.96
Petunia (*Petunia hybrida*), 6.113
  chloroplast DNA, 6.120
  linkage map, 6.113
Petunia hybrida (petunia), 6.113
  chloroplast DNA, 6.120
  linkage map, 6.113
Phycomyces blakesleeanus, 3.84
Pig (*Sus scrofa domestica* L.), 4.110
Pisum sativum (garden pea), 6.106
Podospora anserina, 3.58
Poeciliid fish, 4.160
Poeciliopsis, 4.160
Polyoma virus, 1.90
Pongo pygameus (orangutan), 4.137
Primates, 4.134, *See also Homo sapiens*
  Aotus trivirgatus (owl monkey), 4.148
  Cebus capucinus (capuchin monkey), 4.142
  Cercopithecus aethiops (African green monkey), 4.141
  Gorilla gorilla (gorilla), 4.136
  Hylobates (Nomascus) concolor (gibbon), 4.138
  Macaca mulatta (rhesus monkey), 4.139
  Microcebus murinus (mouse lemur), 4.143
  Pan troglodytes (chimpanzee), 4.135
  Papio papio, *hamadryas*, *cynocephalus* (baboon), 4.140
  Pongo pygmaeus (orangutan), 4.137
  Saquinus oedipus (cotton-topped marmoset), 4.147
Proteus mirabilis, 2.82
Proteus morganii, 2.88
Pr-RSV (*src*), 1.170
Pseudomonas aeruginosa, 2.71
Pseudomonas putida, 2.79

Rabbit (*Oryctolagus cuniculus*), 4.114
Ra-MSV (*ras*), 1.173
Rana pipiens (leopard frog), 4.164
Rat (*Rattus norvegicus*), 4.80
Rattus norvegicus (rat), 4.80
RD114, 1.181
Retroviruses, animal. *See* Type C retroviruses, 1.168
Rhesus monkey (*Macaca mulatta*), 4.139

Rhizobium meliloti and *Rhizobium leguminosarum* biovars:
phaseoli, *trifolii*, *viciae*, 2.104
Rye (*Secale cereale* L.), 6.136

Saccharomyces cerevisiae, 3.30
mitochondrial DNA, 3.50
nuclear genes, 3.30
Salmo (salmonid fishes), 4.151
Salmonella typhimurium, 2.3
Salmonid fishes: *Salvelinus, Salmo, Oncorhynchus*, 4.151
Salvelinus (salmonid fishes), 4.151
Saquinus oedipus (cotton-topped marmoset), 4.147
Schizophyllum commune, 3.90
Secale cereale L. (rye), 6.136
Sheep (*Ovis aries*), 4.118
Simian immunodeficiency virus (SIV), 1.149, 1.163
Simian virus 40, 1.84
SIV, 1.149, 1.163
SK770 (*ski*), 1.174
Slime mold, cellular (*Dictyostelium discoideum*), 3.3
Sordaria macrospora, 3.68
Soybean (*Glycine max* L. Merr.), 6.68
SSV (*sis*), 1.171
Staphylococcus aureus, 2.22
Sus scrofa domestica L. (pig), 4.110
Syrian hamster (*Mesocricetus auratus*), 4.100

T4, 1.24
Tetrahymena thermophila, 2.132
Tomato (*Lycopersicon esculentum*), 6.3
Triticum aestivum (wheat), 6.16
biochemical/molecular loci, 6.28
linkage map, 6.16
Type C retroviruses, 1.168
Ab-MLV (*abl*), 1.173
AEV (*erbB*), 1.178
AKV, 1.168
AMV (*myb*), 1.177
ARV (*rel*), 1.181
BaEV, 1.181
E26 (*ets*), 1.176
FBJ-OV (*fos*), 1.180
FBR, 1.180
FeSV (*fms*), 1.175
Fr-SFFVp, 1.174
FuSV (*fps*), 1.177
GA-FeSV (*fes*), 1.179
GR-FeSV (*fgr*), 1.179
MC29 (*myc*), 1.178
Mo-MLV, 1.169
Mo-MSV (*mos*), 1.172
MSV-3611 (*raf*), 1.175
Pr-RSV (*src*), 1.170
Ra-MSV (*ras*), 1.173
RD114, 1.181
SK770 (*ski*), 1.174
SSV (*sis*), 1.171
Y73 (*yes*), 1.176

Ungulate lentiviruses, 1.166
bovine immunodeficiency virus (BIV), 1.166
caprine arthritis encephalitis virus (CAEV)-CO, 75G63, 1.166
equine infectious anemia virus (EIAV), 1.167
ovine pulmonary adenomatosis virus (OPAV), 1.166
visna, 1.166
Ustilago maydis, 3.92
Ustilago violacea (anther smut fungus), 3.94

Vaccinia virus, 1.138
Visna virus, 1.166

Wasp, parasitic (*Nasonia vitripennis*), 3.198
Wheat (*Triticum aestivum*), 6.16
biochemical/molecular loci, 6.28
linkage map, 6.16

Xiphophorus (fish), 4.160

Y73 (*yes*), 1.176
Yeast (*Saccharomyces cerevisiae*), 3.30
mitochondrial DNA, 3.50
nuclear genes, 3.30

Zea mays L. (corn), 6.39